Plumbing Installation and Design

plumbing
INSTALLATION AND DESIGN

L. V. RIPKA

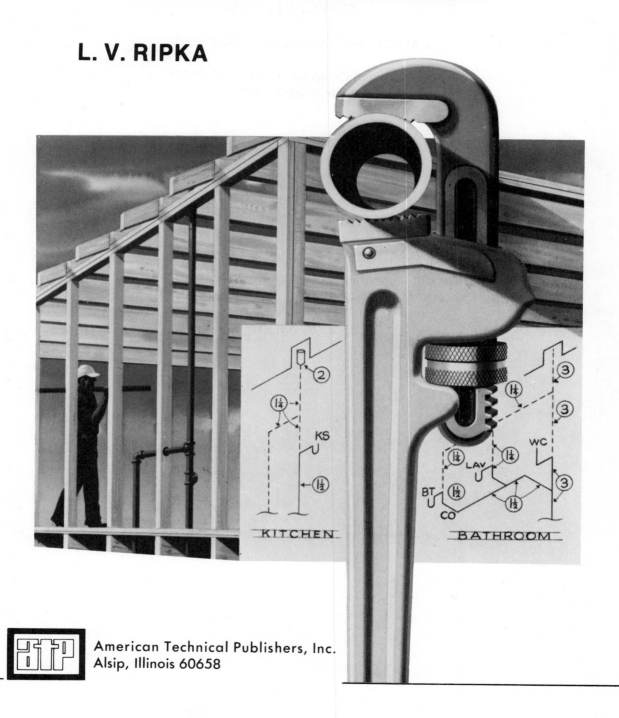

American Technical Publishers, Inc.
Alsip, Illinois 60658

COPYRIGHT © 1978

BY AMERICAN TECHNICAL PUBLISHERS, INC.

Library of Congress Number: 77-73227
ISBN: 0-8269-0600-1

123456789-78-987

PRINTED IN THE UNITED STATES OF AMERICA

PREFACE

PLUMBING INSTALLATION AND DESIGN is a complete learning program planned to help the beginning student of the plumbing trade develop and apply the technical knowledge necessary to attain job-entry skills. The author has attempted to produce an innovative, forward-looking program that stresses the practical, hands-on aspects of the trade. Of necessity, many parts of this text are similar to those found in other successful plumbing texts, this is inevitable in any book that teaches a skill.

According to the DICTIONARY OF OCCUPATIONAL TITLES, a plumber is a person who "assembles and installs air, gas, water, and waste disposal systems; cuts openings in walls for pipe; bends pipe over blocks by hand; cuts, reams, and threads pipe; caulks joints; wipes joints by pouring molten solder over the joints and spreads the solder with a cloth to shape it; tests joints and pipe systems for leaks by filling them with water under pressure and installs gas, water, and sanitary fixtures and equipment with their supports, hangers or foundations."

The objective of PLUMBING INSTALLATION AND DESIGN IS TO DEVELOP THE BASIC SKILLS OF BEGINNING STUDENTS IN THE USE OF PLUMBING MATERIALS, TOOLS, AND EQUIPMENT TOGETHER WITH THE BASIC TECHNICAL KNOWLEDGE REQUIRED TO ENTER THE PLUMBING TRADE. The students learn how to make all the common joints and to install the systems common to the plumbing trade as well as how to install plumbing fixtures and appliances.

The Publishers

CONTENTS

Chapter   **Page**

1 PLUMBING AS A TRADE **1**
   Structure of the Trade 3
   Apprenticeship Standards 4
   Blueprint Reading and Specifications 5
   Plumbing Systems of a Building 8
   Review Questions 12

2 JOB SAFETY **13**
   General Safety on the Job 14
   Electricity 17
   Safe Work in Trenches 17
   Oxyacetylene Welding and Cutting Safety 26
   Review Questions 27

3 PLUMBING MATERIALS **29**
   Cast Iron Soil Pipe and Fittings 29
   Galvanized Steel Pipe and Threaded Fittings 33
   Copper Tubing, Solder Joint, and Flare Joint Fittings 38
   Plastic Pipe and Fittings 42
   Uses of Piping Materials 45
   How to Identify Plumbing Fittings 46
   Plumbing Valves 47
   Water Meters 54
   Review Questions 56

4 PLUMBING TOOLS **57**
   Cast Iron Soil Pipe Tools 57
   Galvanized Steel Pipe Tools 61
   Copper Tubing Tools 72
   Plastic Pipe Tools 76
   Finishing Tools 77
   Layout and Measuring Tools 81
   Cutting and Boring Tools 82
   Concrete Drilling Tools 83
   Testing Tools 86
   Miscellaneous Tools 87
   Review Questions 91

5 JOINING, INSTALLING, AND SUPPORTING PIPE **91**
   Cast Iron Soil Pipe Joints 92
   Galvanized Steel Threaded Joints 100
   Copper Tubing Joints 102
   Plastic Pipe Joints 108
   Installing and Supporting Pipe 112
   Review Questions 118

6 SANITARY DRAINAGE, VENT AND STORM DRAINAGE PIPING **119**
   Sanitary Drainage Piping 119
   Drainage Piping Installation 128
   Venting the Building Sanitary Drainage System 131
   Storm Water Drainage Principles 146
   Review Questions 152

Chapter   **Page**

7   SIZING OF SANITARY DRAINAGE AND VENT PIPING **153**

Single Family Home 154
Two-Story Single Family Home 157
Duplex Residence 159
Apartment Building Bathroom Stack 159
Apartment Building Kitchen Sink Waste Stack 160
Multistory Building Bathroom Stack 162
Two-Story Industrial Building 163
Two-Story Office Building 165
Review Questions 166

8   THE PLUMBING TRAP **167**

Types of Traps 168
Prohibited Traps 173
Trap Seal Loss 175
Review Questions 180

9   SIZING WATER SUPPLY PIPING **181**

Sizing Water Supplies 187
Review Questions 199

10   PLUMBING FIXTURES AND APPLIANCES **201**

Water Closets 201
Urinals 210
Lavatories 214
Bathtubs 218
Shower Baths 220
Bidets 222
Kitchen Sinks 222
Garbage Disposals 224
Domestic Dishwashers 225
Laundry Trays 226
Floor Drains 227
Drinking Fountains and Water Coolers 229
Service Sinks and Mop Basins 230
Water Softeners 232
Water Heaters 235
Installation of Plumbing Fixtures and Appliances 240
Review Questions 242

11   TESTING AND INSPECTING THE PLUMBING SYSTEM **243**

Plumbing System Tests 244
How to Apply a Plumbing Test 254
Review Questions 256

12   PLUMBING A HOUSE **257**

Installation of the Rough Plumbing 281
Installation of the Building Sewer and Water Service 303
Installation of the Finish Plumbing 308
Review Questions 312

HOUSE PLANS FOR CHAPTER 12 .... **313**
GLOSSARY OF TERMS **331**
METRIC INFORMATION FOR PLUMBERS **345**
INDEX **351**

CHAPTER 1

Plumbing as a Trade

What are a plumber's responsibilities? The plumber is responsible for the public health and sanitation. To meet this responsibility, he must provide a safe or potable supply of water into a building, distribute this water supply to the various plumbing fixtures, and remove the waste water and water-borne waste materials. The plumber is also responsible for the maintenance and repair of these water supply and waste pipes. In the process of supplying water to the building and removing the water-borne waste materials, the plumber constructs the plumbing systems of a building. A typical plumbing code might use the following definition of plumbing and a plumbing system:

>*Plumbing.* Plumbing is the art of installing in buildings, the pipes, fixtures, and other apparatus for bringing in the water supply and removing waste water and water-carried waste.

>*Plumbing System.* The plumbing system of a building includes the water supply distributing pipes; the fixtures and fixture traps; the soil, waste, and vent pipes; the building drain and building sewer; the storm water drainage; with their devices, appurtenances, and connections within the building and outside the building within the property line.

The plumber is usually guided in his job of supplying a building with a safe water supply and removing the water-borne waste materials by state and/or municipal plumbing codes. Figure 1-1 shows the 23 basic plumbing principles on which a typical state plumbing code is based.

The plumbing trade has an old and interesting history. Prehistoric man of a hundred thousand years ago left indications of sanitation and plumbing skill. Crude as these devices were, they offered proof that even these primitive people realized the consequences of poor plumbing. The rulers of Egypt, Greece, and Rome, thousands of years before Christ, advocated sanitary facilities of one kind or another. Bathtubs that were mere holes in the ground lined with tile, and water-conveying aqueducts constructed of terra cotta and brick and terminating in a reservoir, were some of these historic accomplishments.

The individual who worked in the sanitary field in ancient Rome was called a *plumbarius,* taken from the Latin word *plumbum,* meaning lead.

Basic Plumbing Principles. This code is founded upon certain basic principles of environmental sanitation and safety through properly designed, acceptably installed and adequately maintained plumbing systems. Some of the details of plumbing construction may vary but the basic sanitary and safety principles desirable and necessary to protect the health of the people are the same everywhere. As interpretations may be required, and as unforeseen situations arise which are not specifically covered in this code, the twenty three principles which follow shall be used to define the intent.

(a) All premises intended for human habitation, occupancy, or use shall be provided with a potable water supply which meets the requirements of the Minnesota State Board of Health. Such water supply shall not be connected with unsafe water sources nor shall it be subject to the hazards of backflow or back-siphonage.

(b) Plumbing fixtures, devices, and appurtenances shall be supplied with water in sufficient volume and at pressures adequate to enable them to function properly and without undue noise under normal conditions of use.

(c) Plumbing fixtures shall be designed and adjusted to use the minimum quantity of water consistent with proper performance and cleaning. Hot water shall be supplied to all plumbing fixtures which normally need or require hot water for their proper use and function.

(d) Devices for heating water and storing it shall be designed and installed to prevent all dangers from explosion and over heating.

(e) Every building with installed plumbing fixtures and intended for human habitation, occupancy or use when located on premises where a public sewer is available within a reasonable distance shall be connected to the sewer.

(f) Each family dwelling unit shall have at least one water-closet, one lavatory, one kitchen type sink, and one bathtub or shower to meet the basic requirements of sanitation and personal hygiene. All other structures for human habitation shall be equipped with sufficient sanitary facilities.

(g) Plumbing fixtures shall be made of durable, smooth, non-absorbent and corrosion resistant material and shall be free from concealed fouling surfaces.

(h) The drainage system shall be designed, constructed, and maintained to conduct the waste water with velocities which will prevent fouling, deposition of solids and clogging.

(i) The piping of the plumbing system shall be of durable material free from defective workmanship and so designed and constructed as to give satisfactory service for its reasonable expected life.

(j) The drainage system shall be provided with an adequate number of cleanouts so arranged that in case of stoppage the pipes may be readily cleaned.

(k) Each fixture shall be provided with a separate, accessible, self-scouring, reliable water-seal trap placed as near to the fixture as possible.

(l) The building drainage system shall be designed to provide adequate circulation of air in all pipes with no danger of siphonage, aspiration or forcing of trap seals under conditions of ordinary use.

(m) Each vent terminal shall extend to the outer air and be so installed as to minimize the possibilities of clogging and the return of foul air to the building.

(n) The plumbing system shall be subjected to adequate tests and to inspections in a manner that will disclose all leaks and defects in the work or the material.

(o) No substance which will clog or accentuate clogging of pipes, produce explosive mixtures, destroy the pipes or their joints, or interfere unduly with the sewage-disposal process shall be allowed to enter the drainage system.

(p) Proper protection shall be provided to prevent contamination of food, water, sterile goods, and similar materials by backflow of sewage. When necessary, the fixtures, device, or appliance shall be connected indirectly with the building drainage system.

(q) No water-closet or similar fixture shall be located in a room or compartment which is not properly lighted and ventilated.

(r) If water-closets or other plumbing fixtures are installed in a building where there is no sewer within a reasonable distance, suitable provision shall be made for disposing of the building sewage by methods of disposal which meets the requirements of the Minnesota State Board of Health and the Minnesota Pollution Control Agency.

(s) Where a building-drainage system may be subjected to back flow of sewage, suitable provision shall be made to prevent its overflow in the building.

(t) Plumbing systems shall be maintained in a safe and serviceable condition from the standpoint of both mechanics and health.

(u) All plumbing fixtures shall be so installed with regard to spacing as to be accessible for their intended use and cleansing.

(v) Plumbing shall be installed with due regard to preservation of the strength of structural members and prevention of damage to the walls and other surfaces through fixture usage.

(w) Sewage or other waste shall not be discharged into surface or sub-surface water unless it first has been subjected to an acceptable form of treatment.

Figure 1-1. Basic plumbing principles on which plumbing codes are based. (Minnesota Plumbing Code)

Because his work consisted of shaping lead, this name seemed to fit him well. It is interesting to note that until just recently much lead was still used for waste and water supply, and after two thousand years the sanitarian is still called a plumber. Evidence of the skill of these artisans can be seen in the aqueducts they built, some of which are still in use today.

During the period known as the Dark Ages (A.D. 400 to 1400) the culture of the early Romans deteriorated. Disease was rampant, and unsanitary conditions were responsible for destroying at least one quarter of the population of ancient Europe. In the fight for supremacy during this period, the Goths, Christians, and other invaders destroyed what remained of Roman culture. Europe was dormant for almost ten centuries.

During the Renaissance, a gradual upbuilding of plumbing again began. Early in the 17th century the first plumbing apprentice laws were passed in England. France began building public water service installations in the 18th century. In general, Europe was in a period of building, including in this progress the art of sanitary science.

In the United States, which was largely devoted to agriculture, very little progress in plumbing was made up to the year 1800. The kitchen sink and portable bathtub were the first two indoor plumbing fixtures. The outside privy was a common means of disposing of waste matter. Water closets, imported from England (where they were first patented), were in use in a few instances, but it is doubtful whether scientific principles were applied in installations of that day.

After the Civil War, plumbing improvements came slowly but steadily. Patents were issued on traps and methods of ventilation. Public water supply and sewage disposal systems became more evident, and plumbing came to be regarded as a necessity rather than the luxury it was considered twenty years before. Up to 1900 very few homes in urban localities provided more than a hydrant and a slop hopper for the disposal of waste. After the turn of the century, plumbing progressed more rapidly. Water closets of the hopper and washout varieties as well as sinks, wash basins, and bathtubs were provided within

the walls of a building. Scientific methods were becoming used in constructing plumbing installations.

Fixture traps were ventilated, and hot and cold running water was introduced. The siphon washdown closet appeared during this period, and states were developing legislation for the control of sanitation. The greatest progress in plumbing took place after the year 1910, which is rather recent for a trade that has a background of thousands of years. Modern manufacturing methods provided materials and equipment that could be scientifically incorporated into a plumbing system. Buildings became larger, and the people who occupied them demanded more sanitary facilities.

 ## STRUCTURE OF THE TRADE

The plumbing trade is structured into various levels of craftsman: apprentice, journeyman, foreman, superintendent.

Apprenticeship or learning usually lasts from four to five years. From man's earliest history apprentices in the various crafts were *indentured* (a contract binding one person to work for another for a given length of time to learn a trade) to a master craftsman, most often a contractor, for a number of years to learn the trade.

In many cases, the apprentice's father had to pay the master a fee to get him to teach his son the trade. From the medieval days down through most of the 19th century, the apprentice would live with the master and would get room and board plus some clothing. However, he was a virtual slave to the master, subject to his every wish.

Today in the plumbing trade, apprentices are protected by federal and state laws, and the local JATC (Joint Apprenticeship and Training Committee) in regard to hours of work, wages, and conditions of employment. Furthermore, there is no control over the apprentices outside of the working hours. Also, apprentices are now selected from applicants who meet the standards of the local JATC. The apprentice is indentured to the JATC, and they assign him to a contractor. If the contractor runs out of work the JATC will place him with another contractor. This permits the JATC to control the training and handle the federal, state, and veterans' paperwork.

The *journeyman*, or experienced craftsman, is one who has completed an apprenticeship in the trade. He is now a free agent and can work for any contractor he pleases. He may travel from place to place, going where the work is to be found.

The *foreman* (or supervisor) is a journeyman who has been placed in the job of supervising a group of men. He is given this job because of his ability as a craftsman and his knowledge of how to supervise other craftsmen. He is responsible for laying out the work for the journeyman and apprentices on the job and seeing that they have enough tools and material to work with. On larger jobs there will sometimes be a general foreman who supervises the foremen.

The *superintendent* is usually a foreman who has been promoted to this important position. He is in charge of all the work in the field for his contractor and supervises the work of the foremen and general foremen.

In the construction industry, the foremen, general foremen, and superintendent keep their union membership. Many of the smaller contractors are permitted to retain their union cards in some unions. Plumbers, too, permit this in some areas.

Large plumbing contractors will employ an *estimator* who works in their offices to estimate the cost of the jobs the contractor wants to bid on. Working from the blueprints and specifications, he *takes off* (measures and/or counts) the plumbing fixtures and equipment, and measures the footage of piping required to install the given job. He does this to figure his costs for material and labor so that he may arrive at a fair bidding figure or estimate for the contractor to make a fair profit on the job. The estimator must be skilled in mathematics, blueprint reading, trade practices, and the cost of labor and materials.

The last person in this team of workers, supervisors, and estimator is the *contractor*. He must know all phases of the business. He must know all the regulations governing the construction industry, be licensed to install plumbing in the area in which he intends to work, and be able to

provide the money for payrolls and materials. The livelihood of his employees depends on his overall ability to run the business successfully.

 ## JOINT APPRENTICESHIP AND TRAINING COMMITTEE APPRENTICESHIP STANDARDS

The plumbing industry has, in cooperation with the U. S. Department of Labor, Bureau of Apprenticeship and Training, set up National Standards of Apprenticeship. These standards define what the term *apprentice* in the trade shall mean. The standards set forth age limits, educational requirements, length of apprenticeship, ratio of apprentices to journeymen, hours of work, and wages.

The Joint Apprenticeship and Training Committee, commonly known as the JATC, is composed of equal representation from labor and management, with consultants from the Bureau of Apprenticeship and the state or local board of education attending as nonvoting advisors.

The JATC has the delegated power to set the local standards consistent with the basic requirements established by the national committee. These local standards are particularly important in the plumbing industry; the apprentice plumber must learn the local plumbing codes and ordinances so that upon the completion of his apprenticeship training he will be able to pass a plumbing license examination based upon them. Figure 1-2 shows a typical set of local standards for a four-year program. Note that the standards also include the wage rate structure the apprentice will receive during the term of his apprenticeship.

The apprentice, when he signs the indenture agreement, agrees to live up to all its provisions and in turn is protected by its rules and regulations. Some state Bureaus of Apprenticeship and Training will then issue the apprentice an identification card. This card will be the apprentice's personal identification when on the job site, since he will not yet have a journeyman plumber's license.

In addition to supervising the on-the-job train-

PLUMBER QUALIFICATIONS

Age—At least 17 years of age

Education—High School graduation or equivalent

Physical Examination—Physical examination by a doctor may be required by the JATC.

Other—Applicants must be citizens of the United States, or in the process of naturalization.

Term of Apprenticeship

The term of apprenticeship shall be not less than four years, to be divided as follows: 7024 hours of work experience, and not less than 800 hours of related instruction (200 hours per year); shall be considered to be the minimum requirements for the development of a journeyman.

Work Processes Covered During Training

Installation of piping for waste, soil sewerage, vent, and leader pipes	1760 hours
Installation of piping for hot and cold water for domestic purposes	640 hours
Installation of tin pipe, lead pipe, sheet lead	80 hours
Assembly and connection of fixtures and appliances	1600 hours
Welding	320 hours
Maintenance and repair of plumbing	1200 hours
Installation of other work usually performed by plumbers	1424 hours
Total	7024 hours

Apprentice Wages

1st 1756 hours	50%	
2nd 1756 hours	60%	of
3rd 1756 hours	70%	journeyman wage rate
4th 1756 hours	80%	

Figure 1-2. Typical local apprenticeship standards.

ing the apprentice receives, the JATC also establishes the curriculum for the related instruction the apprentice receives. This related instruction consists of both classroom and shop classes.

During the period of his four-year apprenticeship the apprentice plumber will receive classroom instruction in the following areas: plumbing theory, natural gas piping, blueprint reading, and plumbing codes. Typical plumbing shop classes are: gas welding, arc welding, soldering and silver brazing, and lead working.

When an apprentice completes his classroom training and the required number of hours of on-the-job training, the JATC notifies the Bureau of Apprenticeship and Training, and this agency issues a completion certificate. In addition, upon passing his plumbing license examination(s) the apprentice plumber will be issued his state and/or local plumbing licenses. These licenses are important because local plumbing ordinances usually require that a plumber be licensed in the area before he can work there, and the ordinances usually do not permit apprentices (who are not licensed) to work alone on jobs.

 BLUEPRINT READING AND SPECIFICATIONS

In addition to practical training with tools and materials, the apprentice plumber will have to devote much of his time to learning to read and understand blueprints and specifications. Blueprints and specifications are the working drawings and the written instructions that tell the various crafts how the architect and the various engineers he is working with (electrical, mechanical, and structural) want the building constructed.

STANDARD SYMBOLS FOR PLUMBING, PIPING AND VALVES

PLUMBING

Corner Bath	
Recessed Bath	
Roll Rim Bath	
Sitz Bath	SB
Foot Bath	FB
Bidet	B
Shower Stall	
Shower Head	(Plan) (Elev.)
Overhead Gang Shower	(Plan) (Elev.)
Pedestal Lavatory	PL
Wall Lavatory	WL
Corner Lavatory	LAV
Manicure Lavatory Medical Lavatory	ML
Dental Lavatory	DENTAL LAV
Plain Kitchen Sink	S
Kitchen Sink, R & L Drain Board	
Kitchen Sink, L H Drain Board	
Combination Sink & Dishwasher	
Combination Sink & Laundry Tray	S & T
Service Sink	SS
Wash Sink (Wall Type)	
Wash Sink	
Laundry Tray	L T
Water Closet (Low Tank)	
Water Closet (No Tank)	
Urinal (Pedestal Type)	
Urinal (Wall Type)	
Urinal (Corner Type)	
Urinal (Stall Type)	
Urinal (Trough Type)	TU
Drinking Fountain (Pedestal Type)	DF
Drinking Fountain (Wall Type)	DF

PLUMBING (continued)

Drinking Fountain (Trough Type)	DF
Hot Water Tank	HWT
Water Heater	WH
Meter	M
Hose Rack	HR
Hose Bibb	HB
Gas Outlet	G
Vacuum Outlet	
Drain	D
Grease Separator	
Oil Separator	
Cleanout	
Garage Drain	
Floor Drain With Backwater Valve	
Roof Sump	

PIPING

Soil and Waste	
Soil and Waste, Underground	
Vent	
Cold Water	
Hot Water	
Hot Water Return	
Fire Line	—F——F—
Gas	—G——G—
Acid Waste	ACID
Drinking Water Supply	
Drinking Water Return	
Vacuum Cleaning	—V——V—
Compressed Air	—A—

PIPE FITTINGS

For Welded or Soldered Fittings, use joint indication shown in Diagram A	Screwed	Bell and Spigot
Joint		
Elbow - 90 deg		
Elbow - 45 deg		
Elbow - Turned Up		
Elbow - Turned Down		

PIPE FITTINGS (continued)

For Welded or Soldered Fittings, use joint indication shown in Diagram A	Screwed	Bell and Spigot
Elbow - Long Radius		
Side Outlet Elbow - Outlet Down		
Side Outlet Elbow - Outlet Up		
Base Elbow		
Double Branch Elbow		
Single Sweep Tee		
Double Sweep Tee		
Reducing Elbow		
Tee		
Tee - Outlet Up		
Tee - Outlet Down		
Side Outlet Tee Outlet Up		
Side Outlet Tee Outlet Down		
Cross		
Reducer		
Eccentric Reducer		
Lateral		
Expansion Joint Flanged		

VALVES

For Welded or Soldered Fittings, use joint indication shown in Diagram A	Screwed	Bell and Spigot
Gate Valve		
Globe Valve		
Angle Globe Valve		
Angle Gate Valve		
Check Valve		
Angle Check Valve		
Stop Cock		
Safety Valve		
Quick Opening Valve		
Float Opening Valve		
Motor Operated Gate Valve		

Figure 1-3. Symbols used for plumbing fixtures, piping, fittings, and valves. (American National Standards Institute)

The working drawings are called blueprints because in years past the common building drawing was a blue background sheet with white lines. However, in most cases today, they are white prints (white background with blue or black lines). The name blueprint is still commonly used to describe these drawings.

The blueprints for most larger buildings are divided into three sets:

1. *Structural blueprints* show the supporting structure of the building. This includes the necessary pilings, footings, foundation walls, columns, beams, floor slabs, and roof.

2. *Architectural blueprints* are the complete building plan (except for structural and mechanical details). Architectural blueprints show framing, walls, partitions, wall finish schedules, trim, cabinets, and all the measurements for walls and partitions.

3. *Mechanical blueprints* show the plumbing, heating, and electrical systems of the building. The mechanical blueprints are an outline of the architectural blueprints, but in the case of the plumbing systems, give a complete drawing of the plumbing fixture installation and piping.

On smaller buildings and residential construction, the structural and mechanical blueprints will quite often be incorporated in the architectural blueprints.

Symbols. Piping symbols are used by architects and mechanical engineers on the blueprints to represent the various plumbing fixtures and piping systems as well as the pipe fittings and valves used to construct these systems. Figure 1-3 illustrates the standard symbols used for plumbing fixtures, piping, fittings, and valves that the apprentice will encounter on blueprints.

Plan Views. On the mechanical blueprints, the apprentice will find plan views of the plumbing fixtures and piping as they are to be installed as well as schematic and isometric piping drawings.

A plan view is simply drawn as though the viewer were looking down into the rooms from above. Figure 1-4 illustrates a plan view of the plumbing fixture installation in a bathroom.

Schematics. A schematic, or diagrammatic, piping drawing is a drawing of an entire piping system without regard to either scale or the

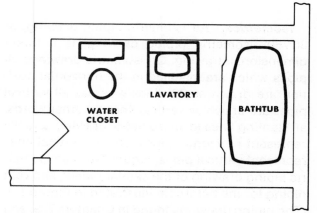

Figure 1-4. Plan view of the plumbing fixtures in a bathroom.

exact location of the items shown on the drawing. Figure 1-5 is a schematic piping drawing of the sanitary drainage and vent piping for the bathroom illustrated in Figure 1-4. (Figures 1-7, 1-8, and 1-9 are also schematic drawings.

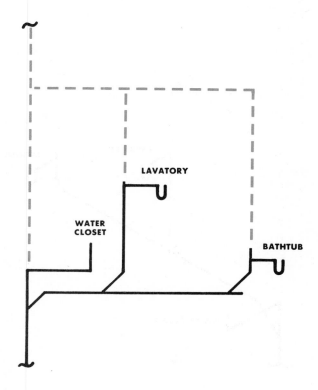

Figure 1-5. A schematic piping drawing of sanitary drainage and vent piping for the bathroom pictured in Figure 1-4.

Isometrics. An isometric piping drawing, or 30°/60° isometric piping drawing, is a three-dimensional drawing. On isometric drawings all pipes which are installed in the horizontal position are drawn as 30° lines whereas all vertical pipes are drawn as vertical lines. In other words, all slanting lines in an isometric drawing actually represent horizontal pipes, and all vertical lines represent vertical pipes. Figure 1-6 is an isometric piping drawing of the sanitary waste and vent piping for the bathroom illustrated in Figure 1-4. (The piping drawings found in Chapters 7, 9, and 12 of this text are nearly all isometric piping drawings.)

On many smaller jobs, the blueprints will not show any piping drawings. The only information provided for the plumber will be the architectural plan views, which show where the plumbing fixtures are to be installed. On these jobs, it will be necessary for the plumber to make his own schematic and isometric piping drawings. For this reason, the apprentice will have to spend a considerable amount of time making schematic and isometric drawings of the different piping systems so that he may acquire a "feeling" for the layout of the different piping systems located in buildings.

Specifications. Specifications are the written instructions from the architect and his engineers that amplify and supplement the working drawings. Depending on the size of the job, the specifications may consist of a few notations printed on the blueprints, a few sheets of paper, or even elaborate books covering hundreds of pages. The specifications give information that cannot be adequately shown on the working drawings. They also include information on legal responsibilities, insurance, quality of workmanship, and other necessary details such as brands and types of plumbing fixtures and equipment.

It is not the intent of this text to teach blueprint reading. The chapters covering the installation of piping systems in buildings will illustrate the architectural and mechanical prints or piping drawings for these buildings. To further your knowledge of blueprint reading you should obtain some basic blueprint reading texts. From these manuals you can learn about floor plans, elevations, building cross sections, and details. You might consider making your own sketches of the piping systems of buildings to appreciate how the building's piping systems fit within the building framework.

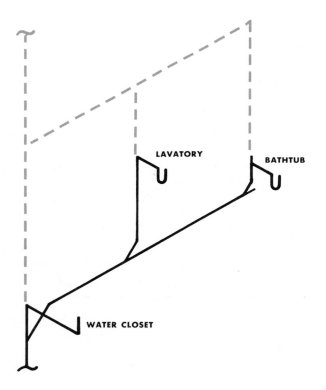

Figure 1-6. An isometric piping drawing of sanitary drainage and vent piping for the bathroom pictured in Figure 1-4.

 PLUMBING SYSTEMS OF A BUILDING

At the beginning of this chapter, it was stated that the plumber constructed the plumbing systems of a building to supply water to that building and remove the liquid and water-borne waste materials. A plumbing code's definition of plumbing systems was also given. In this text the apprentice will be concerned with the 3 most basic plumbing systems of a building:

1. The Potable Water Supply System.
2. The Sanitary Drainage and Vent Piping System.
3. The Storm Water Drainage System.

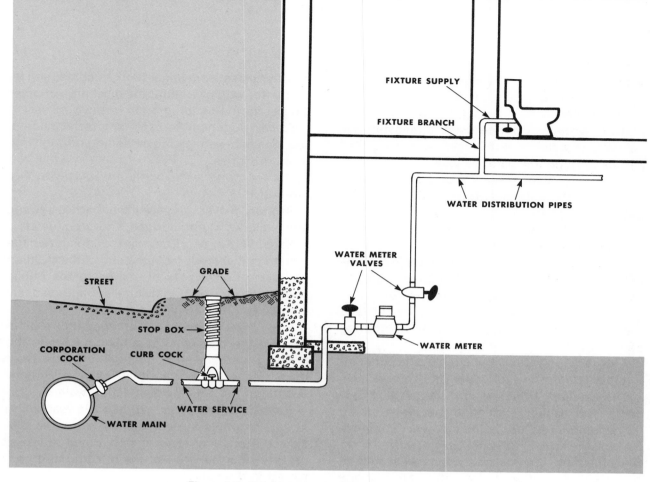

Figure 1-7. Potable water supply system of a building.

Included with the following explanations of each of these 3 systems are a schematic drawing of each system and a list of terms and definitions pertinent to each system. As an apprentice, you will need to know the meaning of these terms and definitions in order to understand the material presented in the later chapters of this text.

The Potable Water Supply System. The potable water supply system of a building is illustrated in Figure 1-7. This supply system supplies and distributes potable water to the points of use within the building.

The following terms (most of which are shown in Figure 1-7) relate to the potable water supply system.

Potable Water. Water free from impurities present in amounts sufficient to cause disease or harmful physiological effects. Its bacteriological and chemical quality shall conform to the re-

quirement of the state board of health. (Minnesota Plumbing Code.)

Potable Water Supply System. The water service pipe, the water distributing pipes, and the necessary connecting pipes, fittings, control valves, and all appurtenances within the building or outside the building within the property lines.

Water Main. The pipe that conveys potable water for public or community use from the municipal water supply source.

Corporation Cock or Corporation Stop. A valve placed on the water main to which the building water service is connected.

Water Service. The pipe from the water main or other source of water supply to the water distributing system of the building.

Curb Cock or Curb Stop. A valve placed on the water service usually near the curb line.

Stop Box or Curb Box. An adjustable cast iron box that is brought up to grade with a removable iron cover. By inserting a shutoff rod down into the stop box it is possible to turn off the curb cock.

Water Meter. A device used to measure the amount of water in cubic feet or gallons that passes through the water service.

Water Distributing Pipe. A pipe that conveys water from the water service pipe to the point of use.

Main. The principal pipe artery to which branches may be connected.

Riser. A water supply pipe that extends vertically one full story or more to convey water to fixture branches or to a group of fixtures.

Fixture Branch. A water supply pipe between the fixture supply pipe and a water distributing pipe.

Fixture Supply. A water supply pipe connecting the fixture with the fixture branch pipe.

The Sanitary Drainage and Vent Piping System. The sanitary drainage and vent piping systems are installed by the plumber to remove the waste water and water-borne wastes from the plumbing fixtures and appliances, and to provide a circulation of air within the drainage piping. A sanitary drainage and vent piping system is illustrated in Figure 1-8. The following terms relate to sanitary drainage and vent piping systems *in general:*

Sanitary Drainage Pipe. Pipes installed to remove the waste water and water-borne wastes from plumbing fixtures and convey these wastes to the sanitary sewer or other point of disposal.

Vent Pipe. A pipe installed to ventilate a building drainage system and to prevent trap siphonage and back pressure.

Sewage. Any liquid waste containing animal or vegetable matter in suspension or solution. It may include liquids containing chemicals in solution. (Minnesota Plumbing Code.)

Sewer Gas. The mixture of vapors, odors, and gases found in sewers.

Cleanout. A fitting with a removable plate or plug that is placed in plumbing drainage pipe lines to afford access to the pipes for the purpose of cleaning the interior of the pipes.

Waste Pipe. A pipe that conveys only liquid waste free from fecal material.

Soil Pipe. A pipe that conveys the discharge of water closets or similar fixtures containing fecal matter with or without the discharge of other fixtures to the building drain or building sewer.

Stack. A general term for any vertical line of soil, waste, or vent piping extending through one or more stories.

The following terms apply *specifically* to Figure 1-8:

Sanitary Sewer. A sewer that carries sewage and excludes storm, surface, and groundwater.

Building Sewer. That part of the drainage system that extends from the end of the building drain and conveys its discharge to the public sewer, private sewer, individual sewage-disposal system, or other point of disposal.

Front Main Cleanout. A plugged fitting located near the front wall of a building where the building drain leaves the building. The front main cleanout may be either inside or directly outside of the building foundation wall.

Building Drain. That part of the lowest piping of the drainage system that receives the discharge from soil, waste, and other drainage pipes inside the walls of the building and conveys it to the building sewer.

Building Drain Branch. A soil or waste pipe that extends horizontally from the building drain and receives only the discharge from fixtures on the same floor as the branch.

Stack Cleanout. A plugged fitting located at the base of all soil or waste stacks.

Waste Stack. A vertical line of piping that extends one or more floors and receives the discharge of fixtures other than water closets and urinals.

Soil Stack. A vertical line of piping that extends one or more floors and receives the discharge of water closets, urinals, and similar fixtures. It may also receive the discharge from other fixtures.

Horizontal Branch. A soil or waste pipe that extends horizontally from a stack which receives only the discharge from fixtures on the same floor as the branch.

Fixture Drain. The drain from the trap of a fixture to the junction of that drain with any other drain pipe.

Fixture Trap. A fitting or device that provides, when properly vented, a liquid seal to prevent

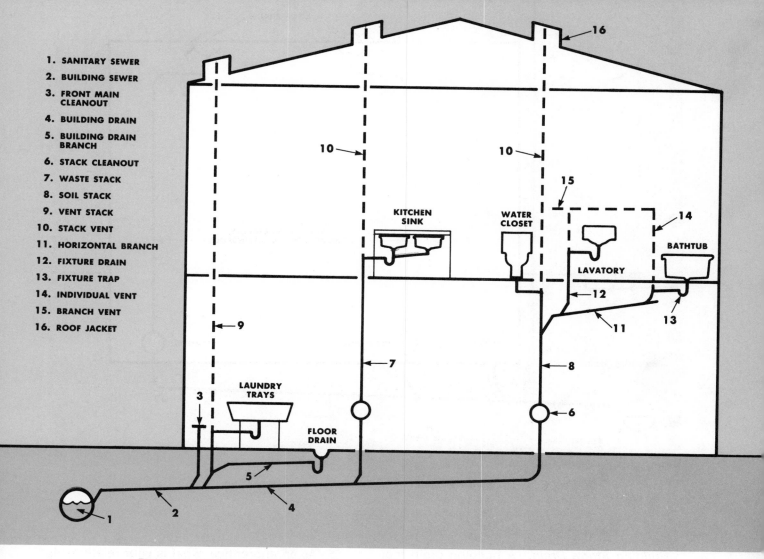

1. SANITARY SEWER
2. BUILDING SEWER
3. FRONT MAIN CLEANOUT
4. BUILDING DRAIN
5. BUILDING DRAIN BRANCH
6. STACK CLEANOUT
7. WASTE STACK
8. SOIL STACK
9. VENT STACK
10. STACK VENT
11. HORIZONTAL BRANCH
12. FIXTURE DRAIN
13. FIXTURE TRAP
14. INDIVIDUAL VENT
15. BRANCH VENT
16. ROOF JACKET

KITCHEN SINK

WATER CLOSET

LAVATORY

BATHTUB

LAUNDRY TRAYS

FLOOR DRAIN

Figure 1-8. Sanitary drainage and vent piping system of a building. (Ralph R. Lichliter)

the emission of sewer gases without materially affecting the flow of sewage or waste water through it.

Individual Vent. A pipe installed to vent an individual fixture trap. It may terminate either into a branch vent, a vent stack, a stack vent, or the open air.

Branch Vent. A vent pipe connecting two or more individual vents with either a vent stack or a stack vent.

Stack Vent. The extension of a soil or waste stack above the highest horizontal drain connected to the stack.

Vent Stack. A vertical pipe installed to provide circulation of air to and from the drainage system.

Roof Jacket or Flange. A jacket or flange installed on the roof terminals of vent stacks and stack vents to seal this opening to prevent rainwater from entering into the building around the vent pipe.

The Storm Water Drainage System. The storm water drainage system, illustrated in Figure 1-9, is the piping system used for conveying rainwater or other precipitation to the storm sewer or other place of disposal.

The following terms apply to Figure 1-9:

Storm Sewer. A sewer used for conveying groundwater, rainwater, surface water, or similar nonpollutional wastes.

Building Storm Sewer. A building sewer that conveys storm water but no sewage.

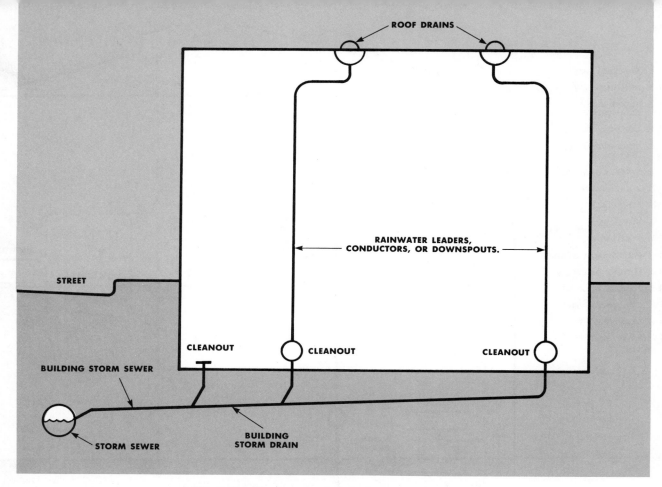

Figure 1-9. Storm water drainage system of a building.

Building Storm Drain. A building drain that conveys storm water but no sewage.

Rainwater Leader, Conductor, or Downspout. A pipe inside the building that conveys storm water from the roof to a storm drain.

Roof Drain. A drain installed to receive water collecting on the surface of a roof and to discharge it into a rainwater leader, conductor, or downspout.

REVIEW QUESTIONS

1. Define the terms *plumbing* and *plumbing system.*

2. What article in the plumbing field received the first patent? In what country?

3. In what period of United States history were the earliest patents issued on plumbing equipment?

4. Define the term *indenture.*

5. As an apprentice, what is your relationship to the JATC?

6. In what way does a plumber's preparation to enter employment differ from that of most other apprentices? What credentials must a plumber hold?

7. How much knowledge and ability must a plumber have in the field of drafting?

8. What are the specifications for a job?

9. Name the 3 plumbing systems of piping you will learn to install and maintain as a plumber.

10. Probably the most important single device in the plumbing system is the fixture trap. Explain what a fixture trap is and why it is used in the system.

11. Why must the drainage system of a building be vented?

CHAPTER 2

Job Safety

Training in accident prevention or job safety is probably one of the most important phases of the apprentice plumber's education. Because the building industry has a comparatively high accident rate, the apprentice plumber must become aware of the hazards associated with the plumbing trade as well as those of all the other trades working on the job site. The apprentice has a responsibility to his fellow workers, to his family, and to himself to prevent accidents. Accidents on the job site can cause temporary disability, permanent disability, and even death.

In addition to serious accidents, the careless worker exposes himself to many small cuts, bruises, and burns, and the pain and temporary handicap that go with them. If he is injured or disabled on the job, the apprentice will probably miss time from work, with the resulting loss in wages causing an additional hardship on his family and himself. The apprentice plumber must become safety conscious to help reduce the accidents on the job to a minimum. He must learn to think of the safety of his fellow workers and himself in every act he performs on the job.

"Safety first" is a slogan adopted on a national scale by all branches of industry. Safety laws and regulations governing employees in the construction industry (and other industries as well)

have been enacted by the federal government under the form of the national Occupational Safety and Health Act of 1971 (OSHA). The law is administered by the Department of Labor and enforced by the Occupational Safety and Health Review Commission through job site inspections by OSHA inspectors. The National Safety Council called the new law "perhaps the single most important event in the history of safety movement." This law guarantees the worker, as an employee, a place to work free from "recognized (safety or health) hazards that are causing or are likely to cause death or serious physical harm." OSHA also requires each employee to comply with occupational safety and health standards as well as all rules, regulations, and orders issued under the act that apply to his or her actions and conduct.

Education in job safety has become an important phase of the instruction of every apprentice plumber. Safety rules important to the plumbing trade (many of which are taken directly from the OSHA safety and health regulations for construction) will be outlined briefly. For your own safety on the job and the safety of coworkers, you must study, know, and practice these safety rules. Training for safety is every bit as important as learning to be a skilled craftsman.

In the normal daily performance of his work, the plumber will work with materials, hand tools, and machinery that can cause serious injury if improperly handled. If an injury should occur, seek first aid no matter how slight the injury and report it to your supervisor. Blood poisoning may result from an insignificant scratch. Blindness may result from a particle left in your eye. Every job is required by OSHA to have an easily accessible, continuously stocked first-aid kit. It is advisable that every apprentice take a first-aid course at the first opportunity; this course is a part of many apprentice plumbing training programs.

General Safety on the Job

Safety is a combination of knowledge and awareness: *knowledge* and *skill* in the use and care of your tools and *awareness* on the job of the particular hazards and safety procedures involved. Tool skills may be learned; awareness, however, depends on attitude. An attitude of care and concern while on the job will help prevent injuries not only to yourself, but also to your fellow workers. Always be alert while on the job and follow recommended safety procedures. *If in doubt, ask questions.*

1. Wrestling, throwing objects, and other forms of horseplay should be avoided. Serious injuries may be the result.

2. Provide a place for everything, and keep everything in its place.

3. Plumbing fittings and materials should be stored where they are accessible but clear of stairways and passageways. Small fittings should be stored in bins by sizes and type. Large fittings will probably have to be stacked in piles. The fittings should not be scattered around the work area or under the work bench where someone could trip or fall over them.

4. Pipe should be stockpiled by size and type in an accessible location. It should be stacked in such a manner that it cannot roll on someone. When removing stockpiled pipe from the pile, do it in a way that will not disturb the entire pile and cause it to shift or move quickly.

5. Keep all work areas clear of pipe scraps, tools, and other materials. Things left scattered about the floor may cause stumbling or tripping

National Safety Council

and result in serious injury from a fall. Remove waste material from the work area and dispose of it properly at regular intervals.

6. Never place articles on windowsills, stepladders, or other high places where they may fall and cause injuries. Check scaffolds and ladders for articles before they are moved.

7. Oil, water, and other slippery substances left on the floor may cause a serious accident. Clean or wipe them up.

8. Keep your arms and body as nearly straight as possible when lifting heavy objects. Place your feet close to the object. Bend your knees, squat, and keep your back as straight as possible. Lift with the legs—not with the back. If the object is too heavy or too bulky, get help.

9. Use the proper tools when moving, raising, or installing heavy pipe or equipment. Do not overload the tools. Plan the moving and raising operations out in advance. Seek advice if you are not sure of the proper method.

10. Use the proper rigging methods and the proper rope, sling, or choker when lifting pipe or other materials and equipment. Inspect all rigging equipment before using it to see that it is in good condition and free from flaws. Do not use any defective rigging equipment. Do not overload the rigging equipment.

11. When raising equipment or pipe into the air, avoid standing below the load. Never stand below a load being raised by a crane.

12. Mark or tag all defective or faulty tools and equipment (or render them inoperable) and have them removed from the job site to prevent someone from using them.

13. Work only in adequately lighted areas.

14. Work with your employer to develop a proper fire protection program and have the proper fire protection equipment on hand. Do not burn yourself out of a job.

15. Notify your immediate supervisor of any known violations of safety rules or of conditions you think may be dangerous.

16. Immediately report all accidents, no matter how slight, to your superior, and report for first aid treatment.

17. Don't take chances.

Clothing and Personal Protective Equipment

The plumber must wear the proper clothing and personal protective equipment for the job he is performing.

1. Wear comfortable overalls or coveralls that fit properly and are in good repair. The pant legs should be tailored to eliminate cuffs, as cuffs tend to catch dirt, hot sparks, and protruding objects.

2. A long-sleeved shirt should be worn for arm protection from burns, cuts, and scratches. The shirt cuffs should either be buttoned or rolled past the elbow to prevent their catching on tools, machinery, or protruding objects.

3. Good, sturdy, leather shoes with hard toe-caps should be worn to protect your feet from falling objects and from dampness. Never wear soft-soled shoes; you might step on a nail that would puncture the soft sole and enter your foot.

4. Head protection (hard hat) is required headwear on virtually all construction jobs. The wearing of an approved plastic helmet (hard hat) is required whenever a plumber is working in an area where there is danger from falling or flying objects or from electrical shock and burns. On the average construction job, there is the constant danger of some workman dropping a tool or piece of material from a scaffold from one of the floors above. A hard hat should always

be worn whenever a crane is being used, or whenever you are working in a trench.

5. Hand protection in the form of gloves, of the proper weight for the job being performed, is very essential for a plumber. In addition to keeping your hands relatively clean, gloves will prevent many minor burns, cuts, and blisters on your hands. Always wear gloves when working with hot lead or when soldering or welding.

6. The proper eye and face protection is necessary whenever the work being performed presents a potential hazard to either your eyes or face. Goggles should be worn when chipping, cutting, or drilling in rock, concrete, or brick. A face mask should be worn when grinding. A plumber who normally wears corrective eyeglasses should wear spectacles with protective (hardened or plastic) lenses (OSHA lists specific types of required eye and face protection for many hazardous tasks.)

7. Ear plugs or ear protectors should be worn when working in extremely noisy areas. Cotton is *not* an acceptable form of ear protection.

8. If it is necessary to work in an area where there are hazardous dusts, gases, or vapors the proper type of respirator, gas mask, or breathing apparatus must be worn. Seek advice and get the proper training before using any of these

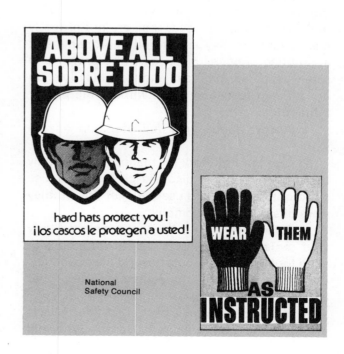

ABOVE ALL
SOBRE TODO

hard hats protect you!
¡los cascos le protegen a usted!

National
Safety Council

WEAR THEM
AS INSTRUCTED

pieces of protective equipment. Do not work alone when the working conditions require the use of a gas mask or breathing device.

Hand Tool Safety

Without the proper hand tools, it would be impossible for a plumber to install the piping systems in a building. Tools are great laborsaving devices. They are also a source of injury if they are improperly used, they are not properly maintained, or they are stored improperly. The apprentice plumber must learn the proper and safe way to use, maintain, and store his tools. Do not attempt to use any tool without the proper training!

1. Always focus your full attention on the work.

2. Use the right tool for the job. Use not only the proper tool, but also the correct size. Use good quality tools and use them for the job they were designed to accomplish.

3. Learn how to use the tool properly. Study your tools—learn the safe way of working with each tool. Never use tools beyond their capacity. Don't be afraid to ask questions on the proper and safe use of a tool.

4. Keep tools in their best condition. Always inspect a tool before using it. Do not use a tool that is in poor or faulty condition. Use only safe tools. Chisels and caulking irons should be sharp; tool handles should be free of cracks and splinters and fastened securely to the working part.

5. Batter-heads of metal tools must be kept ground smooth and square to avoid mushrooming. When the head of a mushroomed tool is struck, bits of metal often break loose, causing serious injuries.

6. Do not strike hardened metal or tools with a hard-faced hammer (such as a carpenter's claw hammer). Chips of metal may break loose and cause injury.

7. Pipe wrenches should be in good condition and have good jaws. Use the right size pipe wrench or chain tongs for the size of pipe being worked on. Always place the pipe wrench jaws in the direction of pulling force (Figure 2-1). Do not use a pipe wrench as a hammer.

8. Always use a handle on a file. Otherwise, the tang may cut into your hand.

9. Soldering and welding torches should be inspected daily to see that they are in good operating condition and do not leak.

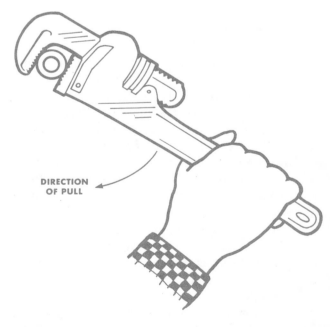

Figure 2-1. Pipe wrench jaws should face the direction of pulling force.

10. Lead furnaces should be placed on a flat, stable surface so that they will not tip and spill the pot of molten lead.

11. When working with lead, preheat any frozen or wet lead before adding it to the molten pot to prevent an explosion. Keep the lead joint free from moisture (wipe dry).

12. Work a safe distance from your fellow workmen to prevent hitting them if you should slip or miss with a hammer blow, or if your wrench should suddenly slip from the pipe.

Electricity

Electricity is the power supply for much of the mechanical equipment used on construction jobs today. Temporary lighting wires are found strung throughout the jobs during construction. Electricity is very dangerous. There is nothing to show that electrical power is flowing through the wires except when a light is burning or a motor is turning. The electricity is there whether it is used or not. Great care must be taken not to touch any bare wire, or to create any condition where the current can flow through your body to a ground. Wet scaffolds, metal pipe, and wet ground and concrete are dangerous, as the wetness or mass of metal improves the grounding condition, permitting a greater flow of current to pass through the body. This causes severe shocks, burns, and possible death.

Never attempt to touch a person who has live current flowing through him or you too may be killed. Try to remove the wire or equipment creating the problem by pushing or lifting it off using a dry piece of wood. Shut off the electricity immediately, if possible; this is the safest method.

1. Do not attempt to use any electrically powered tools or equipment without knowing their principles of operation, methods of use, and general and special safety precautions.

2. Make sure all power tools are either of the approved double insulated type or that they have nondefective, three-wire grounded type cords. Properly made power tools are approved by Underwriters' Laboratories (UL).

3. Use only adequately sized three-wire extension cords to carry the required load. Undersized wire may cause a fire due to overheating or may

Figure 2-2. A three-wire cord and a properly grounded outlet (receptacle).

cause the tool motor to burn out when operating under a heavy load. Inspect all extension cords frequently to see that they are in good condition. Figure 2-2 illustrates an approved three-wire extension cord and a properly grounded outlet (receptacle).

4. Do not leave extension cords lying where they will be pinched in doors or run over and damaged by equipment or scaffold wheels. Do not allow extension cords to kink. Keep them out of water. Extension cords must not be suspended from wire or hung with nails and staples.

5. Extension cords and tool power cords are not to be used to raise and lower the tools from ladders or scaffolding.

6. Never connect electrical equipment to a power source unless the switch is in the OFF position.

7. Report defective power tools and remove them from the job site.

Safe Work in Trenches

A large portion of the pipe that a plumber installs is laid below ground level in trenches. Since he will be working in trench excavations below the ground level, the apprentice plumber

**INSTABILITY
OF THE BOTTOM
OF AN EXCAVATION**

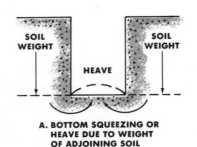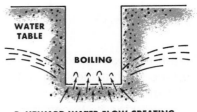

A. BOTTOM SQUEEZING OR
HEAVE DUE TO WEIGHT
OF ADJOINING SOIL

B. UPWARD WATER FLOW CREATING
QUICK CONDITION OF BOTTOM

**METHODS OF FAILURE
IN AN UNBRACED
CUT OR TRENCH**

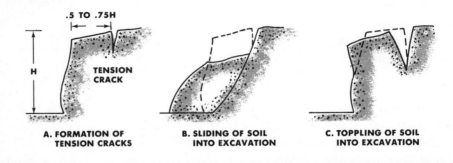

A. FORMATION OF
TENSION CRACKS

B. SLIDING OF SOIL
INTO EXCAVATION

C. TOPPLING OF SOIL
INTO EXCAVATION

**STRESSES AND
DEFORMATION IN AN
OPEN CUT OR TRENCH**

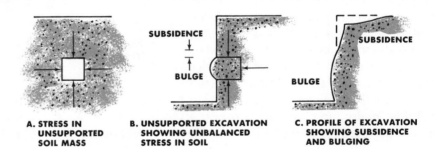

A. STRESS IN
UNSUPPORTED
SOIL MASS

B. UNSUPPORTED EXCAVATION
SHOWING UNBALANCED
STRESS IN SOIL

C. PROFILE OF EXCAVATION
SHOWING SUBSIDENCE
AND BULGING

Figure 2-3. Some common causes of trench failure that proper shoring helps prevent. (National Safety Council)

must recognize a safe trench for his own safety. A trench cave-in can be a plumber's grave!

Trenches can be very dangerous places to work. A trench may fail in a variety of ways: the side walls may topple in, the bottom of the trench walls may slide in, the bottom of the trench may heave up because of pressure from the weight of the side walls, or surface water may boil up into the bottom of the ditch, filling it with mud and water. Figure 2-3 shows the conditions that may lead to a trench failure. Proper sheeting and bracing (shoring) will prevent trench failures.

In addition to a failure of the trench itself, there are dangers from excavated material falling back into the trench. The worker may fall when entering or leaving the trench and may encounter toxic fumes and/or gases in the trench.

To understand the terms associated with trench excavations, the apprentice should study Figure 2-4, a glossary of trenching definitions.* These terms will be easier to understand if the apprentice studies Figures 2-5 (angle of repose), 2-6, and 2-7 (illustrations of trench shoring). Table 2-1 lists the minimum sizes of shoring materials used with the trench shoring methods

*Definitions taken from the National Safety Council pamphlet "Trench Excavations."

DEFINITIONS

Angle of Repose—The greatest angle above the horizontal plane at which a material will lay without sliding.

Cleats (Scabs)—Pieces of wood that solidly connect the crosspieces to the horizontal members (wales).

Excavation—Any man-made cavity or depression in the earth's surface, including its sides, walls or faces, formed by earth removal and producing unsupported earth conditions by reasons of the excavation. If installed forms or similar structures reduce the depth-to-width relationship, an excavation may become a trench. (Put another way, a trench is always an excavation, but an excavation is not necessarily a trench.)

Sheeting—Material (wood, steel, or concrete, which may form a continuous line) placed in close contact and providing a wall to resist the lateral pressure of water, adjacent earth, or other materials.

Spoil—The material resulting from an excavation.

Struts (Braces) — The horizontal members of the shoring system whose ends bear against the uprights or stringers.

Tight Sheeting—Sheeting that is butted close together to form a continuous solid wall to resist the lateral pressure of earth, water, or other material.

Trench — A narrow excavation made below the surface of the ground. In general, the depth is greater than the width, but the width of a trench is not greater than 15 feet.

Trench Shield—A shoring system composed of steel plate and bracing, welded or bolted together, which support the walls of a trench from the ground level to the trench bottom, and which can be moved along as the work progresses.

Wales (Stringers)—The horizontal members of a shoring system whose sides bear against the uprights or earth.

Figure 2-4. Definitions that apply to trench excavations. (National Safety Council)

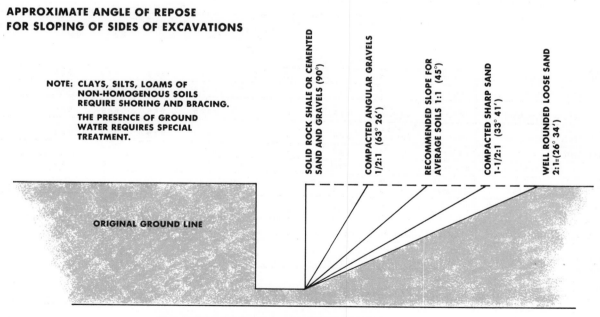

APPROXIMATE ANGLE OF REPOSE FOR SLOPING OF SIDES OF EXCAVATIONS

NOTE: CLAYS, SILTS, LOAMS OF NON-HOMOGENOUS SOILS REQUIRE SHORING AND BRACING.

THE PRESENCE OF GROUND WATER REQUIRES SPECIAL TREATMENT.

SOLID ROCK SHALE OR CEMENTED SAND AND GRAVELS (90°)

COMPACTED ANGULAR GRAVELS 1/2:1 (63° 26')

RECOMMENDED SLOPE FOR AVERAGE SOILS 1:1 (45°)

COMPACTED SHARP SAND 1-1/2:1 (33° 41')

WELL ROUNDED LOOSE SAND 2:1 (26° 34')

ORIGINAL GROUND LINE

Figure 2-5. Angle of repose. (National Safety Council)

illustrated in Figures 2-6 and 2-7. It is not necessary to memorize either Table 2-1 or the trench illustrations shown here, but the apprentice should know they exist for his reference when working in trenches.

The following are some basic rules for safe work in trenches:

1. Locate all underground utilities before starting to dig.

2. Remove all trees, boulders, or other surface objects that are going to create a hazard to the trench.

3. Do not pile the excavated material on sidewalks, walkways, or within 2 feet of the trench excavation.

4. Do not stand below the load of dirt being removed from the excavation by the digging equipment or below any material being lowered into the trench.

5. Trenches *more* than 5 feet deep in hard or compact soil must be shored or sloped to the angle of repose of the soil (Figure 2-5).

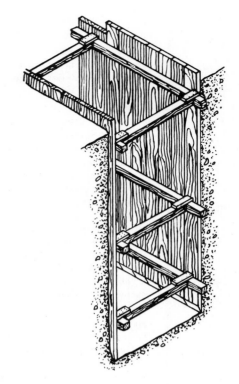

Figure 2-6. An example of shoring at full-depth in soft, sandy soil or in filled ground. (National Safety Council)

TABLE 2-1. MINIMUM SIZES OF TRENCH SHORING LUMBER.

Depth of trench	Kind or condition of earth	Size and spacing of members										
		Uprights		Stringers		Cross braces [1]					Maximum spacing	
						Width of trench						
		Minimum dimension	Maximum spacing	Minimum dimension	Maximum spacing	Up to 3 feet	3 to 6 feet	6 to 9 feet	9 to 12 feet	12 to 15 feet	Vertical	Horizontal
Feet		Inches	Feet	Inches	Feet	Inches	Inches	Inches	Inches	Inches	Feet	Feet
5 to 10	Hard, compact	3 x 4 or 2 x 6	6			2 x 6	4 x 4	4 x 6	6 x 6	6 x 8	4	6
	Likely to crack	3 x 4 or 2 x 6	3	4 x 6	4	2 x 6	4 x 4	4 x 6	6 x 6	6 x 8	4	6
	Soft, sandy, or filled	3 x 4 or 2 x 6	Close sheeting	4 x 6	4	4 x 4	4 x 6	6 x 6	6 x 8	8 x 8	4	6
	Hydrostatic pressure	3 x 4 or 2 x 6	Close sheeting	6 x 8	4	4 x 4	4 x 6	6 x 6	6 x 8	8 x 8	4	6
10 to 15	Hard	3 x 4 or 2 x 6	4	4 x 6	4	4 x 4	4 x 6	6 x 6	6 x 8	8 x 8	4	6
	Likely to crack	3 x 4 or 2 x 6	2	4 x 6	4	4 x 4	4 x 6	6 x 6	6 x 8	8 x 8		6
	Soft, sandy, or filed	3 x 4 or 2 x 6	Close sheeting	4 x 6	4	4 x 6	6 x 6	6 x 8	8 x 8	8 x 10	4	6
	Hydrostatic pressure	3 x 6	Close sheeting	8 x 10	4	4 x 6	6 x 6	6 x 8	8 x 8	8 x 10	4	6
15 to 20	All kinds or conditions	3 x 6	Close sheeting	4 x 12	4	4 x 12	6 x 8	8 x 8	8 x 10	10 x 10	4	6
Over 20	All kinds or conditions	3 x 6	Close sheeting	6 x 8	4	4 x 12	8 x 8	8 x 10	10 x 10	10 x 12	4	6

[1] Trench jacks may be used in lieu of, or in combination with, cross braces.
Shoring is not required in solid rock, hard shale, or hard slag.
Where desirable, steel sheet piling and bracing of equal strength may be substituted for wood.

National Safety Council

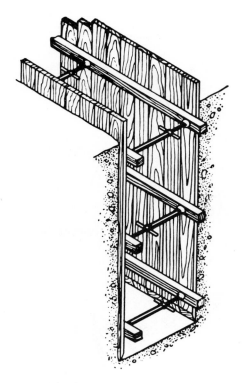

Figure 2-7. A technique of using screw jacks with an example of complete sheet piling. (National Safety Council)

6. Trenches *less* than 5 feet deep in ground where hazardous conditions exist (loose sand, wet soil, old fill, etc.) must also be shored or sloped to the angle of repose of the soil (Figure 2-5).

7. Portable trench boxes (Figure 2-8) or sliding trench shields may be used instead of shoring or sloping the trench if they provide equal or better protection compared to shoring.

8. All trench shoring must be done with good quality material free from defects.

9. All trench shoring and portable trench boxes should be designed by a qualified person.

10. When working in trenches more than 4 feet deep, a ladder must be provided for entering or leaving the trench at every 25 feet of trench.

11. All slopes, shoring, and grading should be inspected daily and after every rainstorm.

12. Never work alone in a trench and when working with others, work at a safe distance from the next worker to prevent injuring one another.

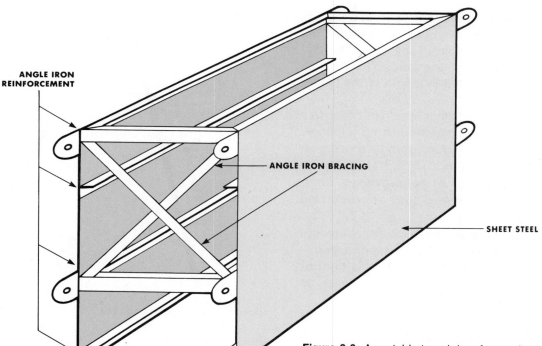

ANGLE IRON REINFORCEMENT

ANGLE IRON BRACING

SHEET STEEL

Figure 2-8. A portable trench box for use in trench excavations instead of shoring. The box is constructed of steel plate with angle iron bracing. The eyes on the ends of the box are for attaching chains so that the excavating machinery may move the box down the trench.

Ladder Safety

The ladder is a very useful tool for a plumber to work from while doing work that cannot be reached from the floor or ground level. At other times the plumber will need a ladder to get onto a roof, down into a ditch, or to get from one level of a building to another when no other means is available. However, the careless use of a ladder can result in a serious injury from falling.

Some basic rules for safe ladder use are as follows:

1. Inspect ladders for cracked, broken, or missing rungs and split side rails. Remove ladders with these defects from the job site because as long as they are on the job there is a chance they will be used.

2. Never paint a wooden ladder because paint can hide cracks and splits in the rungs and side rails.

3. All portable ladders should have safety feet (Figure 2-9, top).

4. When using a ladder, be sure that the bottom rests on solid footing so that it cannot slip.

5. Do not slant the ladder at such an acute angle that the weight of your body will pull the top of the ladder away from the wall. The distance from the base of the ladder to the wall should equal one-fourth the length of the ladder (Figure 2-9, bottom).

6. Keep the ladder steps and your feet free from mud, grease, and oil when using a ladder.

7. Face the ladder when going up or down and grip with both hands.

8. Keep ladders out of passageways, doorways, and driveways unless they are protected by barricades.

9. Tie or block portable ladders to prevent them from moving while they are being used.

10. Ladders used for going from one floor or landing to another must extend at least 36 inches above the next landing (Figure 2-10).

11. Use a ladder of the proper length for the job—neither too long or too short. *Never stand on the top step of a ladder.*

12. Place the ladder in the proper position for the work being performed. If it is too close or too far from the work, move it. Most falls from

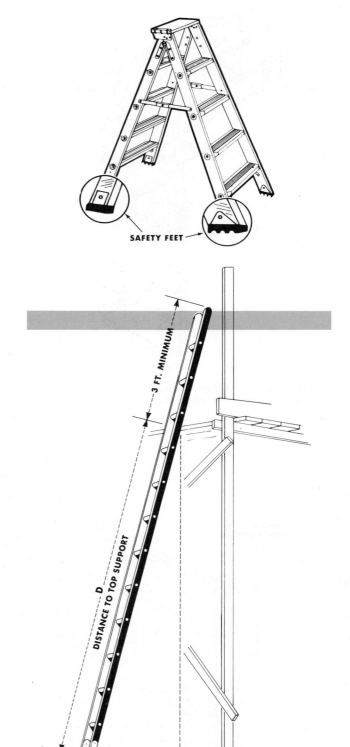

Figure 2-9. Top: Stepladder showing safety feet. Bottom: Wooden ladder showing safety feet and safe ladder angle. (National Safety Council)

LADDER INSPECTION CHECKLIST

General
Item To Be Checked

☐ Loose steps or rungs (considered loose if they can be moved at all with the hand)
☐ Loose nails, screws, bolts, or other metal parts
☐ Cracked, split, or broken uprights, braces, steps, or rungs
☐ Slivers on uprights, rungs, or steps
☐ Damaged or worn nonslip bases

Stepladders

☐ Wobbly (from side strain)
☐ Loose or bent hinge spreaders
☐ Stop on hinge spreaders broken
☐ Broken, split, or worn steps
☐ Loose hinges

Extension Ladders

☐ Loose, broken, or missing extension locks
☐ Defective locks that do not seat properly when the ladder is extended
☐ Deterioration of rope, from exposure to acid or other destructive agents

Trolley Ladders

☐ Worn or missing tires
☐ Wheels that bind
☐ Floor wheel brackets broken or loose
☐ Floor wheels and brackets missing
☐ Ladders binding in guides
☐ Ladder and rail stops broken, loose, or missing
☐ Rail supports broken or section of rail missing
☐ Trolley wheels out of adjustment

Trestle Ladders

☐ Loose hinges
☐ Wobbly
☐ Loose or bent hinge spreaders
☐ Stop on hinge spreader broken
☐ Center section guide for extension out of alignment
☐ Defective locks for extension

Sectional Ladders

☐ Worn or loose metal parts
☐ Wobbly

Fixed Ladders

☐ Loose, worn, or damaged rungs or side rails
☐ Damaged or corroded parts of cage
☐ Corroded bolts and rivet heads on inside of metal stacks
☐ Damaged or corroded handrails or brackets on platforms
☐ Weakened or damaged rungs on brick or concrete slabs
☐ Base of ladder obstructed

Fire Ladders

☐ Markings illegible
☐ Improperly stored
☐ Storage obstructed

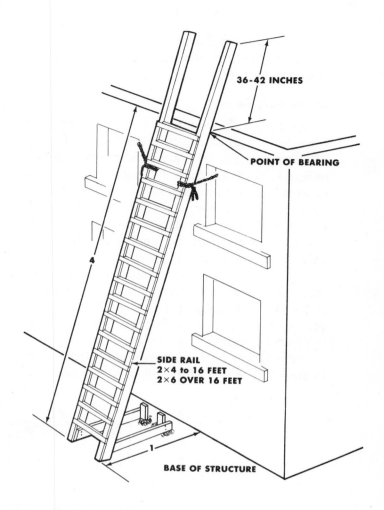

36-42 INCHES

POINT OF BEARING

SIDE RAIL
2×4 to 16 FEET
2×6 OVER 16 FEET

4

1

BASE OF STRUCTURE

Figure 2-10. Typical cleat ladder construction and formula for setting ladders against buildings or scaffolds. A one-to-four ratio (1:4) is used. (National Safety Council)

ladders are caused by loss of balance. Avoid overreaching with your arms either above yourself or too far to one side.

13. Be sure that stepladders are fully opened and locked before climbing them.

14. If a stepladder has a pail or tool shelf, use it as such—it is not a step. Remove tools and materials from this shelf before attempting to move the ladder. They could fall on you!

15. Never allow more than one person to work on a ladder unless the ladder was designed to carry two people.

16. Do not use ladders in the horizontal position as planks, runways, or scaffolding.

17. If it is necessary to use a job-built ladder, it should be built of good quality lumber and as shown in Figure 2-10.

Scaffolding Safety

Plumbers quite frequently find it necessary to work from scaffolding to erect their piping in ceiling areas. Scaffolds are handier than ladders because scaffolds can be safely erected to heights that are impractical to reach from ladders. There is a platform at the working level where tools and fittings can be kept close at hand, and more than one person at a time can safely work on a scaffold platform.

The three main hazards while working on or under scaffolds are falling, dropping tools or materials, and faulty scaffolding. Always watch your step, keep your balance, and handle your tools and equipment carefully.

The majority of a plumber's work from scaffolding will be done from tubular steel scaffold-ing; this scaffolding often has wheels for easy movement along the length of the pipe line(s) being installed. Steel scaffolds can be built to any height and are adaptable to all types of job conditions. These scaffolds can be purchased or rented and the supplying companies give technical services on needs and types best suited for each job or condition.

The type usually used for low heights is made up of prefabricated frames and cross braces. (See Figure 2-11.)

A factor of safety of not less then four times the load is required. Care must be used in the erection of the scaffold so that it rests on a firm base and is kept plumb and level as it is assembled. It must be inspected daily. Care must be taken to keep the frames from injury or from rusting so that they do not lose part of their design strength. Safety rules for metal scaffolding, recommended by the Steel Scaffolding and Shoring Institute, are shown on page 25.

Figure 2-11. Prefabricated metal frames and diagonal braces are assembled quickly to provide safe scaffolds.

SCAFFOLDING SAFETY RULES

as Recommended by

SCAFFOLDING AND SHORING INSTITUTE

(SEE SEPARATE SHORING SAFETY RULES)

Following are some common sense rules designed to promote safety in the use of steel scaffolding. These rules are illustrative and suggestive only, and are intended to deal only with some of the many practices and conditions encountered in the use of scaffolding. The rules do not purport to be all-inclusive or to supplant or replace other additional safety and precautionary measures to cover usual or unusual conditions. They are not intended to conflict with, or supersede, any state, local, or federal statute or regulation; reference to such specific provisions should be made by the user. (See Rule II.)

   I. **POST THESE SCAFFOLDING SAFETY RULES** in a conspicuous place and be sure that all persons who erect, dismantle or use scaffolding are aware of them.

  II. **FOLLOW ALL STATE, LOCAL AND FEDERAL CODES, ORDINANCES AND REGULATIONS** pertaining to scaffolding.

 III. **INSPECT ALL EQUIPMENT BEFORE USING**—Never use any equipment that is damaged or deteriorated in any way.

 IV. **KEEP ALL EQUIPMENT IN GOOD REPAIR.** Avoid using rusted equipment—the strength of rusted equipment is not known.

  V. **INSPECT ERECTED SCAFFOLDS REGULARLY** to be sure that they are maintained in safe condition.

 VI. **CONSULT YOUR SCAFFOLDING SUPPLIER WHEN IN DOUBT**—scaffolding is his business, **NEVER TAKE CHANCES.**

A. **PROVIDE ADEQUATE SILLS** for scaffold posts and use base plates.

B. **USE ADJUSTING SCREWS** instead of blocking to adjust to uneven grade conditions.

C. **PLUMB AND LEVEL ALL SCAFFOLDS** as the erection proceeds. Do not force braces to fit—level the scaffold until proper fit can be made easily.

D. **FASTEN ALL BRACES SECURELY.**

E. **DO NOT CLIMB CROSS BRACES.** An access (climbing) ladder, access steps, frame designed to be climbed or equivalent safe access to the scaffold shall be used.

F. **ON WALL SCAFFOLDS PLACE AND MAINTAIN ANCHORS** securely between structure and scaffold at least every 30' of length and 25' of height.

G. **WHEN SCAFFOLDS ARE TO BE PARTIALLY OR FULLY ENCLOSED,** specific precautions must be taken to assure frequency and adequacy of ties attaching the scaffolding to the building due to increased load conditions resulting from effects of wind and weather. The scaffolding components to which the ties are attached must also be checked for additional loads.

H. **FREE STANDING SCAFFOLD TOWERS MUST BE RESTRAINED FROM TIPPING** by guying or other means.

I. **EQUIP ALL PLANKED OR STAGED AREAS** with proper guardrails, midrails and toeboards along all open sides and ends of scaffold platforms.

J. **POWER LINES NEAR SCAFFOLDS** are dangerous—use caution and consult the power service company for advice.

K. **DO NOT USE** ladders or makeshift devices on top of scaffolds to increase the height.

L. **DO NOT OVERLOAD SCAFFOLDS.**

M. **PLANKING:**
1. Use only lumber that is properly inspected and graded as scaffold plank.
2. Planking shall have at least 12" of overlap and extend 6" beyond center of support, or be cleated at both ends to prevent sliding off supports.
3. Fabricated scaffold planks and platforms unless cleated or restrained by hooks shall extend over their end supports not less than 6 inches nor more than 12 inches.
4. Secure plank to scaffold when necessary.

N. **FOR ROLLING SCAFFOLD THE FOLLOWING ADDITIONAL RULES APPLY:**
1. **DO NOT RIDE ROLLING SCAFFOLDS.**
2. **SECURE OR REMOVE ALL MATERIAL AND EQUIPMENT** from platform before moving scaffold.
3. **CASTER BRAKES MUST BE APPLIED** at all times when scaffolds are not being moved.
4. **CASTERS WITH PLAIN STEMS** shall be attached to the panel or adjustment screw by pins or other suitable means.
5. **DO NOT ATTEMPT TO MOVE A ROLLING SCAFFOLD WITHOUT SUFFICIENT HELP**—watch out for holes in floor and overhead obstructions.
6. **DO NOT EXTEND ADJUSTING SCREWS ON ROLLING SCAFFOLDS MORE THAN 12".**
7. **USE HORIZONTAL DIAGONAL BRACING** near the bottom and at 20' intervals measured from the rolling surface.
8. **DO NOT USE BRACKETS ON ROLLING SCAFFOLDS** without consideration of overturning effect.
9. **THE WORKING PLATFORM HEIGHT OF A ROLLING SCAFFOLD** must not exceed four times the smallest base dimension unless guyed or otherwise stabilized.

O. For **"PUTLOGS"** and **"TRUSSES"** the following additional rules apply.
1. **DO NOT CANTILEVER OR EXTEND PUTLOGS/TRUSSES** as side brackets without thorough consideration for loads to be applied.
2. **PUTLOGS/TRUSSES SHOULD EXTEND AT LEAST 6"** beyond point of support.
3. **PLACE PROPER BRACING BETWEEN PUTLOGS/TRUSSES** when the span of putlog/truss is more than 12'.

P. **ALL BRACKETS** shall be seated correctly with side brackets parallel to the frames and end brackets at 90 degrees to the frames. Brackets shall not be bent or twisted from normal position. Brackets (except mobile brackets designed to carry materials) are to be used as work platforms only and shall not be used for storage of material or equipment.

Q. **ALL SCAFFOLDING ACCESSORIES** shall be used and installed in accordance with the manufacturers recommended procedure. Accessories shall not be altered in the field. Scaffolds, frames and their components, manufactured by different companies shall not be intermixed.

Rolling steel scaffolds should be equipped with large, strong wheels provided with locking devices. Never move a rolling scaffold while workers are on the scaffold. Always clean the floor ahead of the move so that the wheels will not be blocked by an obstruction that might cause the scaffold to tip over. All planks on rolling scaffolds should be securely fastened so they cannot slide off while the scaffolds are moved.

Oxyacetylene Welding and Cutting Safety

The oxyacetylene torch is found on almost all large construction jobs and is a very useful tool. However, if care is not taken in storage and movement of the oxygen and acetylene cylinders, they are potential fire and explosion hazards. The careless use of the welding and cutting attachments could cause fires and burns.

1. Store cylinders in an upright position in an area away from extreme heat, fire hazards, and traffic. Chain or rope the cylinders to a column or wall to keep them from tipping. The acetylene cylinder must always be in an upright position except when it is being moved. (Figure 2-12 shows how to move the cylinder.)

2. The cylinder cap should always be on a tank when it is being stored (whether empty or

Figure 2-12. Correct way to move a cylinder.

full) or being moved unless it is on the welding cart.

3. Never move a cylinder by dragging, sliding, or rolling it on its side. Avoid striking it against any object that might create a spark. There may be just enough gas escaping to cause an explosion. To move a cylinder, roll it on its bottom edge. (See Figure 2-12.)

4. When in use, the cylinders should be on a suitable welding cart or chained or tied off in such a way that they cannot fall. (Figure 2-13.)

5. Inspect all hoses, regulators, tips, and tanks daily for obvious defects. Do not attempt to use faulty equipment. Get it fixed. No attempt should ever be made to repair cylinder valves. If the valves do not function properly, or if they leak, the supplier should be notified.

6. Blow the cylinder valve out (open the valve quickly and then close it) to remove any dirt before attaching the regulators.

7. Release the regulator adjusting screws before opening the cylinder valves.

8. Open the valves slowly after the regulators are attached and stand to the side of the regulator.

9. Do not use oil on the regulators or fittings. Oxygen plus oil equals an explosion!

10. Purge the oxygen and acetylene hoses separately before lighting the torch.

11. Never use acetylene gas in the free state at more than 15 pounds pressure. An explosion could result!

12. Oxygen should never be used as a substitute for compressed air to operate pneumatic tools, blow out pipe lines, or dust off clothing. A serious accident may result.

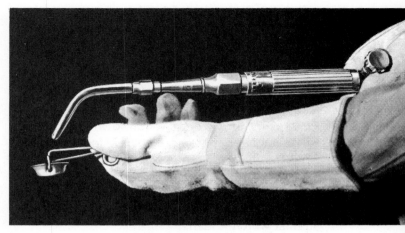

Figure 2-14. Correct method of lighting a welding torch.

13. Light the torch with *only* the acetylene gas turned on.

14. Light the torch *only* with a friction lighter (Figure 2-14). Do not use matches on hot surfaces to light the torch.

15. The cylinders should be turned on only when they are in use.

16. Always wear the proper gloves and welding goggles when doing any cutting or welding.

17. Keep the welding area clear of any debris that could catch on fire.

18. Always have a fire extinguisher at hand when welding or cutting.

Figure 2-13. Welding cylinders fastened to a cart.

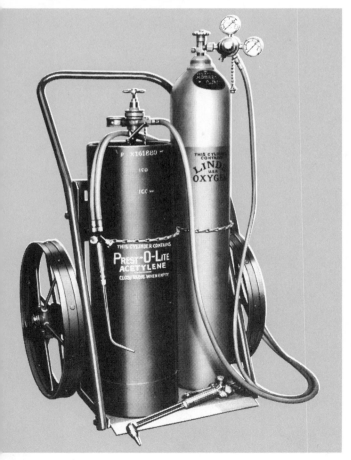

○ REVIEW QUESTIONS

1. What is your responsibility as an apprentice for safety on the job site?

2. What does "safety conscious" mean?

3. Give the basic purpose or purposes of OSHA (Occupational Safety and Health Act of 1971).

4. Write in your own words five important general safety regulations governing construction work.

5. Name and give the reason for using four items of personal protective equipment.

6. Describe or draw a mushroom head on a chisel. Describe or draw a safe head on a chisel.

7. How do you safely add lead to a lead furnace on a frosty, snowy, or wet day?

8. What is meant by a "UL approved" power tool?

9. Explain these terms connected with trench-work in plumbing: angle of repose; expansion jack; spoil; tight sheeting; trench shield.

10. What provision is required when entering or leaving trenches more than 4 feet deep? At what intervals (distance apart) must this provision be made?

11. Give a rule of thumb for how far the foot of a ladder should be from the base of the wall against which it leans.

12. Ladders used to climb from one floor to another should extend how far above the next floor?

13. Name the chief hazards of working on scaffolds.

14. Give the following safety rules for handling oxygen and acetylene cylinders:
   a. how to store them;
   b. how to move them;
   c. the steps in attaching a regulator;
   d. opening the cylinder valve for use after the regulator is attached;
   e. how to light the torch by the approved method.

CHAPTER **3**

Plumbing Materials

The three plumbing systems of a building described in Chapter 1 (the Potable Water Supply System, the Sanitary Drainage and Vent Piping System, and the Storm Water Drainage System) are constructed by plumbers using pipe, fittings, valves, and meters. It is the purpose of this chapter to present to the apprentice some of the more common pipe and fitting materials, as well as plumbing valves and meters, and to describe the uses of these various materials.

The materials that this text will consider for plumbing pipes and fittings are classified into four basic groups:

1. Cast iron soil pipe and fittings;
2. Galvanized steel pipe and threaded fittings;
3. Copper tubing with solder joint and flare joint fittings;
4. Plastic pipe and fittings.

Before going into further detail on these materials, the apprentice must understand that the local plumbing code for his area will specify the type of piping material that may be used for each particular piping system. Local plumbing codes take into consideration such local conditions as soil types, ground conditions, local rainfall, and frost or freezing conditions, all of which can affect the choice of a piping material. For this reason, the local plumbing code must be consulted to see that the material to be used is the code-approved material for the system being piped.

CAST IRON SOIL PIPE AND FITTINGS

Cast iron soil pipe and fittings are manufactured from grey cast iron, a material which is both strong and corrosion resistant. The corrosion resistance of cast iron soil pipe is derived from the metallurgical structure of grey cast iron. During the solidification of the molten cast iron, large graphite flakes form within the pipe and fitting walls to serve as an insulation against corrosion.

After casting, the soil pipe and fittings are coated with coal tar pitch to prevent rust during storage and use, and to improve their appearance.

To ensure a uniform wall thickness, cast iron soil pipe is centrifugally cast. In this process, molten cast iron is poured into a spinning pipe mold, and centrifugal force causes it to form on the sides of the mold where it solidifies to the pipe shape. This method of casting ensures a straight length of soil pipe with a smooth interior surface.

Cast iron soil pipe fittings are cast in permanent metal molds to produce fittings with a uniform wall thickness.

Cast iron soil pipe and fittings have the advantage of being made of a strong material. They neither leak nor absorb water; they are economical and are easily cut and joined, in addition to being corrosion resistant. Cast iron soil pipe and fittings also provide one of the quietest piping systems available because they do not transmit the sound of water draining through the pipe as do some of the other, thinner piping materials.

The main disadvantage to the use of cast iron soil pipe is that it is a heavy material, which has a low tensile strength, and if treated roughly the pipe and fittings crack and break.

Cast iron soil pipe and fittings are available in either the bell and spigot pattern shown in Figure 3-1 or the hubless or no-hub pattern shown in Figure 3-2.

Bell and Spigot Cast Iron Soil Pipe and Fittings

Bell and spigot soil pipe and fittings have a bell or hub cast at the end of the pipe and fittings into which the spigot or plain end of another piece of pipe (or fitting) is inserted to join them together. (See Figure 3-1.) The space between the hub and the plain end of the pipe or fitting is sealed with either a caulked lead and oakum joint or a mechanical compression joint. (Both joining methods are described in Chapter 5.)

Cast iron soil pipe, as pictured at the top of Figure 3-1, is cast in 5-foot and 10-foot lengths of single hub pipe (pipe with a hub on only one end) and 30-inch lengths and 5-foot lengths of double hub pipe (pipe with a hub on each end to minimize waste in cutting short pieces).

At the present time, bell and spigot cast iron

SINGLE HUB, 5' LENGTHS

DIAMS.	2"	3"	4"	5"	6"	8"	10"	12"	15"
SV WTS.	20	30	40	52	65	100	145	190	255
XH WTS.	25	45	60	75	95	150	215	270	375

DOUBLE HUB, 5' LENGTHS

DIAMS.	2"	3"	4"	5"	6"	8"	10"	12"	15"
SV. WTS.	21	31	42	54	68	105	150	200	270
XH. WTS.	26	47	63	78	100	157	225	285	395

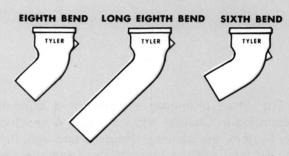

EIGHTH BEND   LONG EIGHTH BEND   SIXTH BEND

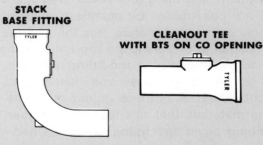

STACK BASE FITTING

CLEANOUT TEE WITH BTS ON CO OPENING

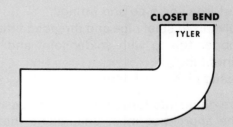

CLOSET BEND

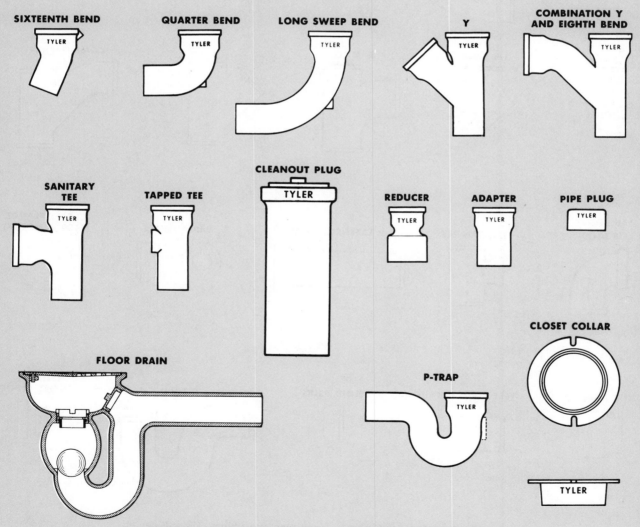

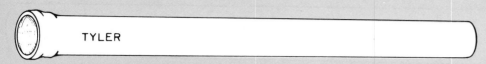

TYLER

SINGLE HUB, 10' LENGTHS

DIAMS	SV WTS.	XH WTS.
2"	38	43
3"	56	83
4"	75	108
5"	98	133
6"	124	160
8"	185	265
10"	270	400
12"	355	480
15"	475	705

TYLER

DOUBLE HUB, 30" LENGTHS
(25" LAYING LENGTHS)

DIAM.	SV WTS.	XH WTS.
2"	11	14
3"	17	26
4"	23	33

SIXTEENTH BEND — TYLER

QUARTER BEND — TYLER

LONG SWEEP BEND — TYLER

Y — TYLER

COMBINATION Y AND EIGHTH BEND — TYLER

SANITARY TEE — TYLER

TAPPED TEE — TYLER

CLEANOUT PLUG — TYLER

REDUCER — TYLER

ADAPTER — TYLER

PIPE PLUG — TYLER

FLOOR DRAIN

P-TRAP — TYLER

CLOSET COLLAR — TYLER

Figure 3-1. Bell and spigot cast iron soil pipe fittings. (Tyler Pipe)

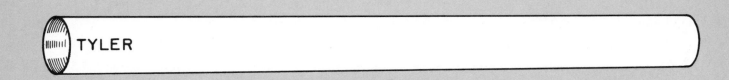

TYLER

NO-HUB PIPE, TEN-FOOT LENGTHS							
DIA.	1-1/2	2	3	4	5	6	8
WT. PER 10'	27	38	54	74	95	118	180

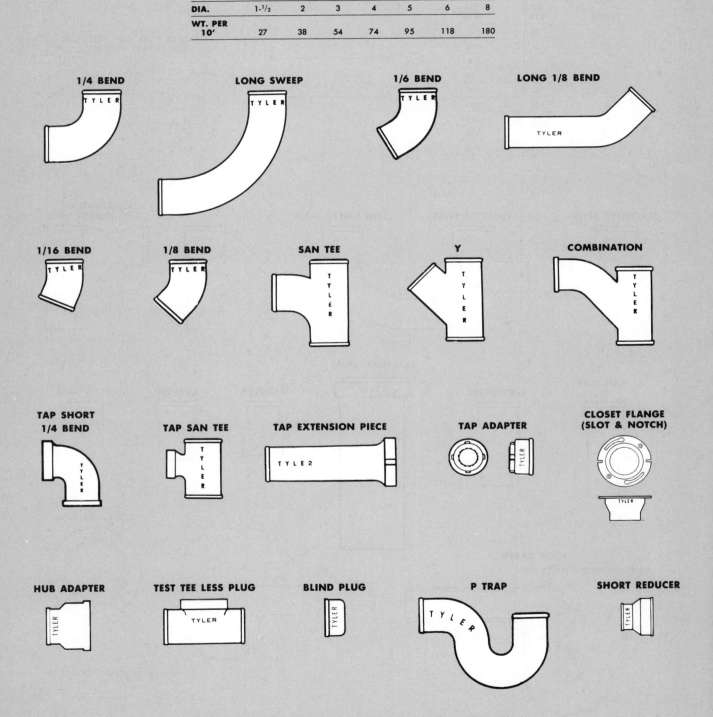

Figure 3-2. No-hub cast iron soil pipe and fittings. (Tyler Pipe)

soil pipe and fittings are cast in two different wall thicknesses or "weights"—service weight (SV) and extra heavy weight (XH)—in sizes from 2-inch to 15-inch inside diameter. For the purpose of comparing these two weights of soil pipe the apprentice should refer to the sizes and weights information located beneath the lengths of bell and spigot soil pipe shown in Figure 3-1. For example, under the illustration for "single hub, 5-foot lengths" he will notice that in the 4-inch size, a 5-foot length of SV (service weight) pipe weighs 40 pounds while the same length of XH (extra heavy) pipe weighs 60 pounds.

As a means of identifying the two weights of soil pipe and fittings, the cast iron industry has adopted the practice of marking the lighter service weight soil pipe and fittings with the letters SV, which are either cast in raised letters or painted on the pipe and fittings. Extra heavy pipe and fittings are marked in the same manner with the letters XH.

Although service weight and extra heavy soil pipe and fittings are identical in appearance, they are not interchangeable within a piping system due to the difference in outside diameters of the pipe and fittings. Service weight pipe and fittings are smaller.

Note: The cast iron soil pipe industry is phasing out the production of the extra heavy weight of pipe and fittings and within the near future only service weight pipe and fittings will be available.

No-Hub Soil Pipe and Fittings

No-hub soil pipe and fittings are a new concept devised by the cast iron soil pipe industry to enable them to compete with a faster method of joining the pipe and fittings. No-hub pipe and fittings (Figure 3-2) are joined together with a mechanical joint composed of a neoprene sleeve gasket into which the plain ends of the pipe and/or fitting are inserted. They are held in place by a stainless steel band with screw clamps which are tightened over the neoprene gasket as detailed in Chapter 5.

No-hub soil pipe and fittings are cast in sizes from 1½-inch to 10-inch inside diameter. The pipe is cast in 10-foot lengths.

The apprentice will find in the sizes and weights information given in Figure 3-2 that the weight of a 10-foot length of 4-inch no-hub pipe is 74 pounds. Referring back to Figure 3-1, notice that the weight of single hub, 10-foot length of SV (service weight), 4-inch pipe is 75 pounds. From this comparison it should be evident that the wall thicknesses (and outside diameters) of service weight soil pipe and no-hub soil pipe are nearly identical.

Uses of Cast Iron Soil Pipe and Fittings. Both bell and spigot and no-hub cast iron soil pipe and fittings are used above and below ground to pipe sanitary drainage, vent, and storm water drainage pipes.

GALVANIZED STEEL PIPE AND THREADED FITTINGS

Galvanized steel pipe is made from mild carbon steel as either a welded pipe or seamless pipe.

Welded pipe, which is also called butt welded and continuous weld pipe, is made by drawing flat strips of steel through a die to form the round shape and then electric butt-welding down the seam. Welded pipe is normally manufactured in 21-foot lengths.

Seamless pipe is made by piercing a red hot, solid, cylindrical billet of steel with a series of mandrels while passing the metal through rollers. Seamless pipe is manufactured in random lengths from 16 to 22 feet long.

Unprotected steel pipe rusts almost immediately upon exposure to the atmosphere and moisture. To protect the pipe, all steel pipe used by plumbers must have a protective coating. The method used most often to protect plumbing pipe is called hot-dip galvanizing. In this process, steel pipe is cleaned and then dipped in a hot (870°F / 465°C) bath of molten zinc.*

The galvanized steel pipe used by plumbers is available in sizes (called nominal pipe sizes) from ⅛ inch to 12 inch, in several different wall thicknesses. Nominal pipe size, which is also called iron pipe size and abbreviated IPS, means

*The *Glossary* gives further information under *Metric* on metric units and conversions. Here, Fahrenheit temperatures are converted to Celsius.

TABLE 3-1. GALVANIZED STEEL PIPE DATA

NOMINAL PIPE SIZE (INCHES)	OUTSIDE DIAMETER (INCHES)	SCHEDULE 40 (standard wall)		SCHEDULE 80 (extra strong wall)	
		WALL THICKNESS (INCHES)	INSIDE DIAMETER (INCHES)	WALL THICKNESS (INCHES)	INSIDE DIAMETER (INCHES)
$1/8$	.405	.068	.269	.095	.215
$1/4$	.540	.088	.364	.119	.302
$3/8$	.675	.091	.493	.126	.423
$1/2$	.840	.109	.622	.147	.546
$3/4$	1.050	.113	.824	.154	.742
1	1.315	.133	1.049	.179	.957
$1^1/4$	1.660	.140	1.380	.191	1.278
$1^1/2$	1.900	.145	1.610	.200	1.500
2	2.375	.154	2.067	.218	1.939
$2^1/2$	2.875	.203	2.469	.276	2.323
3	3.500	.216	3.068	.300	2.900
$3^1/2$	4.000	.226	3.548	.318	3.364
4	4.500	.237	4.026	.337	3.826
5	5.563	.258	5.047	.375	4.813
6	6.625	.280	6.065	.432	5.761
8	8.625	.322	7.981	.500	7.625
10	10.750	.365	10.020	.500 (xs)	9.750
12	12.750	.375 (std) .406 (sch 40)	12.000 11.938	.500 (xs)	11.750

closest to the desired size. There are eighteen nominal pipe sizes from $1/8$ inch to 12 inch as listed in Table 3-1.

The terms used by the pipe industry to describe the different wall thicknesses of galvanized steel pipes are *standard wall, extra strong wall,* and *double extra strong wall.* The terms *schedule 40* and *schedule 80* are also used to describe pipe and wall thicknesses. The terms are somewhat interchangeable in that through 10-inch size pipe, schedule 40 pipe and standard wall pipe have the same wall thickness. Schedule 80 pipe and extra strong wall pipe are identical through 8-inch size. Virtually all of the galvanized steel pipe installed by plumbers is schedule 40 or standard wall thickness.

Table 3-1 also lists the actual outside diameter (OD) of each size of galvanized steel pipe as well as the actual inside diameter (ID) for both schedule 40 and schedule 80 wall thicknesses. The apprentice will notice on this table that the outside diameter is the same in any given size of pipe for both schedule 40 and schedule 80 wall thickness pipe, so that the pipe fittings are interchangeable.

Since nearly all of the galvanized steel pipe used by plumbers is joined with a threaded joint, the pipe is normally supplied with threads on both ends of each length, with a coupling threaded onto one end of the length of pipe. Pipe supplied in this manner is called *T and C* (thread and coupling) by the plumbing industry.

Galvanized steel pipe is an inexpensive, strong, and rugged material, and is not easily damaged by rough handling. It resists both shock and stress. Galvanized steel pipe cuts and threads easily, although this method of joining (because it is somewhat slower in terms of installation time than the joining method of other piping materials) tends to be a rather expensive piping method. To eliminate some of the labor costs associated with cutting and threading galvanized steel pipe, plumbing contractors purchase short lengths of pipe with a thread on each end called *nipples*. These nipples vary in length from *close* (nipples that are virtually all thread) up to 6 inches in length (in $1/2$-inch increments). Nipples are also available in 1-inch increments from 7 to 12 inches in length.

Uses of Galvanized Steel Pipe. Galvanized steel pipe is used only above ground with the appropriate threaded fittings for vent piping, sanitary and storm drainage piping, and potable water supply piping. The threaded fittings used to pipe each of these piping systems are:

1. For *vent piping* — standard cast iron threaded fittings,
2. For *sanitary and storm drainage piping* — cast iron recessed drainage fittings,
3. For *water supply piping* — galvanized malleable iron fittings.

Standard Cast Iron Threaded Fittings

Standard cast iron threaded fittings (see Figure 3-3) are cast from grey cast iron in sand molds. They are a very economical fitting, but they are very brittle and may have imperfections called *sand holes* because of the way they are cast. The standard cast iron fittings are used by plumbers on vent piping in $1\frac{1}{4}$-inch to 8-inch sizes. They are available with either a plain, uncoated finish (referred to as black) or with a galvanized iron coating. However, black fittings are most commonly used on vent piping.

Cast Iron Recessed Drainage Fittings

These fittings are also cast from grey cast iron in sand molds. These drainage fittings (Figure 3-4) are cast with a recessed shoulder to provide a smooth interior surface so that when properly installed there is no obstruction to the flow of waste material through the pipe. Figure 3-5 illustrates a sectional view of a cast iron recessed drainage elbow with the threaded joint properly installed.

Another feature of these drainage fittings is that 90° elbows (ells), short pattern tees (Ts), and long sweep tee-wyes (T-Ys) (Figure 3-4) have the

90° ELBOW

45° ELBOW

TEE

REDUCING TEE

REDUCER

CAP

PLUG

Figure 3-3. Standard cast iron threaded fittings.

22-1/2° ELBOW	45° ELBOW	60° ELBOW	SHORT PATTERN 90° ELBOW
TEE	SHORT PATTERN T-Y	LONG SWEEP 90° ELBOW	WYE
			LONG SWEEP T-Y
TUCKER	TUCKER TEE	TRAP	STREET 45° ELBOW
			STREET 90° ELBOW

Figure 3-4. Cast iron recessed drainage or Durham fittings. (Stockham)

threads tapped at a slight angle so that a horizontal drainage pipe will be pitched at ¼ inch per foot.

Cast iron recessed drainage fittings are available in 1¼-inch to 12-inch sizes for use on sanitary drainage and storm drainage piping. The fittings are available with either a black tarred coating (which is most commonly used) or a galvanized iron coating.

Another name sometimes used for cast iron

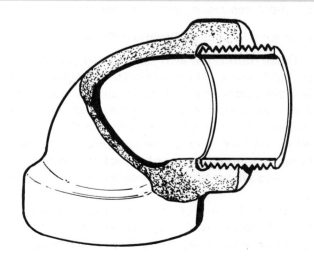

Figure 3-5. A sectional view through a cast iron recessed drainage elbow with the threaded joint properly installed. (Stockham)

recessed drainage fittings is Durham fittings, in honor of the Durham brothers who developed this piping system in New York City about 1880.

Galvanized Malleable Iron Fittings

Galvanized malleable iron fittings (Figure 3-6) are cast from ordinary grey cast iron in a foundry mold, but they are control cooled (heat treated) over a 72-hour period. This controlled cooling produces a change in the grain structure of the iron that makes it a tough, elastic material. These fittings, which plumbers use for water supply piping in $3/8$-inch to 6-inch sizes, are then given a protective galvanized coating. Malleable fittings must never be reheated and/or welded or they revert back to ordinary grey cast iron fittings.

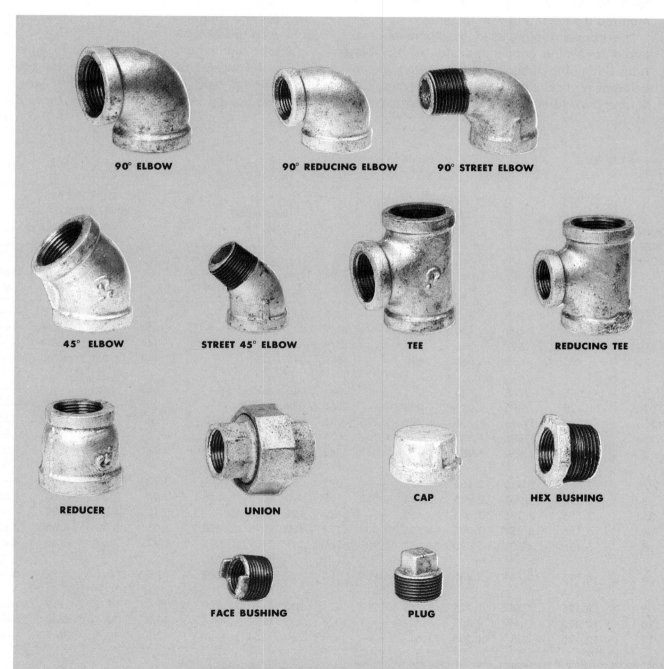

90° ELBOW   90° REDUCING ELBOW   90° STREET ELBOW

45° ELBOW   STREET 45° ELBOW   TEE   REDUCING TEE

REDUCER   UNION   CAP   HEX BUSHING

FACE BUSHING   PLUG

Figure 3-6. Galvanized malleable iron fittings.

COPPER TUBING, SOLDER JOINT, AND FLARE JOINT FITTINGS

Copper tubing, which is manufactured from 99.9% pure copper, is made by drawing a heated billet of copper through a die while a piercing rod or ball passes through the middle of the billet to draw out the copper tubing to the required diameter and wall thickness.

The copper tubing used by plumbers is referred to by the copper industry as *plumbing tube.* Plumbing tube is manufactured in four different wall thicknesses or types—types K, L, M, and DWV (drainage, waste, and vent). Type K

copper tubing has the heaviest wall thickness and type DWV has the thinnest wall.

All four of these types of copper plumbing tubing have the same *outside diameter* which is always 1/8 inch larger than the stated or *nominal* size (which refers to the actual inside diameter). That is, 1/2-inch copper tubing has an actual *outside diameter* of 5/8 inch; 3/4-inch has an outside diameter of 7/8 inch; etc. Table 3-2 lists the sizes and weights of tube which are available in 1/4-inch to 12-inch sizes.

Copper plumbing tube is available in either drawn or annealed tempers. Drawn copper tubing is called *hard* copper by the trade, while annealed copper is called *soft* copper. Types K,

TABLE 3-2. SIZES AND WEIGHTS OF COPPER PLUMBING TUBE

NOMINAL SIZE (INCHES)	OUTSIDE DIAMETER (INCHES)	INSIDE DIAMETER (INCHES)				WALL THICKNESS (INCHES)				POUNDS PER LINEAR FOOT*			
	Types K-L-M-DWV	Type K	Type L	Type M	Type DWV	Type K	Type L	Type M	Type DWV	Type K	Type L	Type M	Type DWV
1/4	.375	.305	.315	.325	——	.035	.030	.025	——	.145	.126	.106	——
3/8	.500	.402	.430	.450	——	.049	.035	.025	——	.269	.198	.145	——
1/2	.625	.527	.545	.569	——	.049	.040	.028	——	.344	.285	.204	——
5/8	.750	.652	.666	.690	——	.049	.042	.030	——	.418	.362	.263	——
3/4	.875	.745	.785	.811	——	.065	.045	.032	——	.641	.455	.328	——
1	1.125	.995	1.025	1.055	——	.065	.050	.035	——	.839	.655	.465	——
1 1/4	1.375	1.245	1.265	1.291	1.295	.065	.055	.042	.040	1.04	.884	.682	.650
1 1/2	1.625	1.481	1.505	1.527	1.541	.072	.060	.049	.042	1.36	1.14	.940	.809
2	2.125	1.959	1.985	2.009	2.041	.083	.070	.058	.042	2.06	1.75	1.46	1.07
2 1/2	2.625	2.435	2.465	2.495	——	.095	.080	.065	——	2.93	2.48	2.03	——
3	3.125	2.907	2.945	2.981	3.035	.109	.090	.072	.045	4.00	3.33	2.68	1.69
3 1/2	3.625	3.385	3.425	3.459	——	.120	.100	.083	——	5.12	4.29	3.58	——
4	4.125	3.857	3.905	3.935	4.009	.134	.110	.095	.058	6.51	5.38	4.66	2.87
5	5.125	4.805	4.875	4.907	4.981	.160	.125	.109	.072	9.67	7.61	6.66	4.43
6	6.125	5.741	5.845	5.881	5.959	.192	.140	.122	.083	13.9	10.2	8.92	6.10
8	8.125	7.583	7.725	7.785	7.907	.271	.200	.170	.109	25.9	19.3	16.5	10.6
10	10.125	9.449	9.625	9.701	——	.338	.250	.212	——	40.3	30.1	25.6	——
12	12.125	11.315	11.565	11.617	——	.405	.280	.254	——	57.8	40.4	36.7	——

*Slight variations from these weights must be expected in practice.

TABLE 3-3. COMMERCIALLY AVAILABLE LENGTHS OF COPPER PLUMBING TUBE

TUBE	DRAWN (hard copper)		ANNEALED (soft copper)	
Type K	Straight Lengths:		Straight Lengths:	
	Up to 8-inch diameter	20 ft	Up to 8-inch diameter	20 ft
	10-inch diameter	18 ft	10-inch diameter	18 ft
	12-inch diameter	12 ft	12-inch diameter	12 ft
			Coils:	
			Up to 1-inch diameter	60 ft
				100 ft
			$1^1/_4$ and $1^1/_2$-inch diameter	60 ft
			2-inch diameter	40 ft
				45 ft
Type L	Straight Lengths:		Straight Lengths:	
	Up to 10-inch diameter	20 ft	Up to 10-inch diameter	20 ft
	12-inch diameter	18 ft	12-inch diameter	18 ft
			Coils:	
			Up to 1-inch diameter	60 ft
				100 ft
			$1^1/_4$ and $1^1/_2$-inch diameter	60 ft
			2-inch diameter	40 ft
				45 ft
Type M	Straight Lengths:		Not available	—
	All diameters	20 ft		
DWV	Straight Lengths:		Not available	—
	All diameters	20 ft		

Copper Development Association, Inc.

L, M, and DWV hard copper tubing are available in straight lengths usually 20 feet long. Soft copper tube is available in types K and L in either 20-foot straight lengths or in long coils. Table 3-3 lists the commercially available lengths of copper plumbing tube of the various types and tempers.

Hard temper copper tubing is coded with a color stripe and lettering which indicates the type of the tubing, the name or trademark of the manufacturer, and the country of origin. The four colors used to identify the four types of hard copper plumbing tube are:

Green (Type K copper tube)

Blue (Type L copper tube)

Red (Type M copper tube)

Yellow (Type DWV copper tube)

In addition to the color coding, the type and manufacturing information is also stamped every 18 inches on the copper tubing.

Soft temper copper tubing is not marked with a color code, but the type and manufacturing information is stamped into the tubing every 18 inches.

Copper tubing is used for piping plumbing systems because it is a lightweight, easily joined, corrosion-resistant material. However, some acids and fumes found in the sanitary drainage system have been known to corrode copper tubing. The ammonia fumes from urine are especially corrosive to copper tubing, and for this reason this material should not be used for the drain pipe from urinals.

Copper Tubing Fittings

These fittings are manufactured of either cast bronze or wrought copper. Cast bronze fittings are cast from molten bronze (an alloy of copper and tin) in sand molds. Wrought copper fittings are formed by a hammering process from copper mill products. There is no difference between the two types of fittings as far as cost or installation practices. The choice between cast bronze and wrought copper fittings is largely a matter of user's preference and fitting availability, as some fittings are not available in both cast bronze and wrought copper.

Copper tubing fittings are joined to the copper tubing by either a soldered joint or a flared joint. Solder joint fittings are available in two patterns: DWV and pressure. Flared joint fittings are a mechanical joint fitting that are manufactured only in cast bronze. (These joining methods will be described in Chapter 5.)

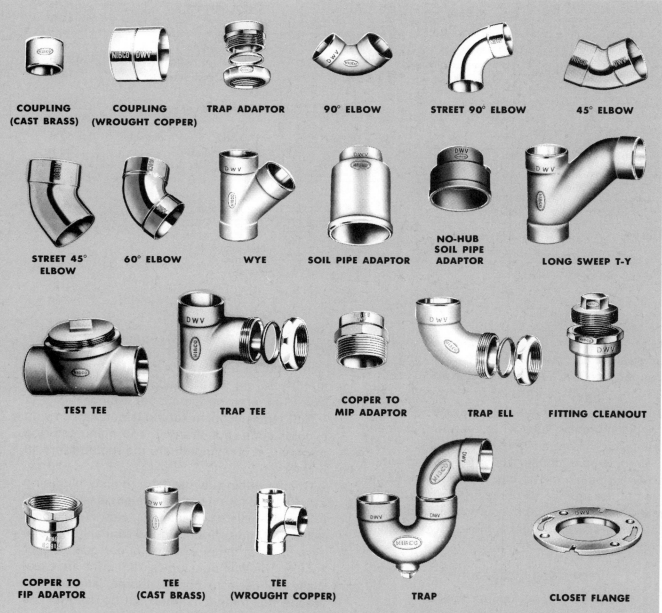

COUPLING (CAST BRASS)   COUPLING (WROUGHT COPPER)   TRAP ADAPTOR   90° ELBOW   STREET 90° ELBOW   45° ELBOW

STREET 45° ELBOW   60° ELBOW   WYE   SOIL PIPE ADAPTOR   NO-HUB SOIL PIPE ADAPTOR   LONG SWEEP T-Y

TEST TEE   TRAP TEE   COPPER TO MIP ADAPTOR   TRAP ELL   FITTING CLEANOUT

COPPER TO FIP ADAPTOR   TEE (CAST BRASS)   TEE (WROUGHT COPPER)   TRAP   CLOSET FLANGE

Figure 3-7. DWV (drainage, waste, and vent) copper solder joint fittings. (NIBCO, Inc.)

DWV Copper Fittings. These solder joint copper fittings, used with type DWV hard temper copper tubing, are shown in Figure 3-7. They are used for drainage, waste and vent fittings, and are characterized by a long radius or sweep in elbows, tees, and other fittings. The long radius allows easy gravity flow of the waste materials draining through them. In addition, these fittings have a shallow solder socket as they are not normally subject to pressure. DWV copper fittings, which are used for sanitary drainage, vent, and storm water piping above ground, are available in 1¼-inch to 8-inch sizes.

Solder Joint Pressure Fittings. Copper solder joint pressure fittings are used above ground with types L and M hard temper copper tubing for water supply piping. Copper pressure fittings (Figure 3-8) have a short radius or sweep on 90° elbows, tees, and other fittings because the water flowing through them is under pressure. As they must contain water under pressure, these fittings have a deeper solder socket than the DWV copper fittings (illustrated in Figure 3-7). Solder joint copper pressure fittings are available in ⅛-inch to 12-inch sizes.

Flared Joint Fittings. These fittings, which

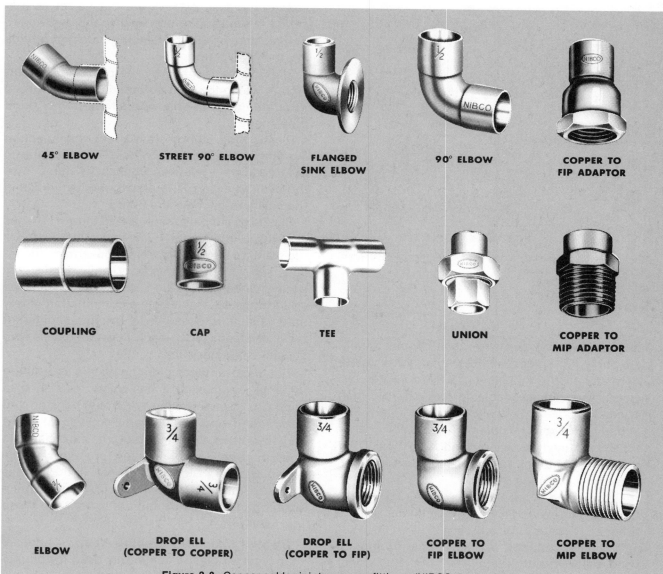

Figure 3-8. Copper solder joint pressure fittings. (NIBCO, Inc.)

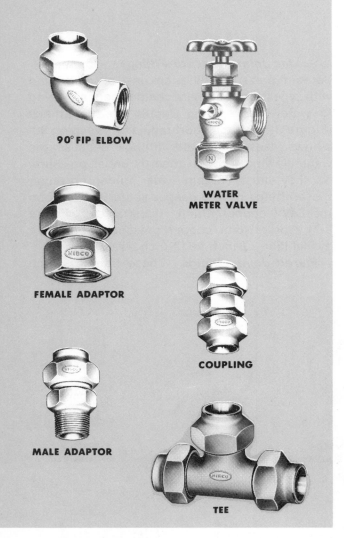

90° FIP ELBOW

WATER METER VALVE

FEMALE ADAPTOR

COUPLING

MALE ADAPTOR

TEE

Figure 3-9. Flare joint fittings for use with soft copper tubing. (NIBCO, Inc.)

are used with types K and L soft copper tubing, are shown in Figure 3-9. Flared joint fittings are normally used only below ground on water services. They are available in ⅜-inch to 3-inch sizes.

Uses of Copper Tubing. Type DWV copper tubing is used above ground with type DWV solder joint fittings for sanitary drainage, vent, and storm drainage piping.

Types L and M hard temper copper tubing are used above ground with solder joint pressure fittings for water supply piping.

Types K and L soft temper copper tubing are used below ground with flare joint fittings for water supply piping.

PLASTIC PIPE AND FITTINGS

Plastics are a family of man-made materials developed from synthetic organic chemicals. The basic raw materials used to make plastics are oil, natural gas, coal, and cellulose (from wood fibers). The raw materials are made into resins, which are classified into two general types of plastics.

Thermosetting plastics are plastic resins that cannot be remelted after they are formed and cured in their final shape.

Thermoplastic plastics are plastic resins that can be heated and reformed over and over again. The plastics used for plumbing pipe and fittings are thermoplastics.

Plastic pipe is made by the process of extrusion in which the heat-softened plastic is forced through a circular-shaped die to form the round pipe shape.

Plastic pipe fittings are made by the injection molding process in which heat-softened plastic is forced into a relatively cool cavity shaped like the desired fitting. The plastic then solidifies in the cavity to become the fitting.

Plastic as a piping material has the advantages of being a lightweight, inexpensive material. It is resistant to corrosion from household chemicals, has very smooth interior walls, and is very easily joined. In addition, because plastic pipe and fittings are made from a combination of raw materials, it is possible to formulate different types of plastic pipe—each of which has its own special characteristics.

The disadvantages of using plastic as a piping material are that (1) it has a low resistance to heat; (2) it has a very high rate of expansion and contraction when heated and cooled; (3) it is a very flexible material that requires extra hangers and supports; (4) it has a low crush resistance (when compared to metal pipes); and (5) it has pressure ratings (the amount of internal pressure it can withstand) that are lower than those for metal pipes.

Plastics Used for Plumbing Pipes. There are four types of plastic commonly used for plumbing pipes and fittings:

1. *Acrylonitrile-Butadiene-Styrene*, or *ABS* as it is more commonly called.
2. *Polyvinyl Chloride*, or *PVC*.
3. *Chlorinated Polyvinyl Chloride*, or *CPVC*.
4. *Polyethylene*, or *PE* plastic tubing.

In the following sections of this chapter the common uses of these four materials will be described. (The apprentice is reminded to check his local plumbing code for restrictions on the uses of plastic piping materials.)

ABS plastic pipe and fittings are the type most commonly used in the plumbing industry. This black plastic pipe (with solvent weld DWV fittings like those pictured in Figure 3-10) is used to construct sanitary drainage and vent piping systems as well as for storm water drainage systems both above and below ground.

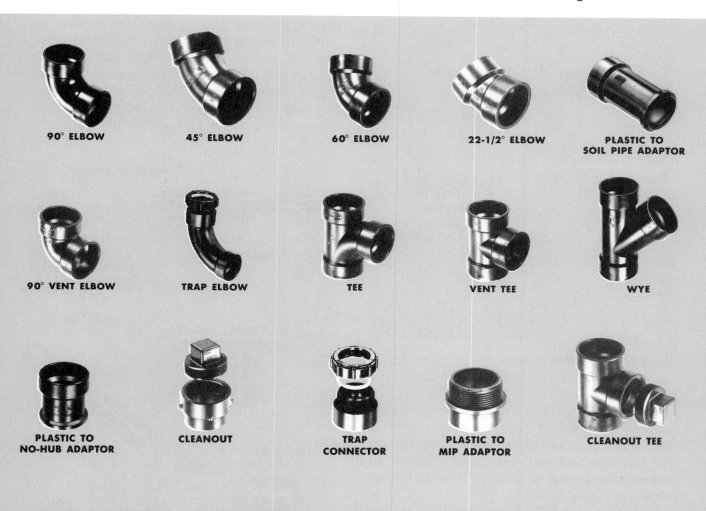

90° ELBOW   45° ELBOW   60° ELBOW   22-1/2° ELBOW   PLASTIC TO SOIL PIPE ADAPTOR

90° VENT ELBOW   TRAP ELBOW   TEE   VENT TEE   WYE

PLASTIC TO NO-HUB ADAPTOR   CLEANOUT   TRAP CONNECTOR   PLASTIC TO MIP ADAPTOR   CLEANOUT TEE

LONG SWEEP T-Y   CLOSET FLANGE   COUPLING   PLASTIC TO FIP ADAPTOR   TRAP

Figure 3-10. An assortment of DWV plastic solvent weld fittings that are available in ABS and PVC plastic. (NIBCO, Inc.)

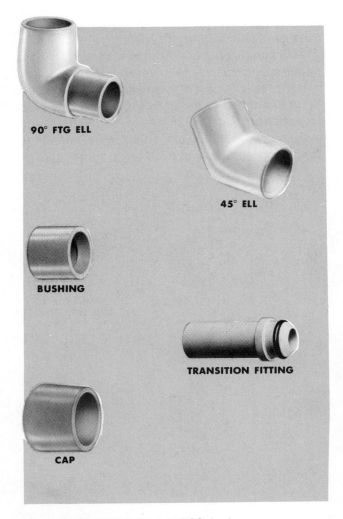

Figure 3-11. CPVC fittings. (NIBCO, Inc.)

Since ABS and PVC plastic pipe and fittings have identical uses, the only reasons for selecting one material over the other for a piping system are either the cost of the material or local plumbing code restrictions. That is, in most instances the least expensive of the two materials will be used unless there is a local plumbing code restriction prohibiting the use of that material.

CPVC plastic pipe is a light or cream colored plastic that has been specially formulated to withstand higher temperatures than other plastics. It is normally rated for 180° F (82° C) at 100 pounds per square inch of pressure with a substantial safety margin. CPVC is used for piping hot and cold water distribution systems in homes. CPVC plastic pipe and fittings, which are

PVC plastic pipe and fittings, which are light colored, are used in the same applications as ABS plastic pipe and fittings. PVC plastic fittings for drainage and vent piping are identical in appearance (except for color) to the ABS fittings shown in Figure 3-10.

Even though ABS and PVC plastic pipe and fittings have identical uses and are identical in appearance (except for color) the two materials are not interchangeable, and each material— ABS and PVC — has its own special solvent cement.

Both ABS and PVC plastic pipe and fittings for drainage and vent piping are available in 1¼-inch to 6-inch sizes. The pipe is available in 10- and 20-foot lengths.

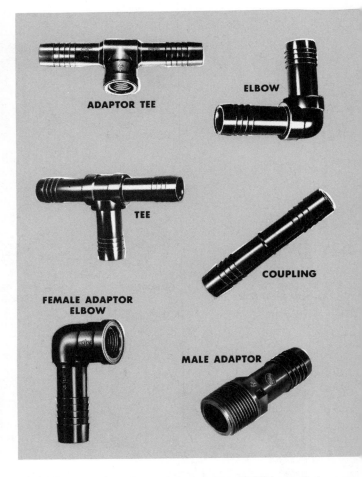

Figure 3-12. Plastic insert fittings for use with PE (polyethylene) plastic tubing. (Celanese Piping Systems, a division of Celanese Plastics Co.)

available in ½- and ¾-inch sizes, are also joined by solvent welding. CPVC pipe is available in 10-foot lengths. An assortment of fittings is illustrated in Figure 3-11.

PE plastic tubing is a flexible plastic tubing, available in ¾-, 1-, 1½-, and 2-inch sizes, that is used for water supply piping below ground in yard areas (the building water service). PE tubing is black and is usually sold in 100-foot coils. It is joined to nylon or plastic insert fittings (Figure 3-12) with hose clamps (as shown in Chapter 5). PE plastic may also be flared by the method shown in Chapter 5 and used with cast bronze flare fittings of the type shown in Figure 3-9 for use with soft copper tubing.

 ## USES OF PIPING MATERIALS

For your future reference use, all of the information regarding the uses of the four basic types of pipe and fittings has been incorporated into Table 3-4, "Uses of Piping Materials."

TABLE 3-4. USES OF PIPING MATERIALS

PIPING MATERIAL	Potable Water Supply Piping: Below Ground	Below Ground in Yard Areas	Above Ground	Sanitary Drainage and Vent Piping: Below Ground	Above Ground	Storm Water Drainage Piping: Below Ground	Above Ground
Bell and Spigot Cast Iron Soil Pipe				•	•	•	•
No-Hub Cast Iron Soil Pipe				•	•	•	•
Galvanized Steel Pipe			•		•		•
Standard Cast Iron Threaded Fittings					•		
Cast Iron Recessed Drainage Fittings					•		•
Malleable Iron Fittings			•				
Copper Tubing:							
Hard Temper—Types L and M			•				
—Type DWV					•		•
Soft Temper—Types K and L	•	•					
DWV Copper Fittings					•		•
Copper Pressure Fittings			•				
Flare Fittings	•	•					
ABS Plastic Pipe and Fittings				•	•	•	•
PVC Plastic Pipe and Fittings				•	•	•	•
CPVC Plastic Pipe and Fittings			•				
PE Plastic Tubing and Insert Fittings or Flare Fittings		•					

HOW TO IDENTIFY PLUMBING FITTINGS

In addition to being able to recognize the various types of plumbing pipes and fittings and knowing the uses of these materials, the apprentice must learn to identify or *read* the more common types of plumbing fittings. To assist the student in identifying plumbing fittings, a variety of the most commonly used plumbing fittings is illustrated on Figure 3-13. In this descriptive illustration each opening of the fitting is marked with a letter that indicates the *sequence* to be followed in identifying the size of the fitting. That

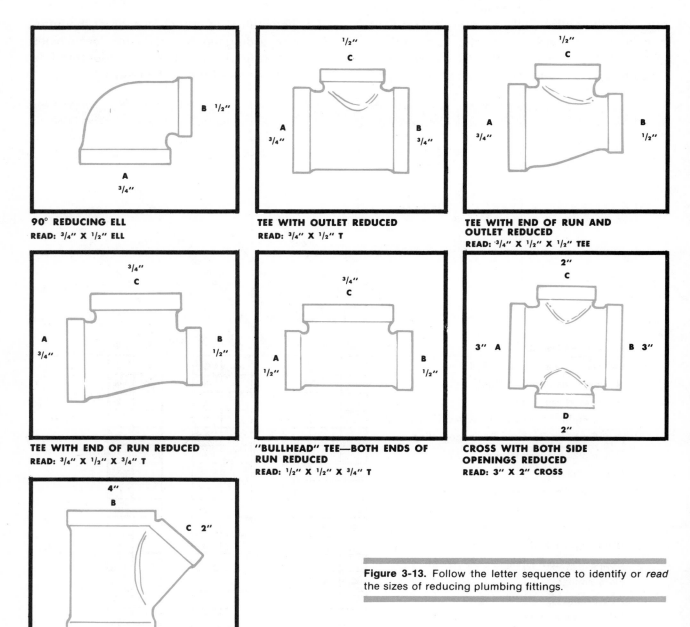

90° REDUCING ELL
READ: 3/4" X 1/2" ELL

TEE WITH OUTLET REDUCED
READ: 3/4" X 1/2" T

TEE WITH END OF RUN AND OUTLET REDUCED
READ: 3/4" X 1/2" X 1/2" TEE

TEE WITH END OF RUN REDUCED
READ: 3/4" X 1/2" X 3/4" T

"BULLHEAD" TEE—BOTH ENDS OF RUN REDUCED
READ: 1/2" X 1/2" X 3/4" T

CROSS WITH BOTH SIDE OPENINGS REDUCED
READ: 3" X 2" CROSS

WYE WITH SIDE OPENING REDUCED
READ: 4" X 2" WYE

Figure 3-13. Follow the letter sequence to identify or *read* the sizes of reducing plumbing fittings.

is, in reading or specifying the outlets of reducing fittings, the openings should be *read* in the order indicated by the sequence of the letters A, B, C, and D.

On reducing elbows and couplings, the largest opening is named first. Example: $3/4 \times 1/2$ elbow. (See Figure 3-13.)

On bushings, the male thread size is named first, and then the female thread.

On tees, wyes (Ys), sanitary tees, and combination wye and $1/8$ bends, the largest run size is named first, then the smaller run, and then the side outlet last. A tee having a side outlet larger than the run size is called a bull head tee. (See Figure 3-13.)

On crosses with two run openings of the same size and the two side openings of the same size, it is common practice to give the run size first and the side outlet size last.

On side outlet reducing fittings (which are not illustrated), the side outlet is named last, with the other sizes given in the previously mentioned order.

Note: Although all of the fittings illustrated in Figure 3-13 are threaded fittings, the same rules apply to the identifying of reducing solder joint, soil pipe (both hub type and no-hub), and other types of plumbing fittings.

⊙ PLUMBING VALVES

A valve is a fitting installed by plumbers on a piping system to control the flow of fluid within that system in one or more of the following ways:

1. To turn the flow on;
2. To turn the flow off;
3. To regulate the flow by permitting flow in one direction only (that is, to prevent backflow), to regulate pressure, or to relieve excess temperature and/or pressure.

To accomplish these methods of control over the fluids within the piping system, plumbers install the following types of valves:

Gate valves	Check valves
Globe valves	Backwater valves
Core cocks	Pressure-reducing
Ball valves	valves
Butterfly valves	Relief valves

These various types of plumbing valves are manufactured from either cast bronze or cast iron. In general, valves sized 2 inches and smaller will be made from cast bronze with bronze internal working parts. Valves $2^{1}/_{2}$-inch size and larger usually have cast iron bodies with bronze internal working parts.

A term sometimes used to describe some types of valves is *rated valve.* A rated valve is a valve that meets an engineering standard for the normal pressure range of the fluids contained within the system it is controlling. The pressure rating for a particular valve is marked in raised letters on the side of the valve body. Most of the valves installed by plumbers will be marked 125 # SWP and 200 # WOG. This marking means that the valve is rated for use at a maximum of 125 pounds steam working pressure (SWP) or 200 pounds water, oil, or gas (WOG) pressure.

Gate Valves. Gate valves are valves that control the flow of fluid moving through the valve by means of a gate-like wedge disk which fits against smooth machined surfaces, called *seats,* within the valve body. Figure 3-14 illustrates the internal working parts of typical bronze gate valves.

When the disk is in the open position, a gate valve is designed to permit a straight, full, and free flow of fluid. For this reason gate valves are referred to as *full-way* valves.

Gate valves, which are the most commonly used type of plumbing valve, should be used only where they will be left either wide open or completely closed.

Gate valves have two different types of wedge disks, as illustrated in Figure 3-14, a solid-wedge disk and a split-wedge disk. The solid-wedge disk gate valve, which is the most often used of the two types, may be installed with the valve stem and handle in any position.

The split-wedge disk, which is the best sealing of the two types of gate valves, is designed so that closing pressure forces the two parallel halves of the wedge disk outward onto the body seat. This type of gate valve will close off the flow of fluids completely even when there is scale trapped on one seat. Split-wedge gate valves should be installed with the valve stem in the vertical position.

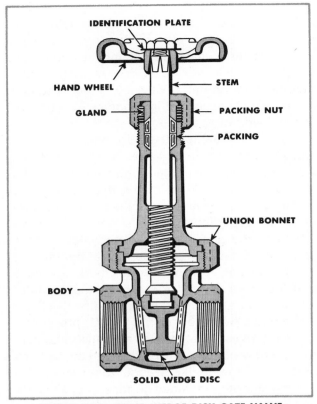

RISING STEM, SOLID WEDGE DISK GATE VALVE

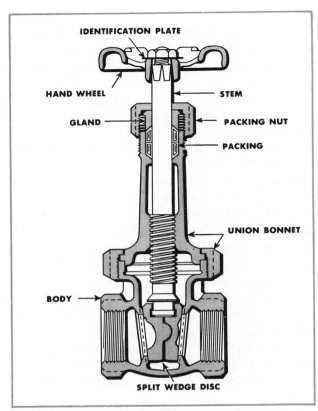

RISING STEM, SPLIT WEDGE DISK GATE VALVE

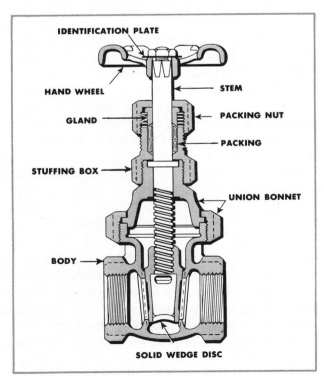

NONRISING STEM GATE VALVE

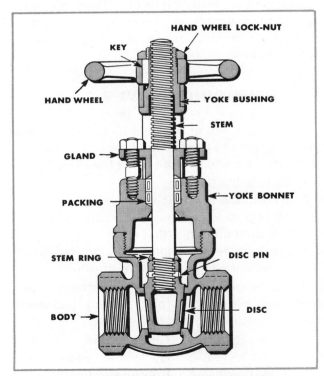

RISING STEM-OUTSIDE SCREW AND YOKE (OS&Y GATE VALVE)

Figure 3-14. Gate valves. (Stockham)

Gate valves are characterized by three different types of stems (Figure 3-14):

1. Rising stem-outside screw and yoke;
2. Rising stem-inside screw;
3. Nonrising stem-inside screw.

On a *rising stem-outside screw and yoke* (OS & Y) *gate valve,* when the hand wheel is turned the stem rises as the yoke bushing engages the stem threads. OS & Y gate valves are used where it is necessary to have a visual indication of whether or not the valve is open or closed.

The *rising stem-inside screw gate valve* is the most common design in bronze gate valves. The stem and hand wheel rise as the valve is opened to indicate the position of the wedge disk. However, because both the hand wheel and stem rise, adequate clearance must be provided for the operation of the valve.

On a *nonrising stem-inside screw gate valve,* neither the hand wheel nor the stem rises when the valve is opened. Nonrising stem gate valves are used where there is not enough clearance for the operation of a rising stem valve. However, since the stem and hand wheel do not rise, there is no way to visually check the open or closed position of this type of valve.

Gate valves are usually not installed in piping systems where they will have to be opened and closed frequently. If the piping system requires the use of a valve that must be opened or closed frequently, plumbers install either a globe valve or one of the other types of full-way valve.

Globe Valves. A globe valve is a compression type valve in which the flow of water is controlled by means of a circular disk that is forced (compressed) onto or withdrawn from an annular ring, known as the seat, which surrounds the opening through which water flows. Figure 3-15 illustrates the internal working parts of a globe valve. The name globe valve is derived from the fact that globe valves have a round or globe-like body. The configuration of a globe valve requires water flowing through it to change direction several times, causing turbulence, resistance to the flow of water, and a considerable pressure drop.

A rated globe valve is one that has a full size seat opening, that is, a 1-inch rated globe valve

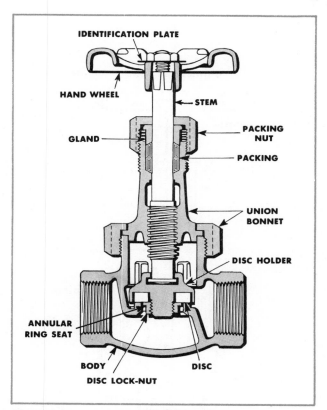

Figure 3-15. A globe valve. (Stockham)

has a 1-inch diameter seat opening. Non-rated globe valves are called *compression stops.*

Unlike gate valves, all globe valves have rising stems. Another feature of most of the globe valves used by plumbers is a replaceable composition disk. These disks, which are also called *washers,* are available made of different materials for use with cold water, hot water, chemicals, etc.

Angle Valves. An angle valve (Figure 3-16) is a globe valve in which the inlet and outlet openings are at a 90° angle to one another. An angle valve offers less resistance to the flow of water through it than a globe valve and the 90° elbow it replaces. The use of an angle valve also reduces the number of joints and thus saves installation time.

Globe valves and angle valves are recommended on installations that require frequent operation, throttling, and/or a positive shutoff

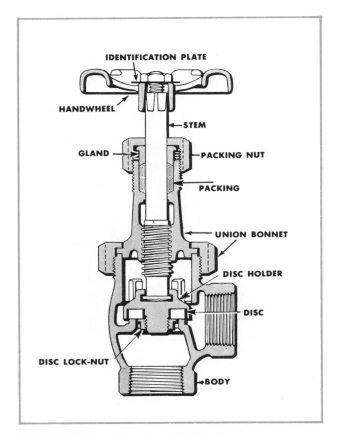

Figure 3-16. An angle valve. (Stockham)

Figure 3-17. Internal working parts of a gas cock.

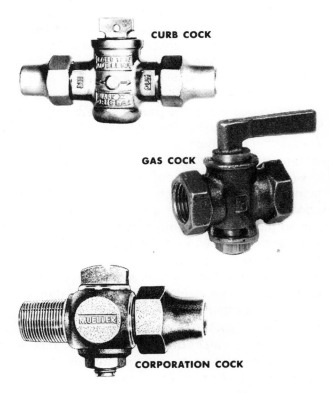

Figure 3-18. Core cocks installed by plumbers: corporation cock (Mueller Co.), curb cock (Mueller Co.), and gas cock.

when closed. On plumbing installations most globe and angle valves are used for individual plumbing fixture control valves.

Core Cocks. Core cocks, or *plug valves* as they are also called, are a type of valve through which the flow of water is controlled by a circular core or plug that fits closely in a machined seat. This core has a port bored through it to serve as a water passageway. Core cocks, like gate valves, are full-way valves. However, to open or close a core cock requires only a 90° turn of the valve handle or key. Figure 3-17 illustrates a sectional view of a gas cock.

Figure 3-18 shows the three types of core cocks most commonly used by plumbers. They are: the corporation cock, the curb cock, and the lever handle gas cock.

Ball Valves. A ball valve is a valve in which the flow of fluid is controlled by a rotating drilled ball that fits tightly against a resilient (flexible)

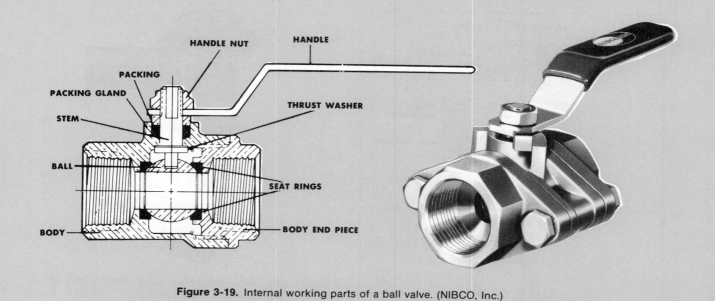

Figure 3-19. Internal working parts of a ball valve. (NIBCO, Inc.)

seat in the valve body. A ball valve (Figure 3-19) is basically a core cock with a spherical core. Like a core cock, a ball valve requires only a 90° rotation of the handle to open or close the valve. Ball valves are full-way valves that can be used for throttling.

Plumbers use ball valves on water supply piping in place of gate valves or globe valves, and on low pressure gas piping in place of the lever handle gas cock.

Butterfly Valves. A butterfly valve, illustrated in Figure 3-20, is a valve with a rotating disk (the butterfly) that fits within the valve body. The rotating disk is controlled by a shaft. In its closed position, the disk seats against a resilient seat. A butterfly valve is opened or closed by only a 90° turn of the handle. It is a full-way valve that may be used for throttling purposes. One of the main advantages to the use of a butterfly valve is its very thin body, which permits it to be installed in spaces where no other valve would fit.

Check Valves. A check valve is a valve that permits the flow of water within the pipe in only one direction and closes automatically to prevent backflow (the flow in a reverse direction). Check valves offer quick, automatic reactions to changes in the direction of the flowing water because the pressure of the flowing water keeps the valve open and any reversal of flow closes it.

Figure 3-20. A butterfly valve. (NIBCO, Inc.)

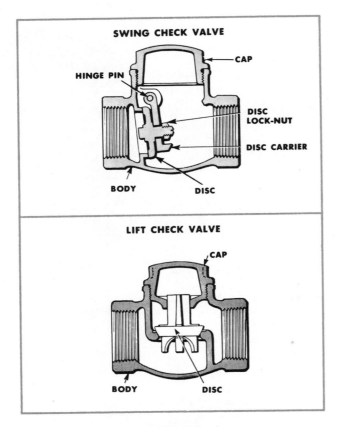

Figure 3-21. Check valves. (Stockham)

Figure 3-22. Backwater valves as an integral part of floor drains. Top: Ball-type backwater valve. Bottom: Swing-check-type backwater valve. (Wade Division, Tyler Pipe)

Check valves, which are illustrated in Figure 3-21, are available in two different styles, swing checks and lift checks. The swing check, in which the water flows straight through the valve, has a hinged disk that is held open by the flowing water. On the other hand, in the lift check, the water pressure forces the disk up from its seat and, in a reversal of flow, the disk drops back onto its seat. In a lift check, the water flow follows a winding course similar to that in a globe valve.

Swing check valves offer very little resistance to the flow of water. They are generally used with gate valves in installations where the water is moving at a low velocity and in which there is seldom a change in direction of flow.

Lift check valves, which have a rather high resistance to the flow of water, are usually used in conjunction with globe and angle valves in installations in which there are frequent changes in the direction of flow.

Backwater Valves. A backwater valve is a type of check valve installed to prevent the backflow of sewage from flooding the basement or lower levels of a building.

Figure 3-22 pictures two types of backwater valves that are an integral part of floor drains, a *ball-type backwater valve* (top) and a *swing-check backwater valve* (bottom). Floor drains with these types of backwater valves are installed in the basements of buildings to prevent flooding in case the building drain or building sewer should become stopped.

Figure 3-23 pictures a swing-check backwater valve. This type of backwater valve is installed on the branches of the building drain serving basement plumbing fixtures in areas where the sewer main in the street is subject to stoppage or flooding.

Figure 3-23. Swing-check backwater valve for building drain piping. (Wade Division, Tyler Pipe)

Figure 3-24. A pressure-reducing valve and strainer. (Watts Regulator Co.)

Figure 3-25. Relief valves for hot water heating equipment. Left: Temperature and pressure (T & P) relief valve. Right: Pressure-only relief valve. (Watts Regulator Co.)

Pressure-Reducing Valve. A pressure-reducing valve is an automatic device used for converting high, fluctuating inlet water pressure to a lower constant pressure. This valve is also called a *pressure-regulating valve.*

A pressure-reducing valve, of the type illustrated in Figure 3-24, is installed by plumbers on the building water service, near the water meter, to reduce excessively high street water main pressures. This pressure-reducing valve has an adjustment screw to adjust the outlet pressure. A strainer must be installed in conjunction with a pressure-reducing valve to keep dirt out of the working mechanism of the valve.

Relief Valves. A relief valve is a safety device that automatically provides protection against excessive temperatures, excessive pressures, or both. Plumbers install relief valves of the type pictured in Figure 3-25 to protect hot water heating equipment from the dangers of overheating and explosion.

Relief valves are available for pressure only or for temperature and pressure (T & P). T & P relief valves, the more commonly used of the two, are used on domestic hot water heaters. Pressure

relief valves are installed on some types of commercial and industrial hot water heating equipment.

Uses of Plumbing Valves

Plumbers install the types of valves just described in the following locations.

1. A corporation cock is installed at the connection of the water service to the water main.

2. A curb cock is located near the curb line of the building property to provide a means of controlling the water service to the building.

3. A full-way valve is installed on each side of the water meter. It may be either a gate valve, a ball valve, or a butterfly valve.

4. If necessary, a pressure-reducing valve would be installed between the water meter valves.

5. A full-way valve is installed on the cold water supply to all hot water heating equipment.

6. A relief valve is installed on all hot water heating equipment.

7. All sill cocks should be provided with a control valve located inside the building.

8. All water closets must have an individual

fixture control valve. It is highly recommended that all other plumbing fixtures and appliances also have individual control valves on their water supplies.

9. In apartment buildings, each family apartment unit must be equipped with shutoff valves controlling the hot and cold water supply to that unit, or the plumbing fixtures and appliances in the unit must each have its own fixture control valve. These valves are required so that repairs may be made in one apartment without interfering with the water supply to other apartments.

10. In buildings other than homes and apartments, shutoff valves must be installed on the water supply piping so that the water supply to all the equipment in each separate room or to each individual fixture can be shut off without interfering with the water supply to other rooms or parts of the building.

In general, all the shutoff valves installed by plumbers on water piping are full-way valves having the same size as the line on which they are installed. The only exception to this is that the fixture control valves for individual plumbing fixtures and appliances are compression stops or globe valves.

Note: One final point the plumber must remember when installing any shutoff valve is to place that valve in an accessible location or it will be worthless.

ⓞ WATER METERS

A water meter is a device used to measure in cubic feet or gallons the amount of water that passes through the water service. A water meter is installed in a building so that the building owner may be charged for the amount of water he uses. The water meter is installed at the end of the water service pipe either directly inside or outside of the building walls in accordance with local plumbing code restrictions. Three types of water meters are installed by plumbers: the *disk-type meter,* the *turbine meter,* and the *compound meter.*

Disk Meters. A disk meter (Figure 3-26) is used for measuring the flow of water through small water services. Disk meters, which are very accurate devices for measuring small flows of water, are available in sizes from 5/8 inch to 2 inch.

The internal working parts of a disk meter are illustrated in Figure 3-27. The water enters the meter and flows into the measuring chamber in which the nutating disk is fitted. As the water passes through the chamber, the disk rotates in a wobbling (nutating) motion and physically displaces all of the water from the chamber. The rotations of the disk are transmitted by way of a gear train to the recording dial.

Disk-type water meters are also called *displacement meters* because during a measuring cycle the water is completely removed or displaced from the measuring chamber with each rotation of the disk.

Turbine Meters. A turbine meter (Figure 3-28) is used in buildings in which water is used in large and constant volumes. In a turbine meter, water strikes the blades of the rotor (or turbine; Figure 3-29) and causes it to turn. This motion is transferred by means of a magnetic coupling to a vertical spindle and then to the gears in the meter's register. Turbine meters of this type are available in 2-, 3-, 4-, and 6-inch sizes. Although turbine meters very accurately measure large flows of water, they are not accurate for small flows.

Compound Meters. A compound meter (Figure 3-30) is a water meter that unites a disk and turbine meter in one body. Compound meters are made in 2- to 10-inch sizes for use in buildings in which there is a large fluctuation in water flow. A typical use is office buildings, which have a large flow of water during regular working hours and very little water use during the evening hours and weekends.

When water enters a compound meter (at a low rate of flow), the disk portion of the meter measures this flow because the *heavy-duty valve* prevents this water from entering the turbine portion of the meter. (These internal working parts are shown in Figure 3-31.) At high flow rates, the water forces this heavy-duty valve open, which closes off the flow of water to the disk portion of the meter and forces the water into the turbine portion of the meter to be measured.

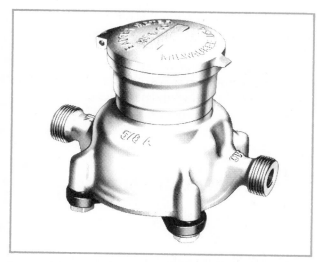

Figure 3-26. A disk-type water meter. (Badger Meter, Inc.)

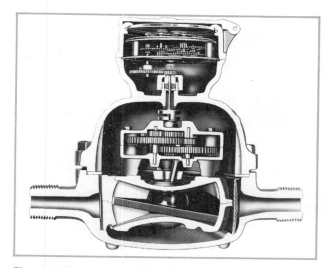

Figure 3-27. Internal working parts of a disk-type meter. (Badger Meter, Inc.)

Figure 3-28. A turbine water meter. (Badger Meter, Inc.)

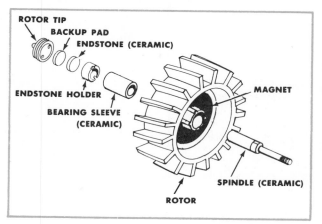

Figure 3-29. Rotor assembly (turbine) found inside of turbine meter pictured in Figure 3-28. (Badger Meter, Inc.)

Figure 3-30. A compound water meter. (Badger Meter, Inc.)

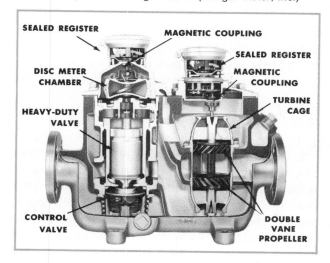

Figure 3-31. Internal working parts of a compound meter. (Badger Meter, Inc.)

Figure 3-32. A disk-type water meter with an outdoor register. (Badger Meter, Inc.)

Outside Registers. An outside register is shown installed on a disk-type water meter in Figure 3-32. Outside registers, which allow the meter reader to read the water meter without entering the building, are also available for turbine and compound meters. Since plumbers install this outdoor register when they install the water meter, the following description of the principle of operation of an outside register is given:

1. A self-contained generator is mounted on the water meter and connected to the outdoor register with a two-conductor wire. The flow of water starts the system operating by causing the meter's disk spindle to rotate. This motion is transferred to a spring-biased, six-pole magnet in the generator by means of a reduction gear train and an escape gear. When the escape gear is released, the biasing springs return the mag-

net to its original position. This action produces a low-voltage pulse in coils located near the magnet.

2. The pulse (approximately 8 volts) is transmitted over the wire to a solenoid in the outdoor register. The outdoor register advances one digit for every pulse received.*

ⓞ REVIEW QUESTIONS

1. Give a use for each of the following kinds of pipe and fittings:

bell and spigot cast iron soil pipe

no-hub cast iron soil pipe

galvanized steel pipe with:
(a) standard cast iron threaded fittings
(b) Durham fittings
(c) galvanized malleable iron fittings

DWV copper pipe and fittings

Types L and M copper and solder joint pressure fittings

Types K and L soft copper and flare fittings

ABS and PVC plastic pipe and fittings

CPVC plastic pipe and fittings

PE plastic tubing and insert fittings

2. Give a use for each of the following types of valves:

gate valve	butterfly valve
globe valve	check valve
angle valve	backwater valve
core cock	pressure-reducing valve
ball valve	relief valve

3. Name the three types of water meters described in the text. Which type would be used in a single family home? In a large office building?

4. What is the advantage to an outside water meter register?

*Courtesy: Badger Meter, Inc.

CHAPTER **4**

Plumbing Tools

The day is gone when a plumber could go out on a construction job with a small hand-tray full of tools and install the plumbing. The price of labor and the new materials used in the plumbing industry have made it necessary to equip today's plumber with many laborsaving and specialized tools which the apprentice plumber must learn to use and care for.

Each type of piping material requires its own special tools. The tools in this chapter will be presented in the same order as the plumbing materials were listed in Chapter 3. This chapter, naturally, will not be able to show *all* the tools of any given type, but will present *examples* of the many types available.

Depending on local union working rules and conditions the plumber may or may not be required to furnish tools. If he is required to furnish any tools they will probably be just basic hand tools.

Basic Plumbing Tools

To start any kind of piping work, the plumber will find it very helpful to have the following basic tools in his pocket:

6-foot folding rule
Soapstone with holder
Pencil
Slip-joint pliers
Torpedo level

These tools are illustrated in Figure 4-1. The 6-foot folding rule is used for measuring, a soapstone for marking soil pipe, a slip-joint pliers is a handy tool for tightening nuts and bolts, and a torpedo level is used for grading pipe.

 CAST IRON SOIL PIPE TOOLS

Cutting Tools. Cast iron soil pipe, either bell and spigot or no-hub, is cut with the same tools. Several different types of cast iron pipe cutters are illustrated in Figure 4-2. The first three cutters with the chain all cut pipe the same way. The chain, which has a cutter wheel at each link, is wrapped around the pipe and then drawn tight by ratchet action, pushing the handles together,

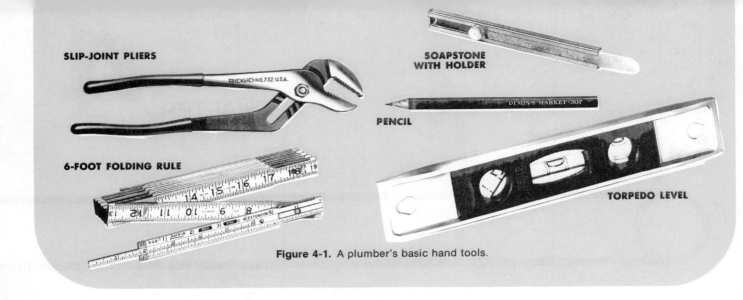

SLIP-JOINT PLIERS

SOAPSTONE WITH HOLDER

PENCIL

6-FOOT FOLDING RULE

TORPEDO LEVEL

Figure 4-1. A plumber's basic hand tools.

or by hydraulic action, until the pipe breaks cleanly beneath the center of the chain.

The *ratchet-type cutter* will cut either service or extra heavy weight soil pipe up to and including 6 inches in size. It has the advantage of being able to make a cut with a minimum amount of effort in close quarters.

The *squeeze-type cutter* is a better choice for no-hub soil pipe. It is quicker and easier to operate on this thin wall pipe than the ratchet-type cutter which has a tendency to crush or crack no-hub pipe and make ragged cuts on it. The squeeze-type cutter can also be used on service or extra heavy weights of bell and spigot soil pipe, but more effort is required to cut the pipe with this type of cutter.

The *hydraulic cutter* pictured is used to cut the larger sizes of soil pipe.

TO HYDRAULIC PUMP ASSEMBLY

HYDRAULIC CUTTER

RATCHET CUTTER

SQUEEZE-TYPE CUTTER

CAULKING HAMMER

COLD CHISEL

Figure 4-2. Cast iron soil pipe cutting tools: squeeze-type cutter (Wheeler Mfg. Corp.), ratchet cutter (Ridgid Tool Co.), hydraulic cutter (Wheeler Mfg. Corp.), caulking hammer (Stanley), cold chisel (Mephisto Tool Co.).

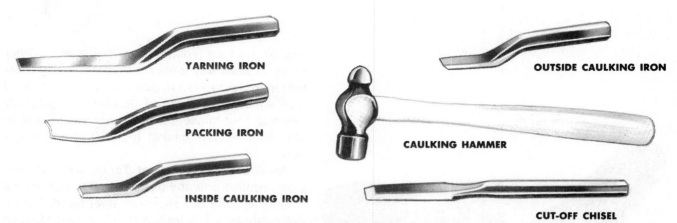

Figure 4-3. Lead and oakum caulking tools: yarning iron, packing iron, inside caulking iron, outside caulking iron, caulking hammer, cut-off chisel.

The *caulking hammer* (2-pound) and *cold chisel* are used to cut the pipe.

Caulk Joint Tools. A *lead and oakum joint* requires the use of *yarning irons, packing irons, inside caulking irons, outside caulking irons,* a *caulking hammer* (12-ounce) and *cut-off chisel.* Figure 4-3 shows these tools.

A *lead furnace* and *lead pot* are necessary to melt the lead, and a *ladle* is required to carry the molten lead to the hub. A *joint runner* or *running rope* is used for lead and oakum joints made in the horizontal position. These tools are shown in Figure 4-4.

Compression Gasket Soil Pipe Joint Tools. A *compression gasket joint* is made by inserting a gasket into the hub of the pipe or fitting and inserting the plain end of the pipe into the gasket until it seats. This can be done with any *assembly*

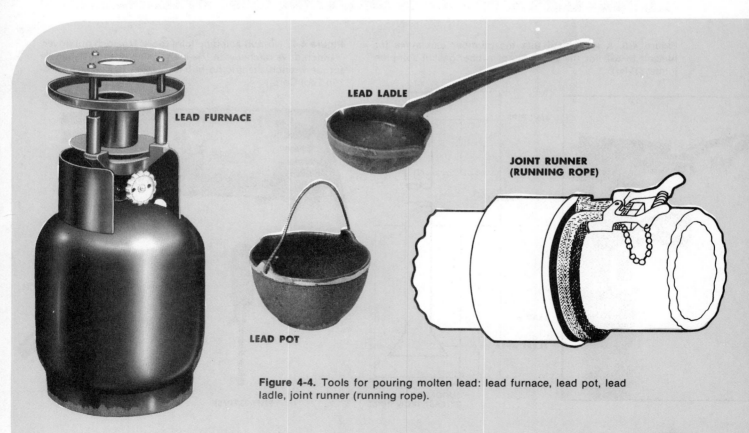

Figure 4-4. Tools for pouring molten lead: lead furnace, lead pot, lead ladle, joint runner (running rope).

Figure 4-5. Compression gasket joint soil pipe assembly tool. (Ridgid Tool Co.)

tool of the type illustrated in Figure 4-5. (Other types of assembly tools are made by many other tool manufacturers.) In addition to the illustrated tool, a *lead maul* (Figure 4-6), which the plumber can make for himself, works quite well to assemble compression gasket joints in some situations.

No-Hub Soil Pipe Joint Tools. No-hub soil pipe requires only the use of a tool to tighten the clamps around the stainless steel band. Figure 4-7 pictures a $5/16$-inch *nut driver*, and $1/4$-*inch drive ratchet* with a $5/16$-inch socket wrench that the plumber will find useful in tightening or loosening the no-hub clamps. The *T-handle torque wrench* tightens the clamp to the recommended 60 inch-pounds of torque and then releases its grip. An electric no-hub nut driver that also releases at 60 inch-pounds is also shown in Figure 4-7. This tool also reverses, allowing couplings to be loosened or removed. It is very useful for prefabrication shops or where many couplings must be made in a close area.

Figure 4-6. A lead maul that the plumber can make for himself to use for the assembly of rubber gasket soil pipe joints. (Tyler)

Figure 4-7. No-hub soil pipe joint tools: $5/16$-inch nut driver, $1/4$-inch drive ratchet with $5/16$-inch socket wrench, T-handle torque wrench, electric no-hub nut driver. (Milwaukee Electric Tool Corp.)

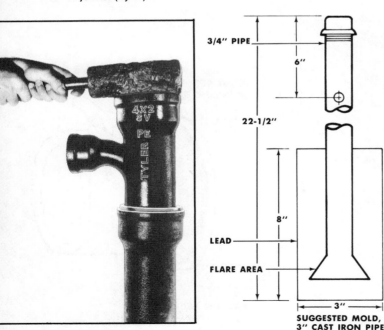

GALVANIZED STEEL PIPE TOOLS

Most of the galvanized steel pipe that plumbers install is threaded. Since very little pipe is installed in full lengths, it will be necessary to cut the pipe to the required length and thread it.

Hand Cutting and Threading Tools

To cut and thread galvanized steel pipe by hand you must have a *pipe vise* (Figure 4-8) to hold the pipe firmly.

A *wheel pipe cutter* (Figure 4-9) is used to make a square cut on the pipe. A square cut is necessary so that the dies will start straight. This type of cutter does not actually remove any metal, but rather the wheel squeezes the metal and forces it ahead of the cutter until the pipe is cut through the wall thickness. Since cutting the pipe in this manner leaves a large ridge on the inside of the pipe which would obstruct the flow, the pipe must be reamed with a *spiral pipe reamer* (Figure 4-10). Pipe is reamed after it is

Figure 4-8. Vise stand for holding pipe for cutting and threading. (Ridgid Tool Co.)

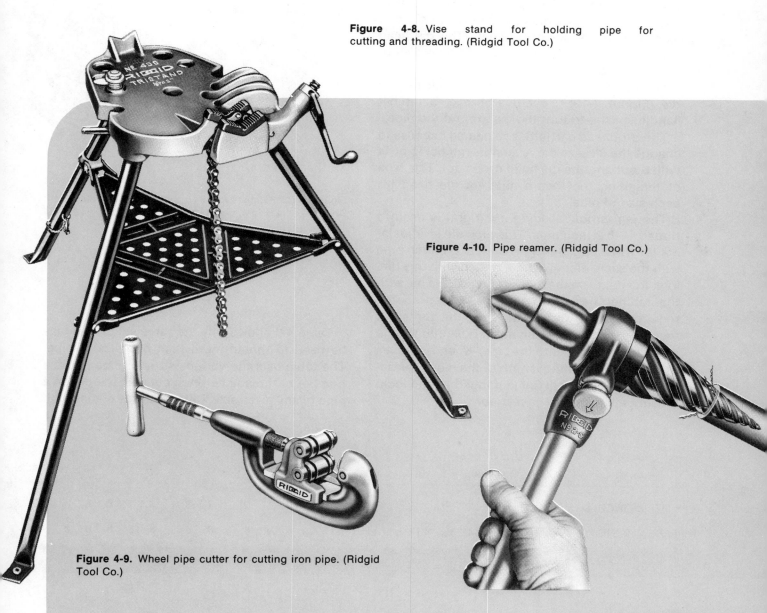

Figure 4-10. Pipe reamer. (Ridgid Tool Co.)

Figure 4-9. Wheel pipe cutter for cutting iron pipe. (Ridgid Tool Co.)

cut, but *before* it is threaded, because the threading operation removes a considerable amount of metal from the front edge of the pipe leaving it quite thin. The cone shape of the reamer tends to enlarge the pipe. The threading operation will remove this enlarged area. If the pipe were threaded first and *then* reamed, this enlarged end of the pipe would be difficult if not impossible to catch in the fitting.

After the pipe is cut and reamed, it may then be threaded with one of the types of dies illustrated in Figures 4-11 and 4-12.

The die shown in Figure 4-11 is the *nonratcheting die*. The handles of this die must be rotated completely around the pipe to turn the die head. This three-way die holds three different sizes of pipe dies in the same die head. The die head either holds $3/8$-, $1/2$-, and $3/4$-inch pipe dies or $1/2$-, $3/4$-, and 1-inch pipe dies.

Another type of hand threader is the *ratchet die* (Figure 4-12). To cut a thread, a ratchet handle is used to turn the die around the pipe.

The *drophead die* (left) is so named because to change the die size the reversible ratchet knob is pulled out and the die head drops out. This type of threading tool has a different die head for each size of pipe.

The *jam-proof receding die* (right) is used to thread 1- to 2-inch pipe. This pipe die is adjustable for threading 1-, $1^{1}/_{4}$-, $1^{1}/_{2}$-, and 2-inch pipe with the same dies. To cut a thread with this die, lock the guide on the pipe and rotate the die with the ratchet handle to cut the thread to the required length. As the die cuts the thread, it recedes on a screw mechanism built into the die head to feed itself onto the pipe. When it reaches the maximum cutting length of the die, it releases so that the die will not jam together and lock; thus it gets the name *jam-proof*.

Figure 4-11. Three-way pipe die for threading three different sizes of pipe. (Ridgid Tool Co.)

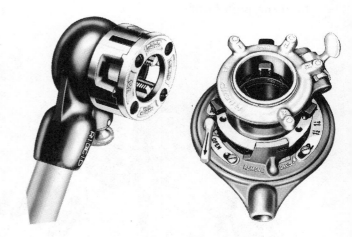

Figure 4-12. Ratchet-type iron pipe threading dies. Left: Drophead die. Right: Jam-proof receding pipe die. (Ridgid Tool Co.)

Table 4-1 shows why the same set of dies may be used to thread more than one size of pipe. The table lists the various sizes of pipe and the number of threads per inch cut with the die on a pipe of any given size. Different sizes of pipe with

TABLE 4-1.   THREADS PER INCH FOR VARIOUS SIZES OF STEEL AND WROUGHT IRON PIPE.

PIPE SIZE IN INCHES	$1/8$	$1/4$	$3/8$	$1/2$	$3/4$	1	$1^{1}/_{4}$	$1^{1}/_{2}$	2	$2^{1}/_{2}$	3	$3^{1}/_{2}$	4	5	6	8	10	12
THREADS PER INCH	27	18	18	14	14	$11^{1}/_{2}$	$11^{1}/_{2}$	$11^{1}/_{2}$	$11^{1}/_{2}$	8	8	8	8	8	8	8	8	8

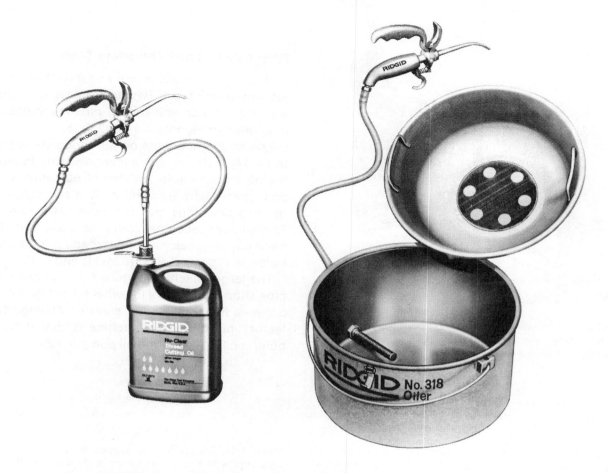

Figure 4-13. Oilers for use in threading iron pipe. Left: Oiler that uses oil can for reservoir. Right: Oiler with reservoir that catches oil drippings. (Ridgid Tool Co.)

the same number of threads per inch can usually be threaded with the same dies in an adjustable die head.

All of the dies illustrated in Figures 4-11 and 4-12 can be used in a pipe machine, but they must be backed off the thread manually to remove the die from the pipe after the thread is cut. This feature makes them awkward to use with a pipe machine because the machine must be turned off and then either the machine reversed or the die head backed off by hand.

All pipe dies must be liberally lubricated while they are cutting the thread. Cutting oil is needed to cool the dies and threads, to speed metal removal, to produce a smooth finish, and to

prevent chips from welding themselves to the dies. Cutting oils are especially formulated for this purpose. To apply the cutting oil, you will need an oiler. Although a plain squirt-type oil can will work for this purpose, Figure 4-13 illustrates oilers that are made for the plumber. The oiler mounted on the *oil container* is simply a pumping mechanism with a hose that uses the oil can for a reservoir. The oiler with the *reservoir* is somewhat neater to use, as the drippings of oil and chips of metal can fall back into the chip pan and the oil can drip through the pan into the reservoir and be reused. An oiler of the reservoir type helps keep the pipe-threading area neat and clean.

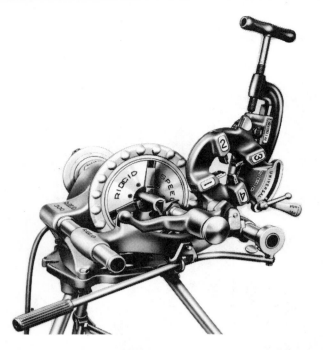

Figure 4-14. Portable pipe-threading machine complete with cutter, reamer, and die head mounted on a tripod stand for threading small diameter (2-inch and under) pipe. (Ridgid Tool Co.)

Power Cutting and Threading Tools

Most iron pipe will be cut and threaded on the job with power tools. Two different types of *small diameter* (2-inch and under) *pipe machines* are illustrated in Figures 4-14 and 4-15.

The smaller machine on the tripod stand (Figure 4-14) is handy where it is necessary to move the pipe machine in and out of buildings or up and down stairs frequently—it is the lighter of the two machines and the more easily carried. The smaller machine could be used with the hand cutting, reaming, and threading tools illustrated in Figures 4-9, 4-10, 4-11, and 4-12.

The larger machine (Figure 4-15) is handy in a pipe shop area or on a job where it can be moved between floors on a lift or elevator. An important feature of the larger machine is that it has a built-in oil reservoir and oil pump to lubricate the

Figure 4-15. Small diameter (2-inch and under) pipe machine mounted on wheel stand. This machine features a cutter, reamer, die head, and built-in oiler assembly. (Ridgid Tool Co.)

Figure 4-16. Adjustable pipe support for use in threading long lengths of pipe to prevent the pipe from overbalancing the machine and tipping it. (Ridgid Tool Co.)

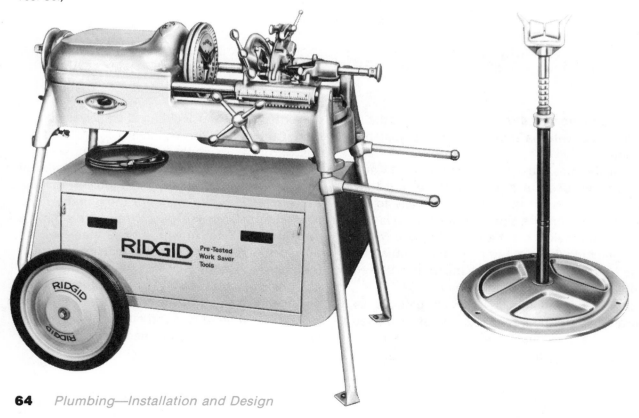

threads while they are being cut. A separate oiler must be carried with the small machine.

Both of these machines, as illustrated, have the cutter, reamer, and a quick-opening die head mounted on the machine.

A *pipe support* (Figure 4-16) should also be used with these machines (especially the smaller machine) to keep long lengths of pipe from overbalancing the machine and tipping it over.

Quick-opening die heads (Figure 4-17) have a lever which is raised to open up the dies when the thread has been cut to the required length; They do not have to be backed off the thread after the thread is cut. Quick-opening die heads, which fit the pipe machines pictured in Figures 4-15 and 4-16, are available in two styles. The *mono head die* is for threading just one size of pipe; a different die head is required for each size of pipe. The *universal die head* is adjustable

and with the same set of dies it is possible to thread 1-, 1$^1/_4$-, 1$^1/_2$- and 2-inch pipe. Two other die sets that fit in the same universal die head permit the threading of $^1/_2$- and $^3/_4$-inch and $^1/_4$- and $^3/_8$-inch pipe.

Figure 4-18 illustrates a *large diameter* (2$^1/_2$- to 4-inch) *pipe machine.* It is basically the same as the pipe machine shown in Figure 4-15, in that it has a self-contained cutter, reamer, oiling system, and quick-opening die heads. It is a very fast way to cut and thread large diameter pipe. One disadvantage of this machine is that it is a large, heavy piece of equipment that is not easily moved around the job site. Therefore, the pipe must be cut and threaded at the machine site, and then carried to the work location. Another disadvantage, on some jobs, is that a machine of this type requires more than 120 volts of electrical current.

Figure 4-17. Quick-opening die heads for use in the machines pictured in Figures 4-14 and 4-15. Top: Mono (one size) die head. Bottom: Universal die head. (Ridgid Tool Co.)

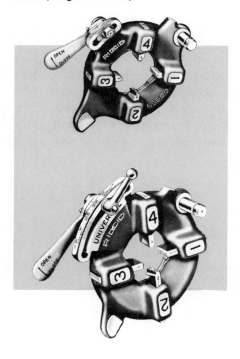

Figure 4-18. Large diameter (2$^1/_2$- to 4-inch) pipe threading machine. Machine shown has built-in oiler, cutter, reamer, and adjustable die head. (Collins Machinery)

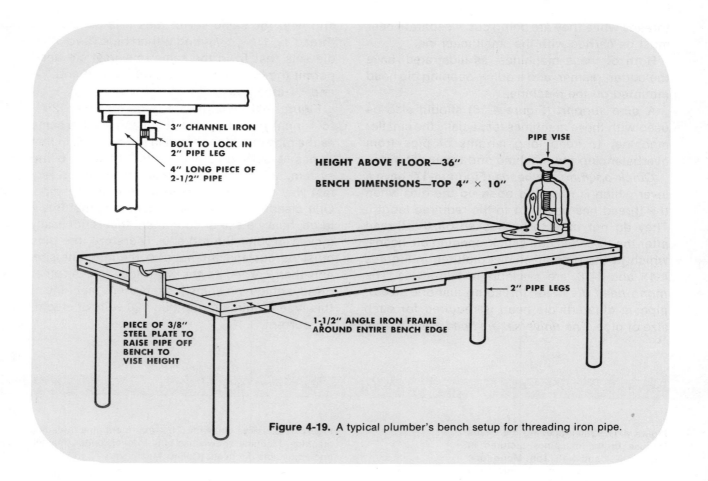

3" CHANNEL IRON

BOLT TO LOCK IN
2" PIPE LEG

4" LONG PIECE OF
2-1/2" PIPE

HEIGHT ABOVE FLOOR—36"

BENCH DIMENSIONS—TOP 4" × 10"

PIPE VISE

2" PIPE LEGS

PIECE OF 3/8"
STEEL PLATE TO
RAISE PIPE OFF
BENCH TO
VISE HEIGHT

1-1/2" ANGLE IRON FRAME
AROUND ENTIRE BENCH EDGE

Figure 4-19. A typical plumber's bench setup for threading iron pipe.

Large diameter pipe may also be cut and threaded at a *bench setup* as illustrated in Figure 4-19, or in a pipe vise stand (Figure 4-8) with the tools illustrated in Figures 4-20 through 4-26.

Figures 4-20 and 4-21 illustrate two different types of *wheeled pipe cutters*. The *single-wheel pipe cutter* (Figure 4-20) is basically the same as those used for cutting smaller diameter pipe.

This cutter must be rotated around the pipe by hand while the pipe is held in a vise. Cutters of this type can cut up to and including 6-inch pipe.

The other large diameter wheel cutter (Figure 4-21) is a *hinged four-wheel pipe cutter*. This cutter does not rotate completely around the pipe to make a cut. It only needs to swing a little more than 90 degrees. It also has the advantage

Figure 4-20. Single-wheel pipe cutter for large diameter (greater than 2-inch) iron pipe. (Ridgid Tool Co.)

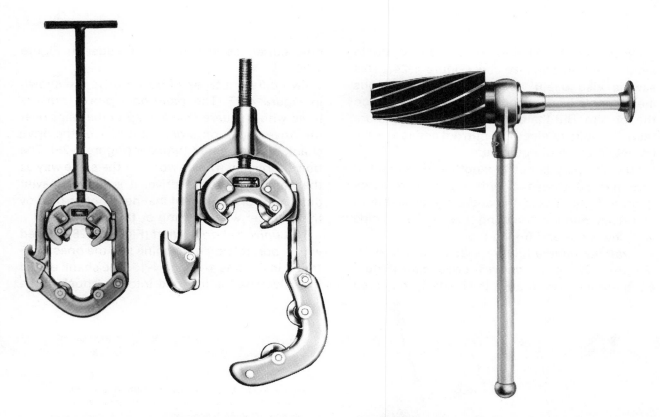

Figure 4-21. Four-wheel hinged pipe cutter for large diameter (greater than 2-inch) iron pipe (Reed).

Figure 4-22. Spiral reamer (2¹/₂- to 4-inch) for removing the ridge left by wheel-type pipe cutters. (Ridgid Tool Co.)

of being hinged so that it may be opened and put over pipe that is already installed and used to make a cut. Hinged cutters of this type are available in sizes to cut up to 12-inch pipe.

A blade-type geared pipe cutter cuts the pipe by actually removing metal from the pipe as it rotates around the outside diameter of the pipe. The cutter is usually power driven. As its four cutting blades rotate around the pipe, the pipe is hand fed into the cutter until the pipe is cut through. Pipe cutters of this type are available for cutting 2¹/₂- to 12-inch pipe. This cutter is made in two sections so that it may be opened

and closed around installed pipe, enabling the plumber to cut the pipe in place. It is usually not necessary to ream the pipe that has been cut with the geared pipe cutter because it cuts the pipe by removing metal rather than by forcing it ahead of the cutter wheel, as do the cutters illustrated in Figures 4-20 and 4-21.

A *large diameter spiral reamer* (Figure 4-22) is available for 2¹/₂- to 4-inch pipe. Pipe larger than 4 inches must be reamed with a *half-round file* (Figure 4-23) because no reamers are made for the larger pipe sizes.

Figure 4-23. Half-round file and handle for reaming iron pipe. (Nicholsen File Co.)

Large diameter pipe may also be quickly cut to length with an *abrasive saw* (Figure 4-24). This saw is also useful for cutting miscellaneous metal (angle iron, channel iron, and sleeves) on the job site. But the abrasive saw requires more than 120 volts of electrical current and is a large, immobile piece of equipment.

After the pipe is cut to length and reamed, it can then be threaded with a *geared threader* (Figure 4-25). These threaders are available in two sizes, one for threading 2¹/₂- to 4-inch pipe and one for 5- and 6-inch pipe.

A *ratchet handle* (Figure 4-26), a *power drive* (Figure 4-27), or a *universal drive shaft* (Figure 4-28) are all used to drive the blade-type geared

pipe cutter, as well as the threader in Figure 4-25.

Two different types of power drive are shown in Figure 4-27. The *hand-held power drive* is fitted with a square shank adaptor that slips over the drive shank on a geared cutter or the drive shank on the geared threader (Figure 4-25). The *power drive on wheels* works in the same way as the hand-held power drive. It is just a heavier duty piece of equipment that need not be held by the plumber when cutting or threading.

The *universal drive shaft* (Figure 4-28) is used with a portable pipe machine like the ones illustrated in Figures 4-14 and 4-15. The shank end of the drive shaft is inserted into the chuck of the

Figure 4-24. Abrasive saw for cutting large diameter iron pipe and other kinds of metal on the job site. (Collins Machinery)

Figure 4-25. Geared threader for threading iron pipe. Available in two sizes: for threading 2¹/₂- to 4-inch pipe; and for threading 5- and 6-inch pipe. (Ridgid Tool Co.)

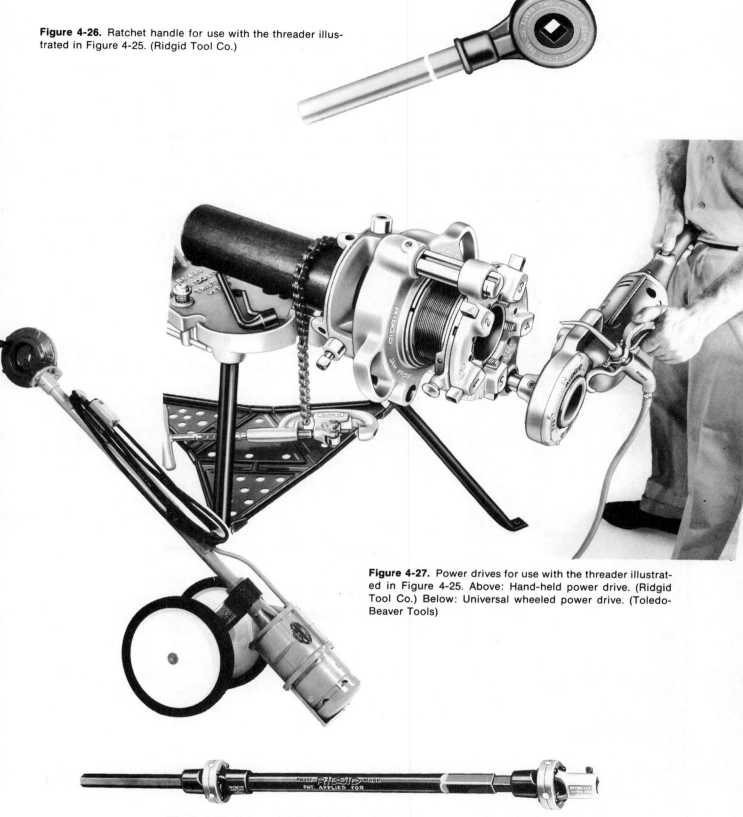

Figure 4-26. Ratchet handle for use with the threader illustrated in Figure 4-25. (Ridgid Tool Co.)

Figure 4-27. Power drives for use with the threader illustrated in Figure 4-25. Above: Hand-held power drive. (Ridgid Tool Co.) Below: Universal wheeled power drive. (Toledo-Beaver Tools)

Figure 4-28. Universal drive shaft that uses either of the pipe machines illustrated in Figures 4-14 or 4-15 as a power source. (Ridgid Tool Co.)

machine, and the end with the square head opening is slipped over the drive pinion of the tool.

Assembly Tools

To assemble threaded pipe, the plumber will need to have on hand a variety of sizes and types of *pipe wrenches*, as illustrated in Figure 4-29.

The size and type of pipe wrench a plumber uses will be determined by the size of pipe he works on, the closeness of the area in which he must work, and on personal preference for a certain type of wrench.

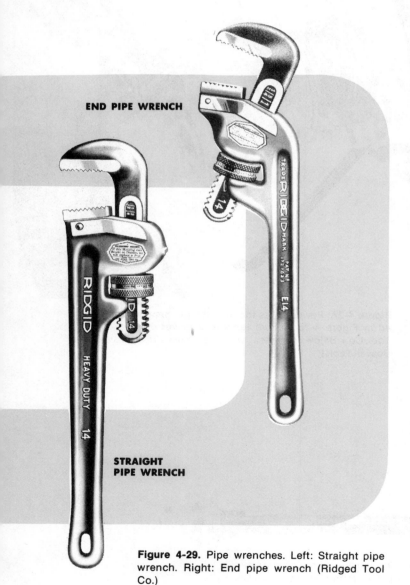

END PIPE WRENCH

STRAIGHT PIPE WRENCH

Figure 4-29. Pipe wrenches. Left: Straight pipe wrench. Right: End pipe wrench (Ridged Tool Co.)

The *straight pipe wrench* (Figure 4-29) is available in sizes from 6 to 60 inches. Straight pipe wrenches made from aluminum rather than malleable iron are available. Aluminum wrenches have the advantage of being some 40 percent lighter than malleable iron wrenches of the same size. Aluminum straight pattern pipe wrenches are made in 10-inch to 48-inch sizes.

The *end pipe wrench* (Figure 4-29) is used in areas where the pipe is hard to get at, such as when it is close to a wall, in tight quarters, or with other closely spaced parallel pipe lines. End pipe wrenches are made in 6- to 36-inch sizes. They are also available in aluminum in three sizes: 14, 18, and 24 inches.

Table 4-2 lists the suggested wrenches to use with the various sizes of pipe. In addition, the table gives the approximate weight of each wrench.

Chain wrenches and *chain tongs* can also be used to tighten threaded iron pipe and fittings. They have the advantage of distributing the biting force evenly around the pipe without crushing it and are handy in close quarters, as the only clearance necessary is enough room to be able to wrap the chain around the pipe or fitting. Figure 4-30 illustrates several different types of chain wrenches and chain tongs. Chain

TABLE 4-2. SUGGESTED WRENCHES TO USE WITH VARIOUS PIPE SIZES.

WRENCH LENGTH (INCHES)	PIPE SIZE (INCHES)	Steel straight and end wrench	Aluminum straight wrench	Aluminum end wrench
		WEIGHT OF WRENCH IN POUNDS		
6	3/4	1/2	——	——
8	1	3/4	——	——
10	1 1/2	1 3/4	1	——
12	2	2 3/4		——
14	2	3 1/2	2 1/3	1 3/4
18	2 1/2	5 3/4	3 2/3	3 1/2
24	3	9 3/4	6	5 3/4
36	5	19	11	——
48	6	34 1/4	18 1/2	——
60	8	51 1/4	——	——

Ridgid Tool Co.

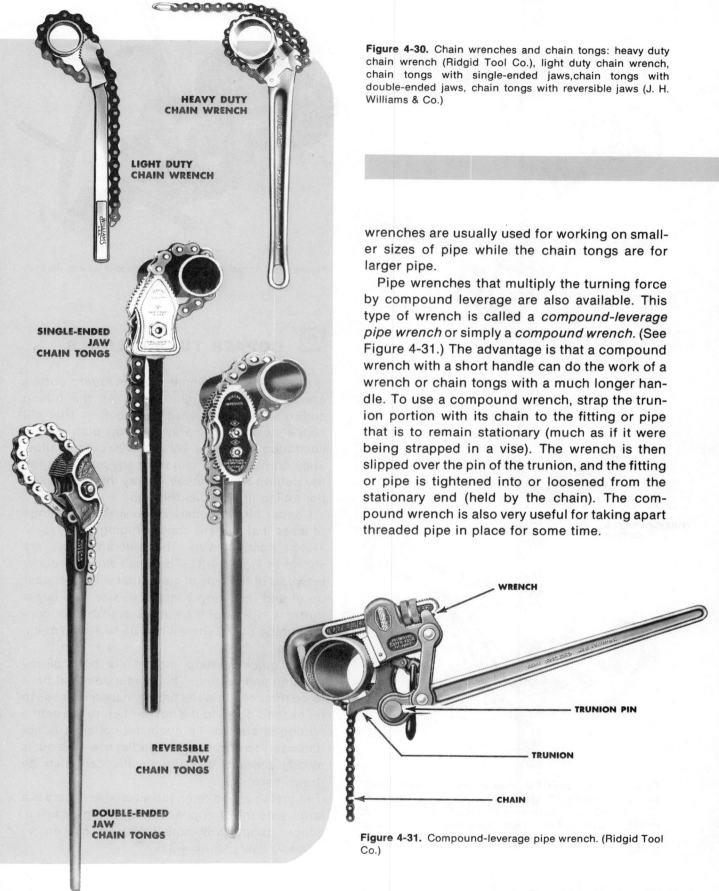

HEAVY DUTY CHAIN WRENCH

LIGHT DUTY CHAIN WRENCH

SINGLE-ENDED JAW CHAIN TONGS

REVERSIBLE JAW CHAIN TONGS

DOUBLE-ENDED JAW CHAIN TONGS

Figure 4-30. Chain wrenches and chain tongs: heavy duty chain wrench (Ridgid Tool Co.), light duty chain wrench, chain tongs with single-ended jaws, chain tongs with double-ended jaws, chain tongs with reversible jaws (J. H. Williams & Co.)

wrenches are usually used for working on smaller sizes of pipe while the chain tongs are for larger pipe.

Pipe wrenches that multiply the turning force by compound leverage are also available. This type of wrench is called a *compound-leverage pipe wrench* or simply a *compound wrench.* (See Figure 4-31.) The advantage is that a compound wrench with a short handle can do the work of a wrench or chain tongs with a much longer handle. To use a compound wrench, strap the trunion portion with its chain to the fitting or pipe that is to remain stationary (much as if it were being strapped in a vise). The wrench is then slipped over the pin of the trunion, and the fitting or pipe is tightened into or loosened from the stationary end (held by the chain). The compound wrench is also very useful for taking apart threaded pipe in place for some time.

WRENCH

TRUNION PIN

TRUNION

CHAIN

Figure 4-31. Compound-leverage pipe wrench. (Ridgid Tool Co.)

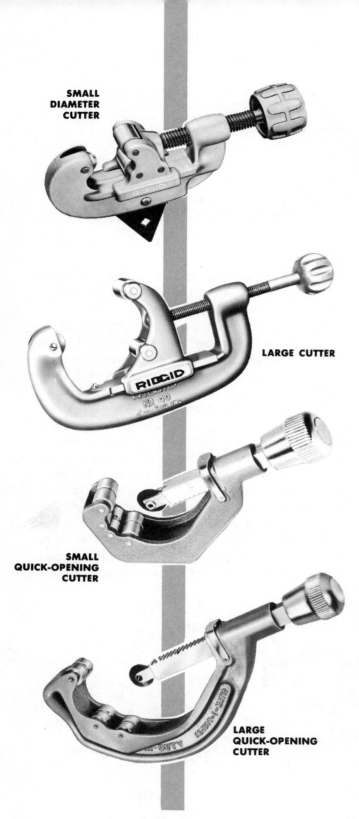

SMALL DIAMETER CUTTER

LARGE CUTTER

SMALL QUICK-OPENING CUTTER

LARGE QUICK-OPENING CUTTER

Figure 4-32. Copper tubing cutters: small diameter (2-inch and under) cutter (Ridgid Tool Co.), 2- to 4-inch cutter (Ridgid Tool Co.), quick-opening cutter for 2½-inch and smaller copper tubing (Imperial-Eastman Corp.), quick-opening cutter for 2- to 4-inch tubing (Imperial-Eastman Corp.).

Figure 4-33. Abrasive saw for cutting copper tubing. (Ridgid)

⊙ COPPER TUBING TOOLS

Cutting Tools. Some *copper tubing cutters* are illustrated in Figure 4-32. All the tubing cutters shown are wheel-type cutters that cut copper in the same way the iron pipe cutters illustrated in Figures 4-9 and 4-20 cut iron pipe. They do not cut by removing any metal; rather, the cutting wheel forces its way into the wall of the tubing until it cuts through.

Copper tubing cutters are available in a range of sizes that will cut copper tubing from ¼- to 4-inch nominal size. The different types are shown in Figure 4-32. The *small diameter cutter* is typical of the type of cutter that will cut copper up to and including 2 inches in size. The *larger cutter* is typical of the cutters available for copper tubing over 2 inches and up to 4-inch nominal size.

The *quick-opening cutters* are both quick-opening and closing. The handle does not have to be threaded into the tubing; it merely needs to be pushed down to the copper before the actual cutting is started. To open the cutter, it is not necessary to unscrew the handle; a release nut is merely pressed and the handle can then be drawn back.

In prefab shops or on job sites where there is a large amount of copper tubing being installed, copper tubing will sometimes be cut on an *abrasive saw* (Figure 4-33).

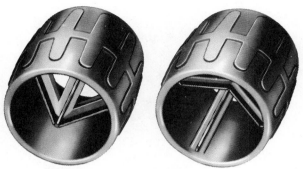

Figure 4-34. Combination inside-outside reaming tool for ¹/₄- to 1¹/₂-inch copper tubing. (This is the same tool seen front and back.) (Ridgid Tool Co.)

Reamers. Being wheel-type, these copper cutters leave a ridge on the inside of the tubing that must be removed or it will cause a restriction in the flow through the pipe. Cutters for the smaller sizes of copper tubing (such as in Figure 4-32) have *reamers* built into the body of the tool. Figure 4-34 illustrates a *combination inside-outside reamer* for copper tubing ¹/₄- through

1¹/₂-inch nominal size. Copper tubing may also be reamed with either a full- or half-round file.

Cleaning Tools. To prepare copper tubing and fittings for solder, the outside of the pipe and the inside of the fittings must be mechanically cleaned bright. This can be done with the tools illustrated in Figure 4-35.

The copper tubing *cleaning tool* is for cleaning inside and outside of ¹/₂- and ³/₄-inch nominal size copper tubing and fittings only. Although it only cleans these two sizes of copper tubing, it is still a very useful tool for residential work where most of the plumbing involves ¹/₂- and ³/₄-inch copper tubing.

Individual copper tubing *cleaning brushes* are available for copper tubing from ¹/₈- to 1-inch nominal size. For cleaning copper tubing larger than 1-inch nominal size (and any of the smaller sizes), the plumber can use *abrasive sandcloth,* which is available in roll form.

Copper tubing *fitting brushes* are available for cleaning the inside of copper fittings from ¹/₈- to 2¹/₂-inch nominal size. Larger copper fittings will have to be cleaned with abrasive sandcloth.

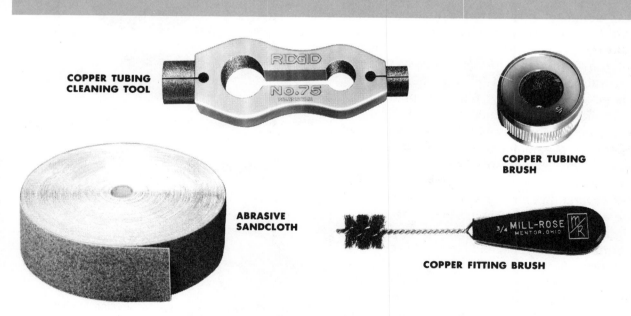

COPPER TUBING CLEANING TOOL

COPPER TUBING BRUSH

ABRASIVE SANDCLOTH

COPPER FITTING BRUSH

Figure 4-35. Copper tubing and fitting cleaning tools: tool for cleaning ¹/₂- and ³/₄-inch copper tubing and fittings (Ridgid Tool Co.), copper tubing brush (Mill-Rose Co.), abrasive sandcloth (Mill-Rose Co.), copper fitting brush (Mill-Rose Co.).

Figure 4-36. Copper-cleaning machine. This machine will clean both the outside of the copper tubing and the inside of the copper fittings as well as deburr the cut end of the tubing. (Ridgid Tool Co.)

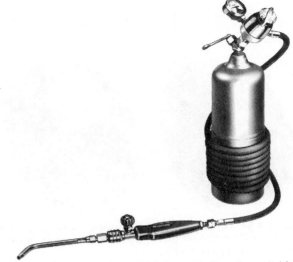

Figure 4-37. Air-acetylene torch for soldering copper tubing. (Ridgid Tool Co.)

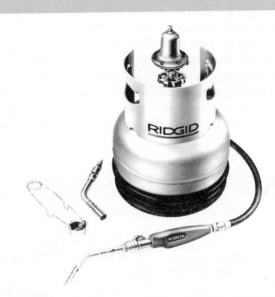

Figure 4-38. A propane torch kit for soldering copper tubing. (Ridgid Tool Co.)

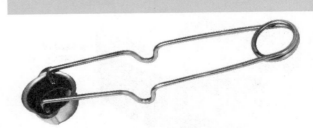

Figure 4-39. A friction ignitor for lighting torches. (Ridgid Tool Co.)

In prefab shops, there will usually be a *copper-cleaning machine* (Figure 4-36) for cleaning the copper tubing and fittings.

Soldering Tools. To make a copper tubing solder joint, heat the joint to the melting point of the solder being used. Use an *air-acetylene torch* (Figure 4-37) or a *propane gas torch* (Figure 4-38). A *friction ignitor* for lighting these torches is illustrated in Figures 4-38 and 4-39.

Flared Copper Tools. Flared copper joints require the use of tools illustrated in Figure 4-40. The *flaring tool* is driven into the end of the copper with a hammer to form the flare. Each size of copper tubing requires a different size flaring tool. The *yoke and screw* flaring tool flares several different sizes of copper with the same tool. To flare with this tool, the copper tubing is inserted into the yoke, flush with the top surface of the yoke, and clamped firmly. The screw portion of the tool is then tightened down into the copper, forcing it to flare out into the yoke. The copper *tubing cutter* illustrated in Figure 4-40 is basically the same as the cutters shown in Figure 4-32. The only difference is that the rollers are grooved so that a flare can be cut off very close to the end to minimize waste.

Bending Tools. There are occasions on the job when it is desirable to bend copper tubing rather than to use a fitting to make a change in direction or an offset. Small diameter (up to 3/4-inch nominal size or 7/8-inch outside diameter) copper tubing may be bent with the tools shown in Figure 4-41. The use of these tools will produce a smooth bend in the tubing without the danger of kinking it. All of the tools illustrated are for bending one size of tubing only. A different size tool is required for each size of tubing.

The *spring bender* is used to bend soft temper copper tubing from 1/4-inch to 7/8-inch outside diameter. The spring is slipped over the end of the tubing to the area of the bend and then the spring and the tubing are bent together. After the bend is completed, the spring is slipped back off the tubing.

The *lever-type bender* is for bending hard or soft temper tubing having a 3/16-inch to 1/2-inch outside diameter. The *geared ratchet-type bender* is for bending hard drawn copper tubing having a 5/8- to 7/8-inch outside diameter. Both of these benders operate in the same fashion. The ratchet bender is geared to make the bending operation easier on the larger sizes of hard drawn tubing. To use both of these tools, insert the tubing into the bending form and then using the handle, bend the tubing to the desired angle.

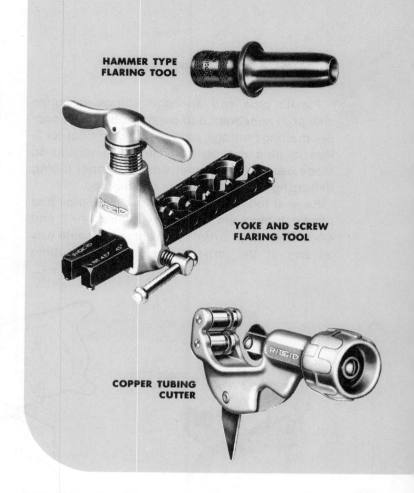

HAMMER TYPE FLARING TOOL

YOKE AND SCREW FLARING TOOL

COPPER TUBING CUTTER

Figure 4-40. Flared copper tools: hammer-type flaring tool, yoke and screw flaring tool, copper tubing cutter for cutting off flared ends of tubing with a minimum amount of waste. (Ridgid Tool Co.)

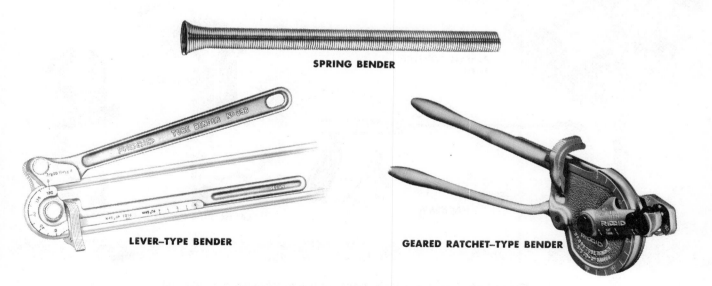

SPRING BENDER

LEVER–TYPE BENDER

GEARED RATCHET–TYPE BENDER

Figure 4-41. Tools for bending copper tubing: spring bender, lever-type bender, geared ratchet-type bender. (Ridgid Tool Co.)

⊙ PLASTIC PIPE TOOLS

Plastic pipe and drainage fittings are quite similar in appearance to copper tubing and copper drainage fittings, as was shown in Chapter 3. Plastic pipe cutting tools are also similar to those used to cut copper tubing. Copper tubing deburring tools can deburr plastic pipe.

Several tools for working on plastic pipe and plastic tubing are shown in Figures 4-42 and 4-43. The *blade cutters* operate in the same way as any of the large diameter copper tubing

cutters illustrated in Figure 4-32, except that instead of a replaceable cutting wheel the cutter has a replaceable cutting blade. Also shown is a *wooden miter box* for holding the plastic pipe when a saw is used to cut the pipe. (Any carpenter's saw will do the job.) A *vise* is used for holding plastic pipe and has covered jaws to protect the plastic.

The last cutting tool pictured in Figure 4-42 is the common *hacksaw*, which is an excellent tool for cutting plastic pipe or tubing as long as you are careful to make a square cut. Any saw cut on

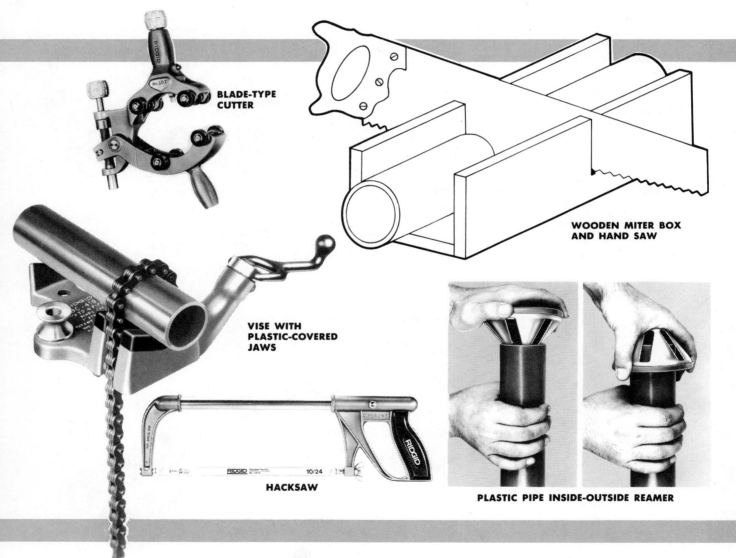

BLADE-TYPE CUTTER

WOODEN MITER BOX AND HAND SAW

VISE WITH PLASTIC-COVERED JAWS

HACKSAW

PLASTIC PIPE INSIDE-OUTSIDE REAMER

Figure 4-42. Plastic pipe tools: blade-type cutter, wooden miter box and hand saw, vise with plastic-covered jaws to prevent marring pipe, hacksaw, plastic pipe inside-outside reamer. (Ridgid Tool Co.)

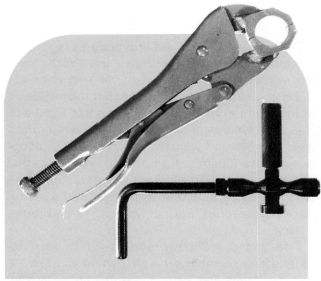

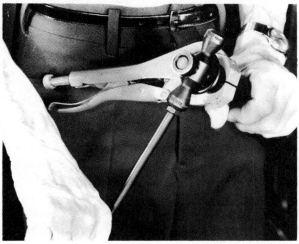

Figure 4-43. Flaring tool for flaring plastic water service tubing. (Reed Mfg. Co.)

plastic pipe tends to leave a very ragged burr that must be removed or it will cause an obstruction in the installed pipe line.

A *reaming tool* (Figure 4-42) is used to remove the inside burrs and the outside ridges from plastic pipe. The tool is available in three sizes for plastic pipe from 1/2- to 4-inch size. In addition, it should be noted that a full- or half-round file or pocket knife will work as well for deburring plastic pipe.

Figure 4-43 illustrates a *flaring tool* for flaring plastic water service tubing when this tubing is used with flare-type fittings. (The use of this tool will be discussed further in Chapter 5.)

◖ FINISHING TOOLS

The tools that have been mentioned previously in this chapter have mainly been for the installation of the plumbing system prior to the final setting of fixtures or *finishing*. The following figures will illustrate a variety of tools used for finishing.

Assembly Tools

Figure 4-44 shows several smooth-jawed tools that are used for tightening the chrome-plated

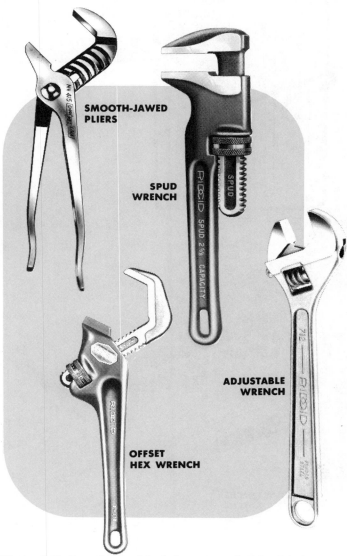

Figure 4-44. Smooth-jawed tools for use on chrome-plated plumbing fixtures: smooth-jawed pliers (Channel Lock), spud wrench (Ridgid Tool Co.), offset hex wrench (Ridgid Tool Co.), adjustable wrench (Ridgid Tool Co.)

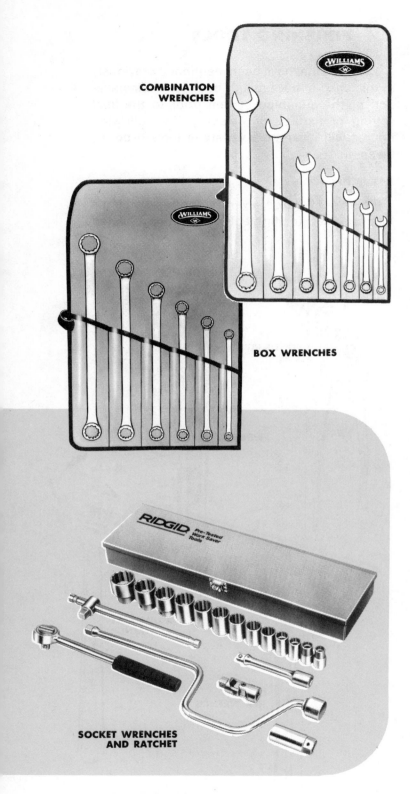

COMBINATION WRENCHES

BOX WRENCHES

SOCKET WRENCHES AND RATCHET

Figure 4-45. Useful wrenches for finishing work: combination wrenches (J. H. Williams & Co.), box wrenches (J. H. Williams & Co.), socket wrenches and ratchet (Ridgid Tool Co.).

nuts that secure plumbing fixtures to their mounting place. They are also used to tighten the various slip nuts that secure trap connections and flush valves to the fixtures without marring the chrome plating. The tools used are the *smooth-jawed pliers, spud wrench, offset hex wrench,* and *adjustable wrenches.* In addition to the illustrated smooth-jawed tools, the plumber will also use an assortment of *combination wrenches, box end wrenches,* and a set of *socket wrenches and a ratchet.* (See Figure 4-45.)

The *basin wrench* (Figure 4-46) is used to extend the plumber's reach behind fixtures to tighten the water supply connections to the faucet and to tighten up slip nuts on fixture traps in tight places. Basin wrenches are available with different sizes of jaws to take nuts from $3/8$-inch to $2^1/2$ inches in size.

The *closet seat wrench* (Figure 4-47) is designed to accept the many different types of nuts that are used to fasten the closet seat to the closet bowl. The wrench ends are adapted for use with a leverage pin, adjustable wrench, or a ratchet. On some closet seats, adjustable wrenches, deep socket wrenches and a ratchet, or even a pliers will work as well.

The *strap wrench* (Figure 4-48) is used for tightening chrome-plated pipe and fittings so that they will not be marred with wrench marks. Powdered rosin is sprinkled on the strap to help it bite on the smooth chrome surface.

The *basket strainer wrenches* (Figure 4-49) are two highly specialized plumbing tools—they are used to tighten the nut that secures the kitchen sink basket strainer assembly to the sink. A large slip-joint pliers or a small pipe wrench may also be used for this purpose.

The *rim wrench* (Figure 4-50) is used to tighten the rim clamps that secure kitchen sinks and lavatories to the counter top. This tool holds the clamp while it is inserted into the rim and being tightened, and then releases when the clamp is tight.

Screwdrivers of both the regular (Figure 4-51) and Phillips (Figure 4-52) type in several sizes are also necessary tools for finishing work, for tightening the various screws and bolts on plumbing fixtures, faucets, etc.

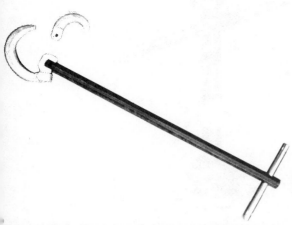

Figure 4-46. Basin wrench with large and small jaws.

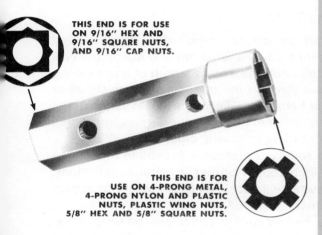

THIS END IS FOR USE ON 9/16" HEX AND 9/16" SQUARE NUTS, AND 9/16" CAP NUTS.

THIS END IS FOR USE ON 4-PRONG METAL, 4-PRONG NYLON AND PLASTIC NUTS, PLASTIC WING NUTS, 5/8" HEX AND 5/8" SQUARE NUTS.

Figure 4-47. Closet seat wrench. (J. A. Sexauer Inc.)

Figure 4-48. Strap wrench for use on chrome-plated pipe. (Ridgid Tool Co.)

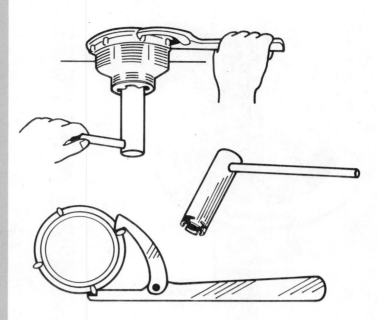

Figure 4-49. Wrenches for tightening basket strainers to a kitchen sink. (Chicago Specialty Mfg. Co.)

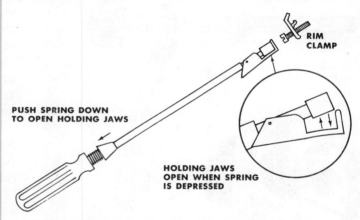

RIM CLAMP

PUSH SPRING DOWN TO OPEN HOLDING JAWS

HOLDING JAWS OPEN WHEN SPRING IS DEPRESSED

Figure 4-50. Rim wrench for tightening rim clamps, which secure kitchen sinks and lavatories to the counter top. (Chicago Specialty Mfg. Co.)

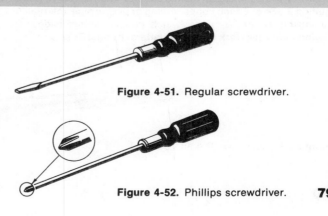

Figure 4-51. Regular screwdriver.

Figure 4-52. Phillips screwdriver.

LARGE INTERNAL CUTTER

SMALL INTERNAL CUTTER

MIDGET TUBING CUTTER

Figure 4-53. Specialized tubing cutters: internal cutter for cutting 2-, 3-, or 4-inch copper tubing or plastic pipe, internal tubing cutter for 1/2- and 3/4-inch copper tubing, midget tubing cutter for work in close quarters. (Ridgid Tool Co.)

Figure 4-54. Internal cutter for cutting 3- and 4-inch cast iron soil pipe. (Capitol Cutter)

Cutting Tools

Some specialized tubing cutters are illustrated in Figure 4-53. The *large internal tubing cutter* is used to cut off the excess length of waste tubing stubbed through the floor for a water closet or shower below the level of the closet flange or the shower strainer. It will cut 2-, 3-, or 4-inch copper tubing or plastic pipe.

The *small internal cutter* is used to trim off 1/2- or 3/4-inch copper or plastic water supply pipes. This cutter will cut closer to the wall or floor than any other.

The *midget tubing cutter* can be used in extremely close quarters on small size copper or plastic tubing. It is available in two sizes to cut tubing up to 15/16-inch outside diameter.

An *internal cutter* is used for trimming off the excess length of cast iron closet bends flush with the floor so that the water closet can be set. (See Figure 4-54.)

The *jab saw* (Figure 4-55) is used to cut off floor-set water closet bolts after the closet has been tightened to the flange. It is also a very useful tool for making hacksaw cuts in tight areas where a regular hacksaw will not fit.

Figure 4-55. Jab saw. (Ridgid Tool Co.)

LAYOUT AND MEASURING TOOLS

Before any pipe is installed on a job, the location of the fixtures and the various water pipe and waste openings must be laid out. To do this, the plumber will need to use the tools illustrated in Figure 4-56.

The *builder's level* is used to get piping elevations and lay out pipe trenches in the open ditch or inside the building when there is no other surface to measure from.

The 50- or 100-foot *steel tape* is used to measure long distances to save time and eliminate the error that can result from using the 6-foot rule for measuring long distances.

The *chalk line and reel* is a timesaving tool for laying out pipe lines on floors and ceilings. It is also used for laying out grid and wall lines on the floor in toilet rooms so that the plumber may accurately install his pipe before the walls are constructed.

The *plumb bob* is probably the plumber's second most useful measuring tool after the 6-foot folding rule. (The folding rule is shown in Figure 4-1.) Its uses as an aid in measuring and layout on the job are almost endless. A plumb bob can be used to center openings for pipes from one floor to the next. It can be suspended through an opening in the floor so that pipe on the floor can be run directly to the center of the opening. A plumb line can also be used to mark an opening so that accurate measurements can be taken to the line for cutting pipe.

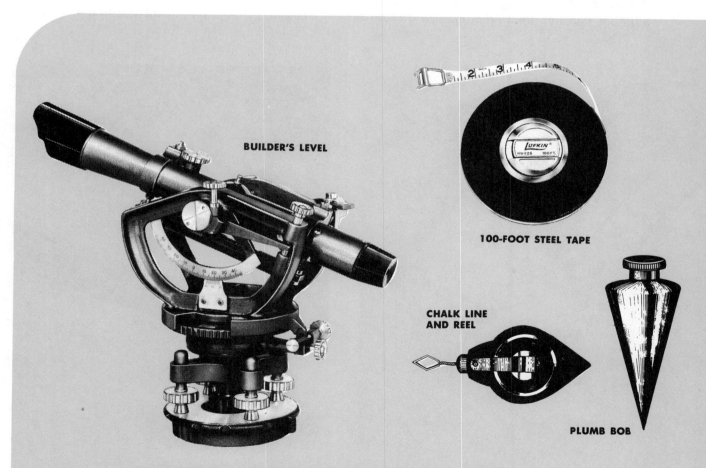

BUILDER'S LEVEL

100-FOOT STEEL TAPE

CHALK LINE AND REEL

PLUMB BOB

Figure 4-56. Plumber's layout and measuring tools: builder's level (David White Instruments, Div. of Realist, Inc.), 100-foot steel tape (Lufkin Rule Co.), chalk line and reel, plumb bob.

CUTTING AND BORING TOOLS

After the locations of the plumbing fixtures and pipes are laid out, the plumber will have to cut holes for his pipe in the building walls and floors if the building is of wood frame construction. Some common woodcutting and boring tools are illustrated in Figures 4-57 through 4-60.

The ½-inch *right-angle drill* (Figure 4-57) and the *electric hacksaw* (Figure 4-58) are almost a necessity for drilling and cutting the holes and openings necessary to install the plumbing pipes in wood frame buildings.

The right-angle-type drill is preferred because the drill will fit between a 16-inch stud or joist space (Figure 4-57) and will drill a hole squarely through the stud or joist. And when holes are drilled squarely through the stud or joist, the plumber can use the minimum size drill bit for the size pipe being installed, thus minimizing the loss of structural strength (a smaller hole causes less structural strength loss). An assortment of drill bits for this drill is also shown in Figure 4-57.

The *electric hacksaw* (Figure 4-58) is used in wood frame construction to cut holes for pipe over 2 inches in size and for cutting openings where pipes pass through the wood construction at angles that make it impossible to drill the proper hole. When fitted with the proper blades, this saw has a variety of uses for the plumber. The saw can also be used to cut off copper tubing, or plastic or steel pipe that is already installed if an opening into an existing pipe line must be cut.

The ¼-inch *drill motor* (Figure 4-59), when used with flat boring wood bits, is a useful tool for drilling small diameter holes (¼- to 1½-inch in diameter) for water pipes in wood frame buildings. The plumber will also find many other

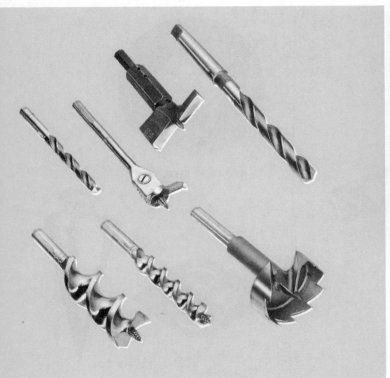

Figure 4-57. Right-angle drill for drilling holes for pipes in studs and joists, and an assortment of wood bits for this drill. (Milwaukee Electric Tool Corp.)

Figure 4-58. Electric hacksaw for cutting large holes for plumbing pipes. (Milwaukee Electric Tool Corp.)

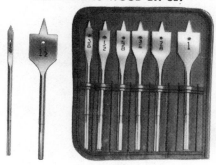

DRILL MOTOR

FLAT BORING WOOD BIT SET

TWIST DRILL SET

Figure 4-59. Common wood-drilling equipment: a ¼-inch drill motor (Milwaukee Electric Tool Corp.), flat boring wood bit set (Milwaukee Electric Tool Corp.), twist drill set.

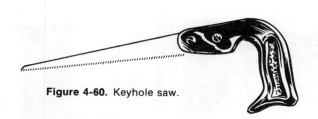

Figure 4-60. Keyhole saw.

uses for this drill when used with either the illustrated wood bit or a set of twist drills in both the rough-in and finishing stages of the job.

The *keyhole saw* (Figure 4-60) is the final wood cutting tool illustrated. The plumber will find a keyhole saw very handy for cutting openings where it is impractical to use an electric hacksaw. To use a keyhole saw, you drill a hole first and then work the saw blade through the hole to make the desired cut.

CONCRETE DRILLING TOOLS

Not all plumbing is installed in buildings of wood frame construction. The majority of the larger buildings are constructed of reinforced concrete. The plumber working in buildings of this type of construction must be equipped to cut and bore holes for his pipe, pipe hangers, and supports in concrete. As a rule in new work in reinforced concrete buildings, the holes through the walls and floors will usually be sleeved in advance and inserts will be placed in the concrete for pipe hangers and supports where this is practical. However, because of additions and/or changes as the job progresses, the plumber will need to make use of concrete cutting and boring tools like those illustrated in Figures 4-61 through 4-64.

The $^3/_8$-inch *hammer drill* (Figure 4-61) is used mainly for drilling small diameter (up to $^9/_{16}$-inch) holes in concrete for pipe anchors. This tool drills concrete more efficiently than a regular drill motor because it "hammers" the bit while rotating it, resulting in faster and cleaner holes. In addition, the drill bits last longer when used in a hammer drill. The *depth guide rod* on the side of the drill is an aid for drilling anchor holes. After the bit is placed in the drill chuck, the rod is adjusted back from the end of the drill bit to the

DEPTH GUIDE ROD

Figure 4-61. A ³/₈-inch hammer drill for drilling small diameter (up to ⁹/₁₆-inch) holes in concrete. (Black and Decker)

proper drilling depth for the type of anchor being used. The hole is then drilled until the depth guide rod reaches the surface of the concrete being drilled. A carbide drill bit for use with this hammer drill is also shown in the illustration.

The 2-inch *rotary hammer* (Figure 4-62) will drill holes in concrete up to a 1¹/₈-inch diameter with solid drill bits and holes up to a 2-inch diameter with "core-" type drill bits. A *solid bit* and a *core-type bit* for use in this drill are also shown. An *adjustable depth guide rod* is mounted on the side of this drill. When fitted with the proper size bit, this rotary hammer can be used to drill holes in concrete for anchors and for small diameter (under a 2-inch outside diameter) pipe lines. Heavy duty rotary hammers are available to drill holes up to a 4-inch diameter in concrete with core-type bits similar to the type shown. A tool of this type is very useful for drilling holes for pipes in concrete where it is not possible to use a "wet" core drill.

The core-type drill bit illustrated in Figure 4-62 is for drilling concrete that does not have heavy steel reinforcing rod buried in it, as it is not designed to cut steel. Because most reinforced concrete has heavy steel reinforcing rod, it is necessary to use a *wet core drill outfit* like the one illustrated in Figure 4-63 to drill holes over 1 inch in diameter. This core drill uses a hollow or core drill bit with industrial diamonds embedded in its cutting surface. These bits are available in

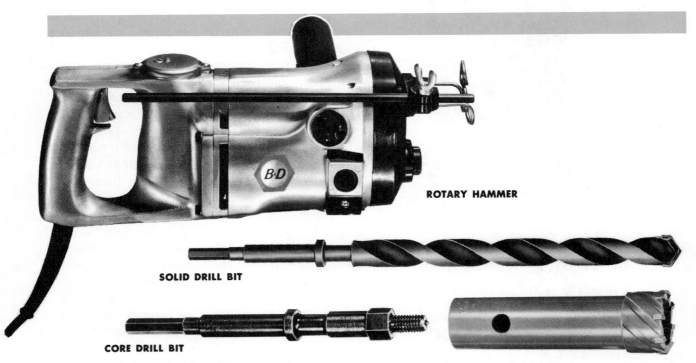

ROTARY HAMMER

SOLID DRILL BIT

CORE DRILL BIT

Figure 4-62. A 2-inch rotary hammer for drilling concrete. Top: Solid drill bit. Bottom: Core drill bit. (Black and Decker)

sizes from 1 to 14 inches in diameter. The diamond-faced cutting surface will drill either concrete or steel reinforcing rod with ease. The bits require a water supply to keep them cool and to flush the cut particles of concrete and steel out of the hole while it is being drilled. A *pressure water tank*, as shown in Figure 4-63, or a garden hose may be used to supply water to the bit through the center shaft of the drill motor.

A core drill must be firmly secured to the surface being drilled. This may be done with a telescoping extension assembly that braces the unit firmly from the ceiling. The drilling rig may also be held in either the horizontal or vertical position by bolting it through the center of its base to an anchor placed in the concrete floor or wall.

The last method of cutting and drilling holes in concrete is the oldest. A variety of *heavy chisels*, *points*, and *star drills* for use with a *heavy hammer* are illustrated in Figure 4-64. A *sledgehammer* for breaking up concrete basement floors is also shown. Eye protection should be worn when using these tools!

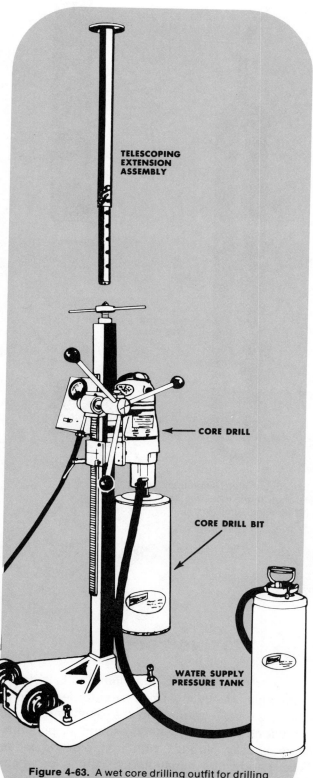

TELESCOPING EXTENSION ASSEMBLY

CORE DRILL

CORE DRILL BIT

WATER SUPPLY PRESSURE TANK

Figure 4-63. A wet core drilling outfit for drilling holes in reinforced concrete. (Milwaukee Electric Tool Corp.)

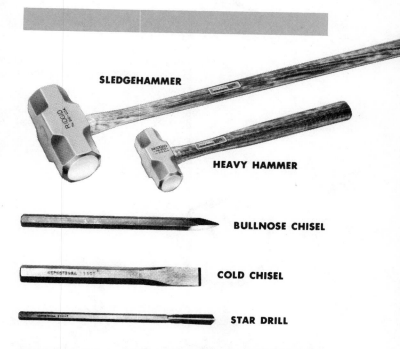

SLEDGEHAMMER

HEAVY HAMMER

BULLNOSE CHISEL

COLD CHISEL

STAR DRILL

Figure 4-64. Hand tools for cutting concrete: sledge hammer, heavy hammer (Ridgid Tool Co.), bullnose chisel, cold chisel, star drill (Mephisto Tool Co.).

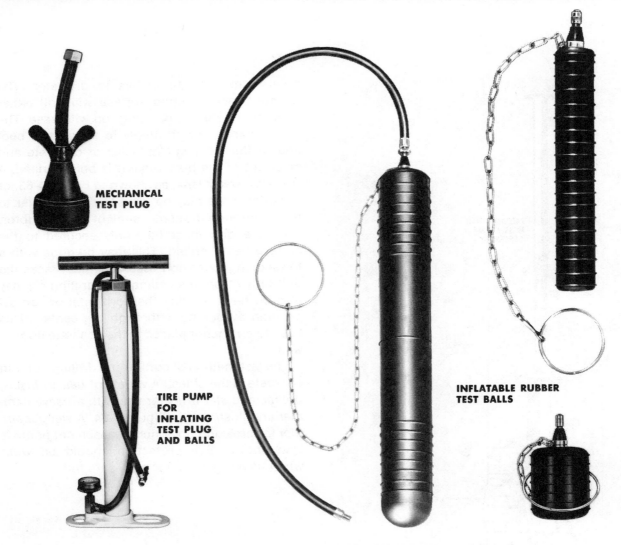

Figure 4-65. Assorted test plugs: mechanical test plug (Richmond Foundry and Mfg. Co.), inflatable rubber test balls (Cherne Industrial, Inc.), tire pump for inflating test plug and balls (Cherne Industrial, Inc.).

Labels within figure:
MECHANICAL TEST PLUG
TIRE PUMP FOR INFLATING TEST PLUG AND BALLS
INFLATABLE RUBBER TEST BALLS

Figure 4-66. Test gauge assembly.

◯ TESTING TOOLS

After each phase of the plumbing is installed, the pipe usually must be tested for leaks. To do this, all the openings in the plumbing system must be plugged. Water, gas, and air pipe lines are usually plugged with iron pipe or copper caps and plugs (depending, of course, on whether the pipe line is copper or iron pipe). To test waste and vent piping, the openings can be plugged with *test plugs,* as shown in Figure 4-65.

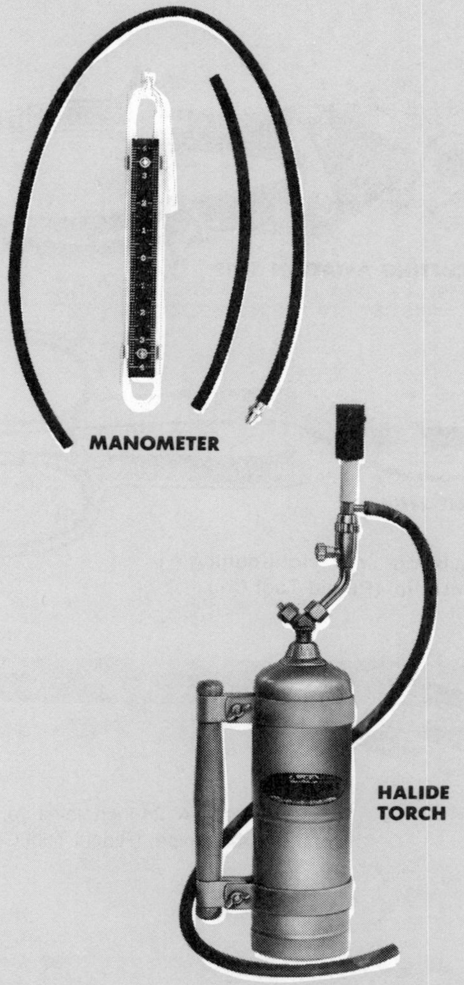

MANOMETER

HALIDE TORCH

Figure 4-67. Manometer (Robinair Manufacturing) and halide torch (Union Carbide Corp., Linde Division)

In addition to the previously mentioned tools, the plumber will need some of the following tools.

To use any of the electric tools discussed in this chapter, the plumber will need a good *three-wire extension cord* (Figure 4-68). An electric power source is rarely close enough to the work site to be reached with the cord on the tool.

3 WIRE GROUNDED

Heavy Duty
Indoor or Outdoor

EXTENSION CORD

50 FT.

Figure 4-68. Three-wire extension cord for electric power tools. (Carol Cable Co.)

A tire pump for inflating the rubber test plugs is also illustrated.

The rough-in test on waste and vent piping can be just a simple air test. A *test gauge assembly,* similar to the one shown in Figure 4-66, must be made by the plumber. A gauge assembly of this type can also be used for testing water, gas, air, and other piping.

When the local plumbing code requires a final air test, a *manometer* and a *halide torch,* as illustrated in Figure 4-67, are required. (The use of all of these testing tools will be discussed more fully in Chapter 12.)

The *aviation snips* and *sheet metal snips* shown in Figure 4-69 are useful in all the phases of a plumbing job. The sheet metal snips make straight cuts. The aviation snips cut right- or left-handed curves. However, the *straight aviation snips* make straight cuts.

A 24-inch *level* (Figure 4-70) is another tool that should be in every plumber's tool box. It is a more accurate leveling tool than the torpedo level shown in Figure 4-1.

Before the plumber can install any below-ground piping, a ditch will have to be dug with a *pick* and *shovel* (Figure 4-71).

On larger jobs or in prefab shops, there will

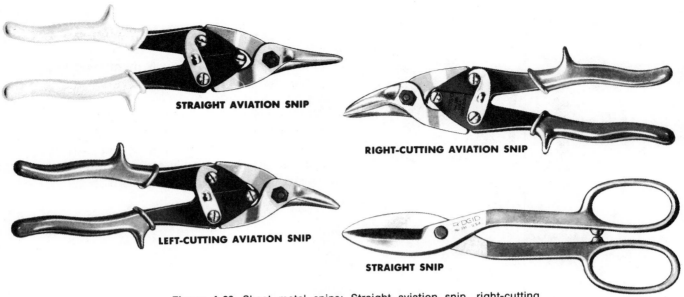

Figure 4-69. Sheet metal snips: Straight aviation snip, right-cutting aviation snip, left-cutting aviation snip, straight snip. (Ridgid Tool Co.)

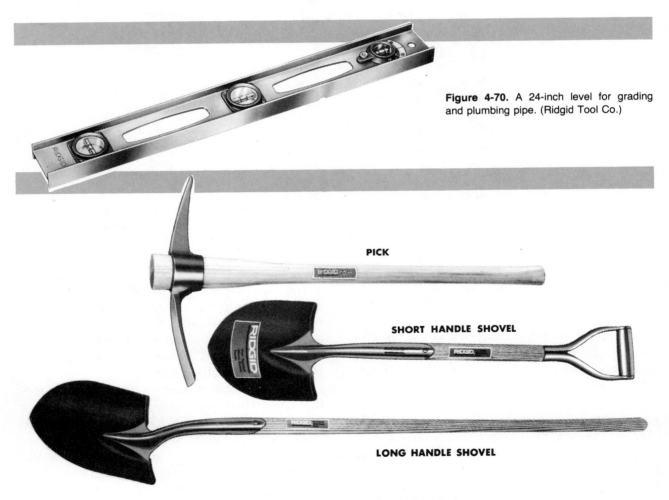

Figure 4-70. A 24-inch level for grading and plumbing pipe. (Ridgid Tool Co.)

Figure 4-71. Pick and shovels. (Ridgid Tool Co.)

also be a need for a *bench grinder* (Figure 4-72), for sharpening tools and grinding the mush-roomed heads from caulking irons, chisels, and points.

Large jobs and prefab shops will also have either a *portable* or *table band saw* (Figure 4-73) for cutting large amounts of hanger rod and hanger iron, and for trimming other metals.

To lift pipe into position will require the use of some of the tools in Figure 4-74. The *ratchet lever hoist* or *come-a-long* and the *hand hoist* or *chain fall* are two very handy tools for this purpose. To lift pipe with either of these hoists, either *nylon lifting slings* or *cable choker* will have to be wrapped around the pipe to provide a lifting place for the hook on the hoist.

The plumber will also need various sizes of

Figure 4-72. Bench grinder for sharpening tools. (Milwaukee Electric Tool Corp.)

stepladders (Figure 4-75) and some metal frame scaffolding to use because very little pipe is installed in locations that can be reached from the floor (see Figure 2-11).

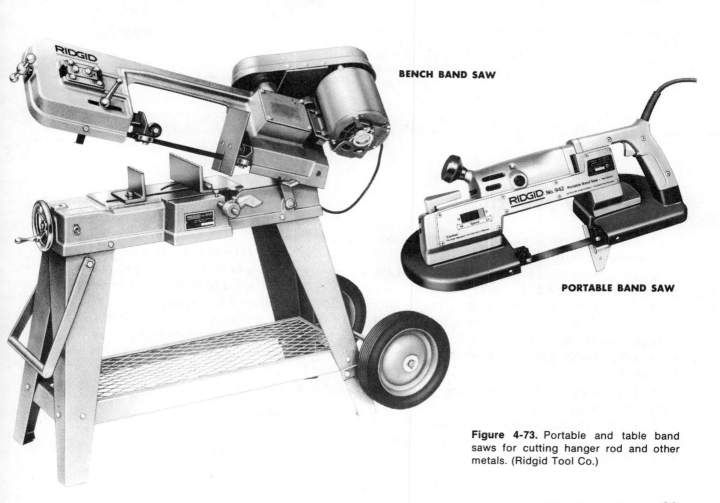

BENCH BAND SAW

PORTABLE BAND SAW

Figure 4-73. Portable and table band saws for cutting hanger rod and other metals. (Ridgid Tool Co.)

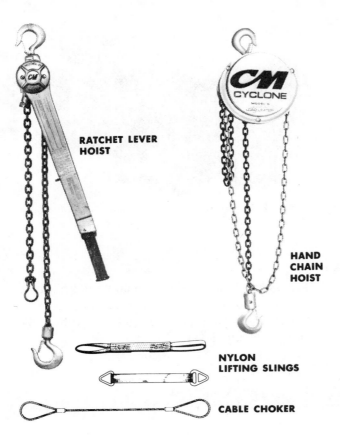

Figure 4-74. Tools for hoisting pipe into position: ratchet lever hoist (Chisholm-Moore Hoist Div. of Columbus McKinnon Corp.), hand chain hoist (Chisholm-Moore Hoist Div. of Columbus McKinnon Corp.), nylon lifting slings, cable choker.

RATCHET LEVER HOIST

HAND CHAIN HOIST

NYLON LIFTING SLINGS

CABLE CHOKER

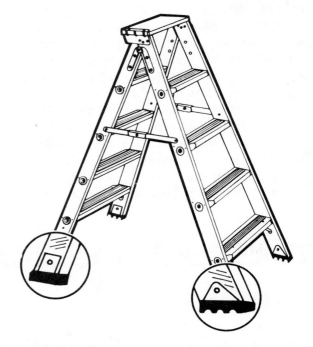

Figure 4-75. Stepladder (National Safety Council)

REVIEW QUESTIONS

1. Name the personal tools every plumber should always carry on the job.

2. Why should pipe cut with a wheel-type pipe cutter be reamed after every cut?

3. Torque wrenches for no-hub pipe joints are set to release at which of these inch-pound figures—30, 45, 50, 60?

4. Why should pipe be reamed before it is threaded?

5. Explain the purposes of using oil on the die when cutting threads.

6. What is the advantage of a compound pipe wrench over a straight pipe wrench? Explain how this advantage works.

7. When should copper tubing be joined with flared joints? When may soldered joints be used?

8. Explain how to prepare a joint for soldering. What tools or materials are required?

9. Why is a blade cutter used to cut plastic pipe rather than a wheel cutter?

10. Make a list of the smooth-jawed tools required to work chromium-plated pipe.

11. How and for what purpose is a rim wrench used?

12. Explain the uses of these special tools:

    Jab saw

    Chalk line

    Plumb bob

CHAPTER 5

Joining, Installing, and Supporting Pipe

A plumber's basic skill is his ability to install and join the various kinds of pipe to complete the plumbing systems of a building. The plumber must know what kind of joint to use for each piping material and be able to make this joint watertight and, in the case of sewer gas, airtight.

Imperfect joints allow the liquid content of pipes to escape. If this happens, property damage can result. Soil and waste pipe lines are often concealed in partitions or under the floor and if a joint fails for any reason, finished walls and ceilings can be damaged.

The plumbing drainage systems contain many noxious sewer gases, which are detrimental to health if they are permitted to enter an occupied building through leaking joints in vent piping. These gases can affect the delicate breathing mechanism of the human body and therefore must be controlled.

A leaking joint on a sewer or drain installed below the surface of the soil would allow sewage to enter the subsoil and could contaminate the drinking water. A leaky joint of this kind might also allow soil to enter the drain and cause stoppage of the drain line. Leaky underground pipe joints in yard areas allow the entrance of fine, fibrous tree roots. Once these roots have entered the pipe, they spread quickly until they fill the entire inside of the pipe and block it. A drain blocked with tree roots usually requires mechanical cleaning of the drain line, and sometimes the pipe can be opened only by digging up the drain and replacing it with a new drain.

For these reasons, plumbing codes require that waste, vent, and rainwater piping systems be pressure tested to ensure that they do not leak. (The testing of the various plumbing systems will be fully explained in Chapter 11.)

The joints of the water supply must be made with precision because these pipes are commonly subjected to high pressures. A defective joint will not only permit a large quantity of water to escape in a few minutes, but it also can cause serious property damage.

In addition to joining the various kinds of pipe, the plumber must install and support it in such a manner that the pipe and its joints will remain leakproof. For this reason, practices for installing, hanging, and supporting the various types of pipe will be covered in the last section of this chapter.

The pipe joints in this chapter will be presented for the various piping materials in the same order as the materials were listed in Chapter 3 and the tools were covered in Chapter 4. Therefore, you may refer back to these two chapters for more detailed pictures of the materials and tools.

Figure 5-1. A strand of oakum and a 25-pound bar of caulking lead.

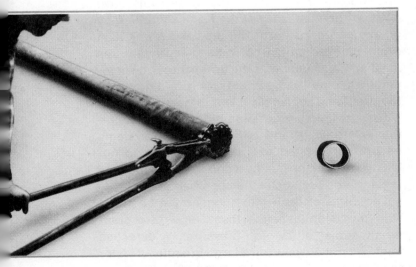

Figure 5-2. Cutting cast iron soil pipe with a squeeze-type cutter.

 # CAST IRON SOIL PIPE JOINTS

There are two different types of cast iron soil pipe: bell and spigot soil pipe and hubless (no-hub) soil pipe. Bell and spigot soil pipe joints are either caulked joints or mechanical joints, made with preformed, molded neoprene rubber compression gaskets. No-hub soil pipe is joined by means of a stainless steel clamp assembly surrounding a sealing sleeve of neoprene rubber.

Caulked Soil Pipe Joint

A caulked joint in soil pipe is made with molten lead and oakum caulked with caulking irons to make the joint watertight. (These tools were shown in Figures 4-3 and 4-4 of Chapter 4.) Oakum is hemp that has been treated with pitch to make it waterproof and resistant to the elements in the waste. Lead is available in 25-pound bars. A bar of lead and a strand of oakum are pictured in Figure 5-1.

The lead is heated on a plumber's lead furnace (Figure 4-4) and poured with the ladle. The lead should be heated until it is hot enough so that it does not stick to the ladle, but not red-hot. Overheating the lead just burns it into slag. Be very careful when adding lead to the molten pot! Any lead that is damp or frozen will explode when added to a lead pot partially full of molten lead. To avoid this, preheat wet or frozen lead on the side of the furnace before adding it to a pot containing any molten lead.

The steps required to make a caulked lead and oakum soil pipe joint are:

1. Cut the soil pipe to the required length with one of the cutters pictured in Figure 4-2 of Chapter 4 or as illustrated in Figure 5-2.

2. After the pipe is cut, wipe the hub (bell) and spigot ends of the pipe dry and clean of foreign material. Moisture in the hub can cause the pipe to split when the molten lead is poured because the trapped moisture may turn to steam and cause a small explosion.

3. Assemble the pipe, aligning and spacing the joint carefully. The spigot end of the pipe must go to the bottom of the hub so that the oakum will not be pushed into the pipe.

Figure 5-3. Yarning oakum into the soil pipe joint.

Figure 5-4. Packing oakum into the soil pipe joint.

Figure 5-5. Pouring molten lead into a vertical soil pipe joint.

Figure 5-6. Pouring molten lead into a horizontal soil pipe joint.

4. Yarn (Figure 5-3) and pack (Figure 5-4) oakum into the hub to a depth of 1 inch from the top. No loose oakum fibers should protrude. Loose fibers may be removed easily by packing them with a yarning or packing iron heated in the molten lead in the lead pot.

5. Pour the molten lead into the vertical soil pipe joint as shown in Figure 5-5, filling the joint in one pour. (The ladle used should be large enough to fill the joint in one pour.) On horizontal soil pipe joints, an asbestos running rope is placed around the pipe and clamped tightly at the top, forming a gate for pouring the lead. The rope should be tapped lightly against the hub with the caulking hammer to make it fit tightly against the hub so that the molten lead will not run out between the hub and the running rope. A wad of oakum is placed under the clamp to retain the lead up to the top of the hub. The lead is then poured into the gate, filling the joint to the top, as shown in Figure 5-6. The running rope is removed after the lead solidifies.

6. After the lead solidifies, in both the vertical and horizontal joints, the lead is driven down in four places around the hub with the caulking hammer and the inside caulking iron to set the joint. On horizontal joints, the neck of excess lead left in the pouring gate is cut off with a hammer and chisel at this time, as shown in Figure 5-7.

7. The soil pipe joint (whether in the horizontal or vertical position) is then caulked around the inside edge with the inside caulking iron. The iron is moved slowly around the joint and struck

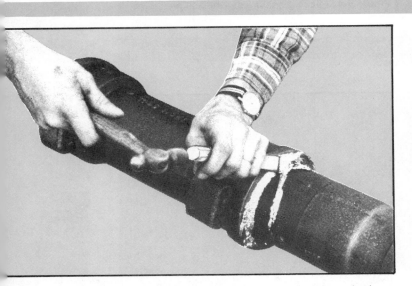

Figure 5-7. Cutting off the excess neck of lead on a horizontal soil pipe joint.

Figure 5-8. Caulking the inside edge of the soil pipe joint.

Figure 5-9. Caulking the outside edge of the soil pipe joint.

Figure 5-10. Sectional view of a lead and oakum soil pipe joint. Notice that the lead is approximately 1 inch thick.

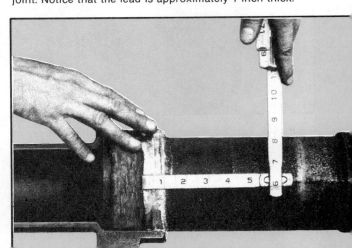

with the caulking hammer. (See Figure 5-8.)

8. After the inside edge of the joint is caulked, the outside edge is caulked in the same manner with the outside caulking iron. (See Figure 5-9.)

Figure 5-10 pictures a cut-away view of a finished caulked lead and oakum joint. The operations shown in Figures 5-3 to 5-10 were performed on a special piece of soil pipe that was designed so that the hub could be split to allow the apprentice to inspect the finished joint.

Compression Gasket Soil Pipe Joint

Compression gaskets are made from neoprene rubber. The gaskets are manufactured in two weights: *service weight,* for service weight soil pipe and fittings, and *extra heavy,* for extra heavy soil pipe and fittings. The two types of gaskets are not interchangeable.

The compression gasket soil pipe joint is made as follows:*

1. Clean the hub and spigot so that they are reasonably free from dirt, mud, gravel, or other foreign material.

2. When using cut pipe, the sharp edge on the cut end of the pipe must be removed because this sharp edge may jam against the gasket. Although the sharp edge will not damage the gasket, it will make it difficult to join the pipe. The sharp edge may be removed by peening with a hammer (Figure 5-11) or by rasping with a file (Figure 5-12) until the sharp outer edge is slightly rounded.

3. Insert the gasket into the cleaned hub. This may be done in two ways: by folding the gasket or by bumping the gasket.

Folded gasket method: Hold the gasket upright with your thumbs at the bottom. Fold the bottom of the gasket up through the top with your thumbs as if it were to be turned inside out. (See Figure 5-13.) Place the gasket in the hub, making sure that the gasket ring is in the groove of the hub. Release the gasket, and it will unfold into the pipe hub, as shown in Figure 5-14. This method is especially recommended for installing the smaller diameter gaskets, because they are more difficult to insert.

*These instructions were supplied by the Tyler Pipe Company for use with Tyler's Ty-Seal Gasket.

Figure 5-11. Peening the cut end of soil pipe with a hammer to prepare the end for inserting into a rubber compression gasket. (Tyler Pipe Co.)

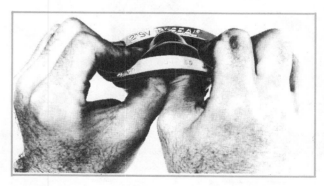

Figure 5-12. Rasping the cut end of soil pipe with a file to prepare the end for inserting into a rubber compression gasket. (Tyler Pipe Co.)

Figure 5-13. Folding the rubber compression gasket before inserting it into the soil pipe hub. (Tyler Pipe Co.)

Figure 5-14. Inserting the folded rubber compression gasket into the soil pipe hub. (Tyler Pipe Co.)

95

Bumping gasket method: Bump the gasket with the heel of your hand (Figure 5-15) or with a board (Figure 5-16). When inserting gaskets into fittings, it may easier to bump the gasket into position by striking the fitting against a board or the floor.

4. Lubricate the gasket by brushing on a smooth, thin coat of lubricant (Figure 5-17) completely around the gasket (Figure 5-18). Be sure to coat the entire circumference of the gasket and give special attention to the inner seal.

CAUTION: Do not overlubricate; no lubricant is required on the spigot end of the pipe.

5. Using the tool of your choice, push or pull the spigot end of the pipe through both seals of the gasket. When using the illustrated assembly tool on pipe, place the yoke of the tool behind the hub of the pipe (Figure 5-19), allowing the tool jaws to slip over the barrel of the pipe (Figure 5-20). Then, push the jaws as far from the hub as possible (Figure 5-21). Next, pull the handle toward the hub (Figure 5-22). Repeat this process until the spigot is through both seals of the gasket. You will feel it hit home. (If using a different tool than the one illustrated, follow the tool manufacturer's instructions.)

Figure 5-15. Bumping the rubber compression gasket into a soil pipe hub with the heel of a hand. (Tyler Pipe Co.)

Figure 5-16. Bumping the rubber compression gasket into a soil pipe hub with a board. (Tyler Pipe Co.)

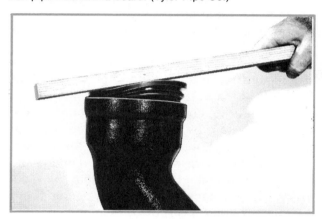

Figure 5-17. A can of lubricant for a rubber compression gasket soil pipe joint. (Tyler Pipe Co.)

Figure 5-18. Applying lubricant to an installed rubber compression soil pipe gasket. (Tyler Pipe Co.)

Figure 5-19. Assembly of a rubber compression gasket soil pipe joint: Assembly tool stirrup placed behind the hub. (Tyler Pipe Co.)

Figure 5-20. Assembly of a rubber compression gasket soil pipe joint: Assembly tool jaws retracted. (Tyler Pipe Co.)

Figure 5-21. Pushing the tool jaws as far from the hub as possible. (Tyler Pipe Co.)

Figure 5-22. Assembly of a rubber compression gasket soil pipe joint: Pull in—pipe joined. (Tyler Pipe Co.)

Joining, Installing, and Supporting Pipe **97**

Figure 5-23. Assembly of a rubber compression gasket soil pipe joint using a lead maul to strike the fitting on the driving lug. (Tyler Pipe Co.)

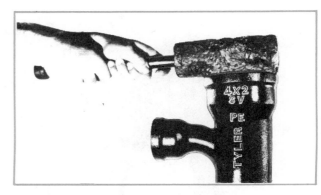

Figure 5-24. Assembly of a rubber compression gasket soil pipe joint using a lead maul to strike the fitting across the hub. (Tyler Pipe Co.)

Figure 5-25. Using a chain tool to assemble fittings when using a rubber compression gasket soil pipe joint. (Tyler Pipe Co.)

Figure 5-26. Sectional view of a properly installed rubber compression gasket soil pipe joint. (Tyler Pipe Co.)

6. *Installing fittings:* To install fittings, force the spigot partially through the first seal and hold it in alignment, then drive it home with a lead maul. Strike the fitting either on the driving lug (Figure 5-23) or across the full hub (Figure 5-24). Hit it hard! The lead will deform without harm to the fitting. When using another device for joining fittings (Figure 5-25), follow the manufacturer's instructions.

Figure 5-26 illustrates a sectional view of a properly installed compression gasket pipe joint.

No-Hub Soil Pipe Joint

The no-hub soil pipe joint is made as follows:

1. Detach the neoprene sleeve from the stainless steel clamp assembly (Figure 5-27).

2. Slide the stainless steel clamp assembly onto the pipe (Figure 5-28).

3. Insert the spigot ends of the pipe or fitting into the neoprene gasket until they butt against the separator ring inside the gasket (Figure 5-29).

4. Slide the stainless steel clamp assembly over the neoprene gasket and tighten the screw clamps with a preset torque wrench (or other suitable tool) to a minimum of 48 inch-pounds of torque (Figure 5-30). The preset torque wrench illustrated releases at 60 inch-pounds of torque. Other tools suitable for tightening the screw clamps are a $5/16$-inch nut driver, a $1/4$-inch ratchet with a $5/16$-inch socket, or an electric nut driver. (These tools are illustrated in Figure 4-7 of Chapter 4.)

Figure 5-27. Assembly of a no-hub soil pipe joint: Detach the neoprene sleeve from the stainless steel clamp assembly.

Figure 5-28. Assembly of a no-hub soil pipe joint: Slide the stainless steel clamp assembly onto the pipe.

Figure 5-29. Assembly of a no-hub soil pipe joint: Insert the spigot ends of the fitting into the neoprene gasket until they butt against the separator ring.

Figure 5-30. Assembly of a no-hub soil pipe joint: Tighten the screw clamps with a preset torque wrench.

⃝ GALVANIZED STEEL THREADED JOINTS

Plumbers join galvanized steel pipe to the water supply, drainage, or vent fittings with a threaded joint. This threaded pipe joint is also used to join other types of pipe, such as, malleable black iron pipe; cast iron pipe; brass and copper pipe; and thick-walled (schedule 80 and 120) plastic pipe, although these materials and their uses are not covered in this text.

The thread used on plumbing pipe and fittings is the American Standard Taper Pipe Thread, or NPT. The NPT is tapered $3/4$ inch per foot of thread length so that the pipe and fittings will thread together *(make up)* tightly to form a leakproof joint. A section of NPT and a table showing its characteristics are illustrated in Figure 5-31 and Table 5-1.

Since all pipe fittings are tapped with the internal (female) pipe thread at the factory when

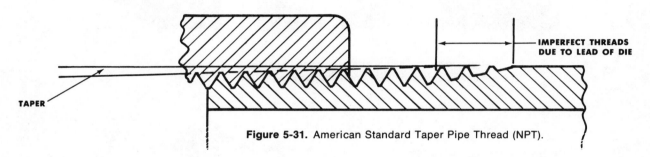

Figure 5-31. American Standard Taper Pipe Thread (NPT).

TABLE 5-1.
CHARACTERISTICS OF THE AMERICAN STANDARD TAPER PIPE THREAD (NPT).*

NOMINAL PIPE SIZE (INCHES)	THREADS PER INCH	APPROXIMATE LENGTH OF THREAD (INCHES)	APPROXIMATE NUMBER OF THREADS TO BE CUT	APPROXIMATE TOTAL THREAD MAKEUP, HAND AND WRENCH (INCHES)
$1/8$	27	$3/8$	10	$1/4$
$1/4$	18	$5/8$	11	$3/8$
$3/8$	18	$5/8$	11	$3/8$
$1/2$	14	$3/4$	10	$7/16$
$3/4$	14	$3/4$	10	$1/2$
1	$11 1/2$	$7/8$	10	$9/16$
$1 1/4$	$11 1/2$	1	11	$9/16$
$1 1/2$	$11 1/2$	1	11	$9/16$
2	$11 1/2$	1	11	$5/8$
$2 1/2$	8	$1 1/2$	12	$7/8$
3	8	$1 1/2$	12	1
$3 1/2$	8	$1 5/8$	13	$1 1/16$
4	8	$1 5/8$	13	$1 1/16$
5	8	$1 3/4$	14	$1 3/16$
6	8	$1 3/4$	14	$1 3/16$
8	8	$1 7/8$	15	$1 5/16$
10	8	2	16	$1 1/2$
12	8	$2 1/8$	17	$1 5/8$

*Dimensions given do not allow for variations in tapping and threading.

they are manufactured, the plumber need only make the external (male) pipe thread. To make an external pipe thread, the plumber should follow these steps:

1. Secure the pipe in the vise (or pipe machine) jaws.

2. Cut the pipe to the required length. (See Figure 5-32.)

3. Ream the inside of the pipe back to the original bore diameter (Figure 5-33) because an unreamed pipe can be reduced in inside diameter by as much as $\frac{1}{4}$ inch.

4. Thread the pipe with the die (Figure 5-34) to the proper length while lubricating the die (Figure 5-35). The correct thread depends on the diameter of the pipe. Table 5-1 illustrates the proper thread length for the various sizes of pipe. As a rule, with a thread of the proper

Figure 5-32. Cutting galvanized iron pipe.

Figure 5-33. Reaming galvanized iron pipe.

Figure 5-34. Threading galvanized iron pipe.

Figure 5-35. Lubricating the die while threading galvanized iron pipe.

Figure 5-36. Applying pipe joint compound to the male galvanized iron pipe thread.

Figure 5-37. Tightening a fitting onto the galvanized iron pipe thread with a pipe wrench.

length, a fitting can be turned onto the thread two-and-one-half to three full turns by hand.

5. Remove the die and wipe the thread clean.

6. Apply pipe joint compound to the male thread. (See Figure 5-36.)

7. Start the fitting onto the pipe thread by hand and tighten the joint with a pipe wrench. (See Figure 5-37.)

Although the illustrated threading and joining operations show the cutting, reaming, and threading of the pipe being done by a pipe threading machine, follow the same sequence of operations for threading pipe by hand using a pipe vise.

◎ COPPER TUBING JOINTS

The three most common methods of joining copper tubing that a plumber will use are the solder joint with capillary fittings; the flared joint; and the compression joint. Solder joints are used on water lines and drainage lines. The flared joint is commonly used on underground water supply tubing, on tubing when it is not practical to use heat to solder, and on joints that have to be disconnected periodically. The com-

pression joint is most commonly used on the exposed water supply tubing to plumbing fixtures.

Solder Joint

Solder joints depend on capillary action drawing free-flowing molten solder into the gap between the fitting and the tube. Figure 5-38 illustrates this capillary space. Flux, applied first, acts as a wetting agent and, when properly applied, permits uniform spreading of the molten solder over the surfaces to be soldered. Table 5-2 estimates the amounts of flux and solder typically required for different sizes of tube.

The selection of the type of solder depends on the operating pressure and temperature of the plumbing line. The two most common types of solder used in plumbing work are 50-50 and 95-5 wire solder. Rolls of 50-50 and 95-5 wire solder are shown in Figure 5-39. The 50-50 solder is composed of 50 percent tin and 50 percent lead, while 95-5 solder is 95 percent tin and 5 percent antimony. The 50-50 tin-lead solder is suitable for moderate pressures and temperatures. For higher pressures, or where greater joint strength is required, the 95-5 tin-antimony solder can be used. The 95-5 solder is more difficult to use because it melts at a slightly higher temperature

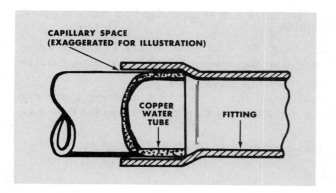

Figure 5-38. Capillary space in a copper tubing solder joint. (Copper Development Assn., Inc.)

Figure 5-39. Rolls of 50-50 and 95-5 wire solder.

than does the 50-50 solder. Also, it has a very narrow *pasty* range. Pasty range is the temperature range between which solder is neither completely solid or completely liquid. It is the working temperature range for a particular type of solder, the range of temperatures between which the solder will flow properly into the joint. The 50-50 solder, which is quite easy to work

with, has a 60°F (33.3°C) pasty range, while 95-5 solder has a pasty range of less than 10°F (5.6°C).

Solder is generally used in wire form, but paste-type solders are also available. Figure 5-40 illustrates a can of 50-50 paste-type solder. These are composed of finely granulated solder in suspension in a paste flux. When using *paste-type* solder, observe these four rules:*

1. Wire solder must be applied in addition to the paste to fill the voids and assist in displacing the flux. Otherwise, the surfaces may be well *tinned,* but a good joint with a continuous bond may not exist.

2. The paste mixture must be thoroughly stirred if it has been standing in the can for more

*Copper Tube Handbook. (Copper Development Assn., Inc.)

TABLE 5-2. TYPICAL CONSUMPTION OF SOLDER AND FLUX FOR 100 JOINTS (CUPS).*

TUBE SIZE (INCHES)	SOLDER REQUIREMENT (POUNDS)	FLUX REQUIREMENT (OUNCES)
1/4	0.25	0.50
3/8	0.25	0.50
1/2	0.40	0.75
5/8	0.45	0.90
3/4	0.50	1.0
1	0.75	1.50
1 1/4	0.90	1.75
1 1/2	1.00	2.00
2	1.25	2.50
2 1/2	1.75	3.50
3	2.25	4.50
3 1/2	2.50	5.00
4	3.25	6.50
5	4.50	9.0
6	8.50	17.0
8	17.50	35.0

*Actual consumption depends on workmanship.

*Copper Development Assn., Inc.

Figure 5-40. A 50-50 paste-type solder.

Figure 5-41. A mildly corrosive soldering flux. (M. W. Dunton Co.)

There are ten simple steps to make a solder joint:*

1. Cut the tube end square to the required length with either a tubing cutter (Figure 5-42) or a hacksaw (Figure 5-43).

2. Ream the cut end of the tubing to remove the small burrs left on the inside of the tubing by

*As recommended by NIBCO, Inc.

than a very short time, because the solder has a tendency to settle rapidly to the bottom.

3. The flux cannot be depended on to clean the tube. Cleaning should be done manually, as is recommended for any other flux and solder.

4. Remove any excess flux. The functions of the soldering flux are to remove residual traces of oxides, to promote wetting, and to protect the surfaces to be soldered from oxidation during heating. The flux should be applied to clean surfaces and only enough should be used to lightly coat the areas to be joined.

An oxide film may re-form quickly on copper after it has been cleaned. Therefore, the flux should be applied as soon as possible after cleaning.

The fluxes best suited to the 50-50 and 95-5 solders are mildly corrosive liquid or petroleum-based pastes containing chlorides of zinc and ammonium. (Figure 5-41 illustrates one such flux.)

Most liquid fluxes are "self-cleaning" fluxes, and their use involves a risk. Some past-type fluxes are also identified by their manufacturers as "self-cleaning," and there is a similar risk in their use. There is no doubt that a strong corrosive flux can remove some oxides and dirt films. However, when highly corrosive fluxes are used this way, there is always an uncertainty as to whether uniform cleaning has been achieved and whether corrosive action continues after the soldering has been completed.

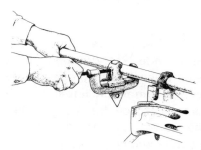

Figure 5-42. Cutting copper water tubing with a roller cutter. Tighten the roller slowly to minimize upsetting the tube. (NIBCO, Inc.)

Figure 5-45. Reaming copper water tubing with a round or half-round file. The outside diameter of the tube should also be cleaned and smoothed. (NIBCO, Inc.)

Figure 5-48. Cleaning a copper fitting with sandcloth. Be sure to clean the bottom of the socket or cup. (NIBCO, Inc.)

the tubing cutter or hacksaw. The tubing may be reamed with either the reamer attached to the tubing cutter (Figure 5-44) or a file (Figure 5-45).

3. Clean the tube end using sandcloth, as shown in Figure 5-46. Steel wool or a wire brush may also be used for cleaning the tube end.

4. Clean the fitting socket with a fitting brush Figure 5-47), steel wool, or sandcloth (Figure 5-48).

5. Apply flux to the tube end (Figure 5-49).
6. Apply flux to the fitting socket (Figure 5-50).
7. Assemble the pipe and fitting.
8. Apply heat to the tubing first, but only momentarily so that the heat will transfer to the end of the tube (Figure 5-51), and then to the fitting (Figure 5-52) until solder melts when placed at the joint of the tube and fitting.
9. Remove the flame and feed solder to the

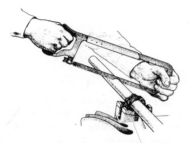

Figure 5-43. Cutting copper water tubing with a hacksaw. Be sure the tube is held firmly in a vise or by some other device. Use a fairly fine-toothed blade. (NIBCO, Inc.)

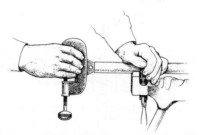

Figure 5-44. Reaming copper water tubing with a flat metal reamer. When this method is used, take care not to expand or flare the tube. Do not overream. Scrape any rough areas off the outside diameter. (NIBCO, Inc.)

Figure 5-46. Cleaning copper water tubing with sandcloth. Use a fine grit. This is a cleaning operation and very little surface needs to be removed. (NIBCO, Inc.)

Figure 5-47. Cleaning a copper fitting with a wire brush. Powered wire brushes are available for use when enough fittings are being cleaned to make a power unit economical. (NIBCO, Inc.)

Figure 5-49. Fluxing copper water tubing. Use enough flux to just barely cover the necessary surface. Do not use more flux than is needed. Brush off any excess. (NIBCO, Inc.)

Figure 5-50. Fluxing a copper fitting. If there are dirty spots on the cleaned surface, they will probably show under the flux. Keep the flux container covered so that the flux does not become contaminated. (NIBCO, Inc.)

Figure 5-51. Heating copper water tubing. The tube is heated momentarily at first so that the heat will transfer to the end of the tube. Point the flame slightly away from the fitting. (NIBCO, Inc.)

Figure 5-52. Heating a copper fitting. Be careful not to overheat; use just enough heat. (NIBCO, Inc.)

Figure 5-53. Applying solder to the heated joint. Do not melt the solder with the torch. Get the joint to the proper temperature and melt the solder by touching it to the tube-fitting juncture with a gentle pushing movement. (NIBCO, Inc.)

Figure 5-54. Solder joints can be wiped with a clean cloth, while they are still hot, to make them even smoother and more attractive. If it is necessary to cool them quickly, the cloth can be damp. (NIBCO, Inc.)

joint at one or two points until a ring of solder appears at the end of the fitting (Figure 5-53). The correct amount of solder is approximately equal to the inside diameter of the fitting—$1/2$ inch of solder for a $1/2$-inch fitting, $3/4$ inch of solder for a $3/4$-inch fitting, etc.

10. Remove excess solder with a cloth while the solder is still pasty (Figure 5-54), leaving a fillet around the end of the fitting as it cools.

Flared Joint

Impact or screw-type tools are used for flaring tube. The procedure for impact-flaring is as follows:*

1. Cut the tube to the required length.

2. Remove all burrs. This is very important to assure metal-to-metal contact.

3. Slip the coupling nut over the end of the tube.

*Copper Tube Handbook. (Copper Development Assn., Inc.)

4. Insert flaring tool into the tube end (Figure 5-55).

5. Drive the flaring tool by hammer strokes, expanding the end of the tube to the desired flare. This requires a few moderately light strokes (Figure 5-56).

6. Assemble the joint by placing the fitting squarely against the flare. Engage the coupling nut with the fitting threads (Figure 5-57). Tighten with two wrenches, one on the nut and one on the fitting.

Using screw-type flaring tools the procedure is as follows:

Steps 1 to 3. Same as for impact-flaring, previously described.

4. Clamp the tube in the flaring block so that the end of the tube is slightly above the face of the block (Figure 5-58).

5. Place the yoke of the flaring tool on the block so that the beveled end of the compression cone is over the tube end.

Figure 5-55. Insert flaring tool into the tube end. (Copper Development Assn., Inc.)

Figure 5-56. Drive the flaring tool with hammer blows. (Copper Development Assn., Inc.)

Figure 5-57. Assembling a flare joint in copper tubing. (Copper Development Assn., Inc.)

Figure 5-58. Flaring copper tubing with a yoke and screw-type flaring tool. Clamp the tube into the flaring block. (Copper Development Assn., Inc.)

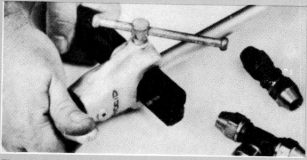

Figure 5-59. Tightening the compressor screw to flare the tubing into the flaring block. (Copper Development Assn., Inc.)

6. Turn the compressor screw down firmly, forming the flare between the chamber in the flaring block and the beveled compression cone (Figure 5-59).

7. Remove the flaring tool. The joint can now be assembled as in Step 6 for impact-flaring.

Copper Compression Joint

A mechanical compression joint consists of the following parts, as illustrated in Figure 5-60:

1. A compression joint fitting.
2. A compression ring.
3. A compression nut.
4. The plain end of the copper tube.

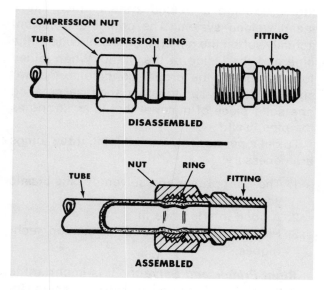

Figure 5-60. Mechanical compression joint for copper tubing.

To assemble a mechanical compression joint for copper tubing, follow these steps:

1. Cut the tube to the required length.
2. Clean the tube end of any foreign material.
3. Slide the compression nut onto the tubing.
4. Slide the compression ring onto the tubing.
5. Slide the mechanical joint fitting over the tube end.
6. Tighten the compression nut onto the fitting with a wrench. By tightening this nut, you compress the compression ring into the tubing and thus seal the joint.

PLASTIC PIPE JOINTS

The various types of plastic pipe and fittings presented in Chapter 3 are joined together with one of three methods, depending on the physical uses to which the particular plastic piping will be put. The three methods used by plumbers to join plastic pipe and fittings are:

1. The solvent weld joint.
2. The insert fitting joint.
3. The flare fitting joint.

Solvent Weld Joint

Solvent weld joints are made by solvent bonding, producing a welded system much like a metal welded system. The primer and solvent actually soften the material on the outside of the pipe and the inside of the fitting. When joined together under the proper conditions, the two surfaces actually run together and fuse, creating one solid piece of material that is as strong as the pipe itself.*

To get a permanent, welded joint, three things are necessary:

1. The right primer and solvent for the plastic you are using.
2. A good interference fit.
3. Proper preparation and installation techniques.

Right Primer and Solvent. It is important to use the right primer and/or solvent cement for each type of plastic. Both should be formulated specifically for the particular type of plastic you are installing. There is no such thing as a totally satisfactory "universal" solvent cement. Different plastics have different molecular structures, which require specific chemicals to penetrate properly. A "universal" solvent may give you a superficial, surface-to-surface bond, but not a properly welded joint.

To assure a good solid weld with PVC and CPVC plastics, *priming is essential,* not optional! Priming cleans the surface, takes away the glaze, and starts the softening-up process. With ABS,

*NIBCO Plastic Piping How-to-and-Why Pocket Handbook. (NIBCO, Inc.)

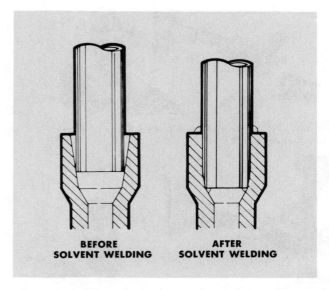

Figure 5-61. Interference fit between solvent weld plastic pipe and fittings. (NIBCO, Inc.)

however, you need only use ABS solvent to weld the joint. No primer is necessary.

Good Interference Fit. Plastic pipe and fittings are precision manufactured to exact specifications to ensure a proper interference fit. Unlike copper fittings, the inside walls of plastic fitting sockets are tapered so the pipe makes contact with the sides of the fitting well before the pipe reaches the seat of the socket. Figure 5-61 illustrates an interference fit.

Before applying any primer or solvent, check for an interference fit—the pipe should fit only about halfway into the fitting socket. Later, after application of primer and/or solvent, you will easily be able to force the pipe all the way in, creating a tight bond with ample surface area to weld together into a single piece.

Preparation and Installation. The following installation techniques have been tested and proven correct for ABS and PVC plastic drainage, waste, and vent systems, and for CPVC pressure systems. By following them carefully, you'll save time and assure a leakproof, permanent joint. These techniques are basically the same for all three types of plastics; differences, when they occur, are noted.

1. *Cut pipe squarely.* Use a miter box or a sharp tube cutter with a special blade for plastic.

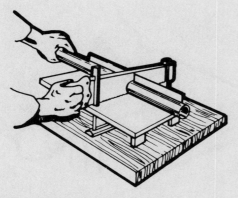

Figure 5-62. Solvent weld plastic pipe joint. Cut the plastic pipe squarely. (NIBCO, Inc.)

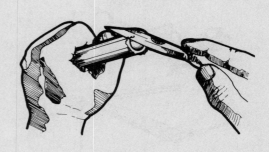

Figure 5-63. Remove the cutting burrs from the cut end(s) of the pipe. (NIBCO, Inc.)

Figure 5-64. Check for interference fit between plastic pipe and fitting. (NIBCO, Inc.)

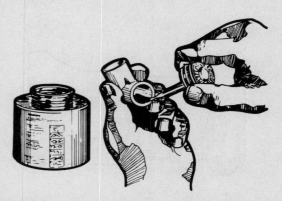

Figure 5-65. Applying primer to CPVC or PVC plastic fitting. (NIBCO, Inc.)

If the end isn't square, the pipe won't seat correctly in the fitting, and a weak joint will result (Figure 5-62).

2. *Smooth the end of the pipe.* Remove the burrs on the end of the pipe after cutting, or the rough edge will scrape away the solvent during assembly. Use a pocket knife or special deburring tool—but make sure the end is perfectly smooth (Figure 5-63).

3. *Check for interference fit.* Try the dry joint. The pipe should go in only about halfway; it *should not* go in all the way to the seat of the fitting. This type of fit is essential to forming a strong, solid joint (Figure 5-64).

4. *Apply primer* (PVC and CPVC only). Be sure surfaces are clean and dry. Use only primer specifically formulated for PVC and/or CPVC. Apply the primer first to the inside of the fitting, then to the outside of the pipe to the depth that will be taken into the fitting when seated. Be careful not to leave a puddle in the bottom of the fitting. Wait 5 to 15 seconds before applying solvent (Figure 5-65).

5. *Apply solvent cement.* While the surfaces are still wet from the primer, brush on a full, even coating of solvent cement to the inside of the fitting. Again, be careful not to form a puddle in the bottom of the fitting. (Applying too heavy a coat or leaving a puddle in the fitting will usually result in some flow restriction.)

Figure 5-66. Applying solvent cement to plastic pipe. (NIBCO, Inc.)

Figure 5-67. Fitting and positioning the plastic pipe and fitting. (NIBCO, Inc.)

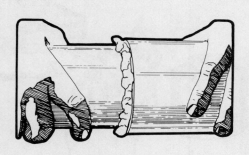

Figure 5-68. Checking for the correct bead of solvent cement. (NIBCO, Inc.)

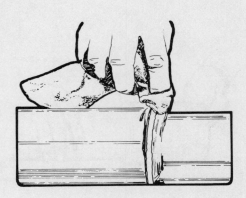

Figure 5-69. Wiping off the excess solvent cement. (NIBCO, Inc.)

Next, apply solvent to the pipe to the same depth as that of the primer. Make certain it's the correct solvent cement for the plastic you're installing! There is no "universal" solvent that will give totally satisfactory welded joints on all plastics. Use the right solvent for each kind of plastic (Figure 5-66).

6. *Fit and position the pipe and its fitting.* Put the pipe and fitting together immediately, before the solvent evaporates. Use enough force to insure that the pipe bottoms in the fitting socket. Give the fitting about a quarter turn as you push it on. This will ensure even distribution and absorption of the solvent. Then hold the joint firmly for about 10 seconds (longer in cold weather) to allow the solvent to start bonding the two surfaces. If you position and release too soon, the interference fit will force the pipe out of the fitting (Figure 5-67).

7. *Check for the correct bead.* Check the ring of cement that has been pushed out during assembly and alignment. If it doesn't go all the way around the joint, it means you haven't used enough cement, and the joint could leak (Figure 5-68).

8. *Wipe off excess cement.* If the bead looks complete, wipe off the excess cement with a clean rag, leaving an even fillet all the way around. This helps the joint cure faster (Figure 5-69).

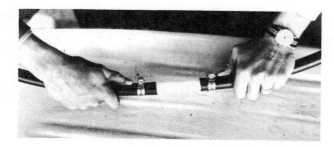

Figure 5-70. Plastic tubing insert fitting joint. Sliding the insert fitting into the tubing. Note that the stainless steel clamps have been placed on the tubing ends.

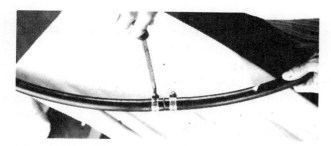

Figure 5-71. Tightening the stainless steel clamps with a screwdriver.

Insert Fitting Joint

Insert fittings for PE plastic tubing, of the type illustrated in Figure 3-12, are simply pushed into the pipe ends and secured with stainless steel clamps.

The following three steps are all that is required to make an insert fitting joint in flexible plastic tubing:

1. Cut the tubing with a plastic tubing cutter, knife, or saw.

2. Join the tubing after slipping the loose stainless steel clamps onto the tubing ends and pushing the insert fitting into the tubing ends up to the fitting shoulder (Figure 5-70).

3. Slide the stainless steel clamps over the serrated portion of the fitting and tighten the clamps with a screwdriver (Figure 5-71).

Flare Fitting Joint

The flare joint for flexible plastic tubing is made by following these steps:

1. Cut the tubing to length, preferably with a plastic tubing cutter. Cut squarely (see Figure 5-72).

2. Place the flare nut on the tubing.

3. Insert the pilot plug of the flaring tool into the fitting, apply the clamping pliers, and rotate the tool 5 to 10 revolutions to complete the flare (see Figure 5-73).

4. Assemble the joint by placing the flared tubing end against the fitting and tightening the flare nut onto the fitting (Figure 5-74). Use two wrenches: one wrench on the fitting and one wrench on the flare nut.

Figure 5-72. Plastic tubing flare joint. Cutting the plastic tubing squarely with a plastic tubing cutter. (Crestline Plastic Pipe Co., Inc.)

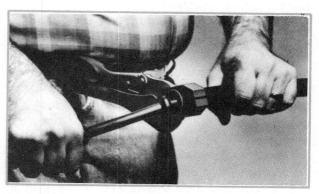

Figure 5-73. Insert the plug of the flaring tool and rotate it 5–10 revolutions. (Crestline Plastic Pipe Co., Inc.)

Figure 5-74. Assembly of the flare joint. (Crestline Pipe Co., Inc.)

⦿ INSTALLING AND SUPPORTING PIPE

As was stated in the first part of this chapter, the plumber must install and support all pipe in such a manner that the pipe and its joints will not leak. Pipe that is not properly installed and supported can sag at the joints, causing them to leak; the pipe may even break or crack between the joints because of stress. In addition, improperly installed and supported drainage piping can sag or shift from its proper grade or pitch, causing portions of drain line to form traps. These trapped portions of the pipe line fill with liquid and solid waste, which will block the pipe.

Installation of Underground Piping

All of the pipe on any given job is installed either above or below ground. The following are some basic rules for the installation of underground piping to be installed on a job.

First, the plumber must make sure it is safe to work in the trench where the pipe is to be installed. (The rules for safe work in trenches were given in Chapter 2.)

Then, after making sure that the trench is safe, the plumber can proceed as follows:

1. Prepare the bottom of the trench for the pipe. The trench bottom should be relatively smooth so that the pipe will lie on a flat surface that is free from large lumps of dirt and rocks. In addition, if the pipe to be installed is a drainage pipe line, the trench bottom must be graded so that the pipe flow will drain by gravity.

2. Lower the pipe carefully into the trench.

3. Install the pipe. Holes should be dug at each coupling or bell so that the pipe will rest on its barrel, not on the pipe coupling or bell (Figure 5-75). At this time, check drainage pipe to see that it has the proper grade. If necessary, raise or lower the pipe by adding or removing dirt from under the pipe. Do *not* use scrap lumber to block the pipe up to grade. If the pipe must be raised by blocking, use concrete blocks and/or bricks. After the pipe is aligned and graded, the joint can be made.

4. When all of the pipe is installed in the trench, *and before it is covered with dirt,* the pipe line must be tested and inspected by the proper

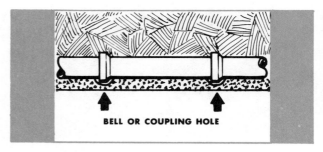

Figure 5-75. Dig bell or coupling holes so that the pipe will rest on the barrel. (National Clay Pipe Institute)

authorities. (The testing of pipe lines will be covered in Chapter 11.)

5. Backfill the trench by hand, tamping the dirt around the pipe to fill any voids under the pipe so that the pipe does not act as a beam holding up the weight of the dirt being used to fill the trench.

It is recommended by the manufacturers of most of the different types of pipe that the trench be backfilled and tamped in stages. The first stage is to backfill to the centerline of the pipe (Figure 5-76). The next stage is to cover the pipe

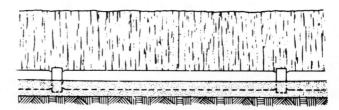

Figure 5-76. First stage of backfilling: backfill and tamp the fill to the centerline of the pipe. (Johns-Manville)

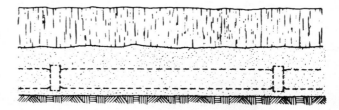

Figure 5-77. Second stage of backfilling: backfill and tamp the fill to at least 1 foot above the pipe. (Johns-Manville)

with at least 1 foot of dirt (Figure 5-77). Backfilling and tamping around the pipe in these two stages should be done by hand; only fine material, such as sand, gravel, loose dirt, etc., should be used. After the pipe is covered by hand with at least 1 foot of dirt, the trench can then be filled with heavier material, such as lumps of dirt, smaller rocks, etc. At this stage, machinery can be used to push the dirt back into the trench to fill it back to the original ground level.

When backfilling with a machine, it is best to backfill and tamp the fill material in layers to avoid having the trench settle at a later date.

Hanging and Supporting Above-Ground Piping

Pipe installed in a trench is supported by the dirt it lies on. Since very little above ground piping can lie on a flat surface for its entire length, the plumber must install the proper pipe hangers, which are anchored to the building construction to keep the pipe in alignment and leakproof.

A variety of pipe hangers are designed to hold and support pipe in either the horizontal or vertical position. Figure 5-78 illustrates some of the more common pipe hangers used by plumbers.

Figure 5-78. Pipe hangers and supports.

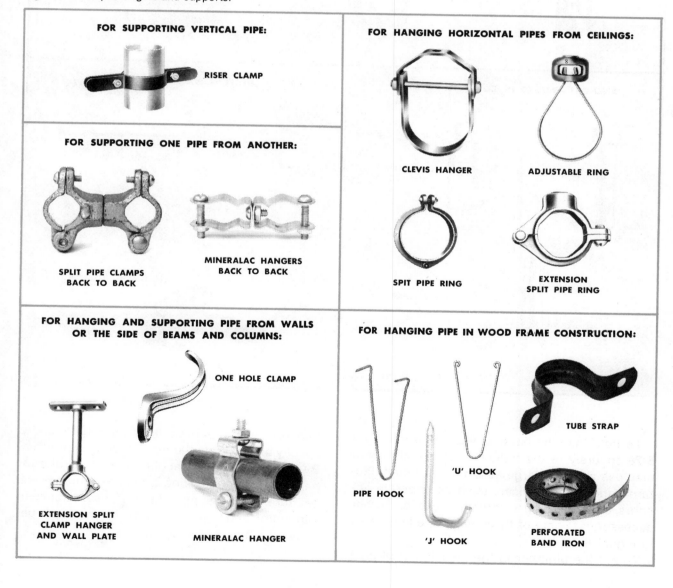

FOR SUPPORTING VERTICAL PIPE:

RISER CLAMP

FOR SUPPORTING ONE PIPE FROM ANOTHER:

SPLIT PIPE CLAMPS BACK TO BACK

MINERALAC HANGERS BACK TO BACK

FOR HANGING HORIZONTAL PIPES FROM CEILINGS:

CLEVIS HANGER

ADJUSTABLE RING

SPIT PIPE RING

EXTENSION SPLIT PIPE RING

FOR HANGING AND SUPPORTING PIPE FROM WALLS OR THE SIDE OF BEAMS AND COLUMNS:

ONE HOLE CLAMP

EXTENSION SPLIT CLAMP HANGER AND WALL PLATE

MINERALAC HANGER

FOR HANGING PIPE IN WOOD FRAME CONSTRUCTION:

PIPE HOOK

'U' HOOK

'J' HOOK

TUBE STRAP

PERFORATED BAND IRON

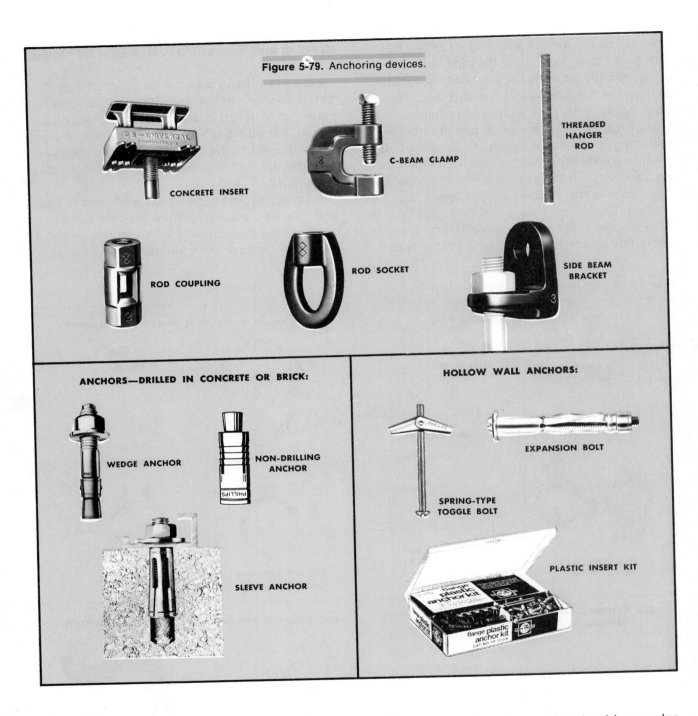

Figure 5-79. Anchoring devices.

CONCRETE INSERT

C-BEAM CLAMP

THREADED HANGER ROD

ROD COUPLING

ROD SOCKET

SIDE BEAM BRACKET

ANCHORS—DRILLED IN CONCRETE OR BRICK:

WEDGE ANCHOR

NON-DRILLING ANCHOR

SLEEVE ANCHOR

HOLLOW WALL ANCHORS:

SPRING-TYPE TOGGLE BOLT

EXPANSION BOLT

PLASTIC INSERT KIT

To hold the pipe hangers illustrated in Figure 5-78 securely to the building construction, you may use any of a large assortment of manufactured anchoring devices, such as concrete anchors, concrete inserts, threaded rods, and other accessories. Some of these items are illustrated in Figure 5-79.

To aid the plumber in the installation of pipe hangers and supports, most plumbing codes specify exactly how the pipe must be supported. By following these local plumbing code rules for hanging and supporting pipe, the plumber will be sure that the pipe he installs remains in alignment and that the joints will not leak because the pipe has sagged.

Supporting Vertical Piping

Vertical piping must be secured at sufficiently close intervals to keep the pipe in alignment. In addition, follow these requirements for vertical support of the types of pipe mentioned below:*

1. *Cast iron soil pipe (both hub-type and no-hub-type):* Support at the base and at each story in height. Rubber gasket joint soil pipe (hub - and no-hub-types) are supported at 5-foot intervals except where 10-foot lengths of pipe are used.

2. *Threaded iron pipe:* Support at every other story in height.

3. *Copper tubing:* Support at each story in height.

4. *Plastic pipe, 1½-inch size and smaller:* Support exposed pipe at 4-foot intervals. Support concealed pipe the same as in requirement No. 5.

5. *Plastic pipe, 2-inch size and larger:* Support at each story.

Figure 5-80 illustrates the use of a *riser clamp* to support a vertical pipe at the floor, while Figures 5-81 and 5-82 illustrate the use of a *bracket* and *one-hole strap* for the support of vertical pipe between floor levels.

*Minnesota Plumbing Code (1975).

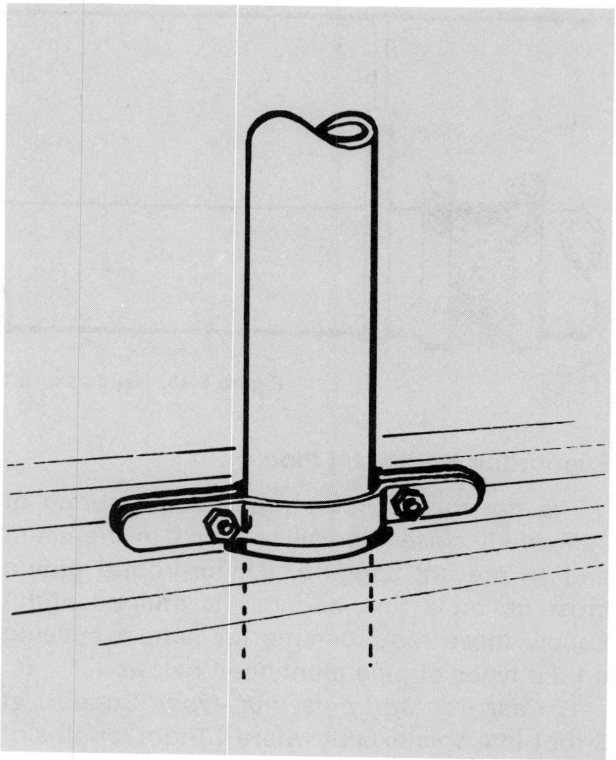

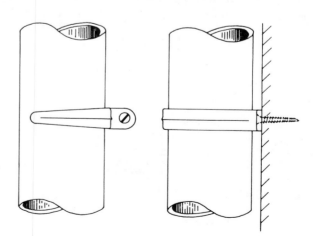

Figure 5-80. A riser clamp is used to support vertical pipe at the floor line.

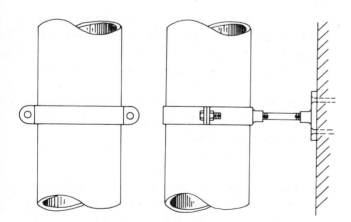

Figure 5-81. Using an extension ring hanger and wall plate to support vertical pipe. (Cast Iron Soil Pipe Research Assn.)

Figure 5-82. A one-hole strap support for vertical pipe. (Cast Iron Soil Pipe Research Assn.)

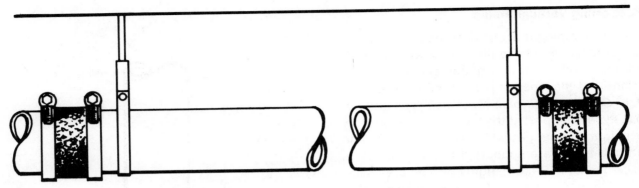

Figure 5-83. Hangers on a 10-foot length of no-hub soil pipe.

Supporting Horizontal Piping

The horizontal piping must be supported at sufficiently close intervals to keep it in alignment and to prevent sagging. (Underground piping must lie on a firm bed for its entire length.) Follow these requirements for hanger spacing on the types of pipe mentioned below:

1. *Cast iron soil pipe (hub-type):* Support at 5-foot intervals except where 10-foot lengths of cast iron soil pipe are used.

2. *Cast iron soil pipe (no-hub-type):* Support at every other joint, except when the developed length between hangers exceeds 4 feet; then, hangers must be provided at each joint. Hangers must be provided at each horizontal branch connection. Hangers must be placed on or immediately adjacent to the no-hub coupling. The no-hub coupling is not designed to support the weight of the pipe. The hangers on a typical length of no-hub pipe suspended from a ceiling would be placed according to Figure 5-83.

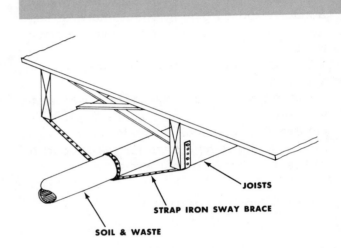

Figure 5-85. A sway brace. (Cast Iron Soil Pipe Research Assn.)

Where no-hub soil pipe is suspended in excess of 18 inches from a ceiling by means of nonrigid hangers, it must be suitably braced against horizontal movement with sway bracing of the type shown in Figures 5-84 and 5-85.

3. *Threaded pipe:* Support at 12-foot intervals.

4. *Copper tubing, 1¼-inch size and smaller:* Support at 6-foot intervals.

5. *Copper tubing, 1½-inch size and larger:* Support at 10-foot intervals.

6. *Plastic pipe:* Support at 32-inch intervals, except where the pipe conveys the waste from dishwashers or similar hot water wastes. In this event, it must be supported on continuous or wood strips for its entire length (Figure 5-86).

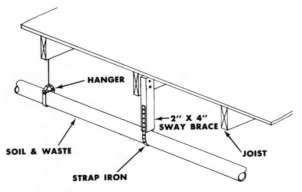

Figure 5-84. Horizontal pipe with a sway brace. (Cast Iron Soil Pipe Research Assn.)

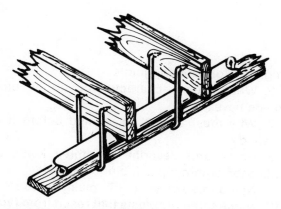

Figure 5-86. Plastic pipe must be supported on continuous wood or metal strips when it conveys hot water waste.

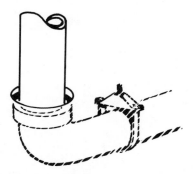

Figure 5-87. A method of supporting a closet bend using a clevis hanger.

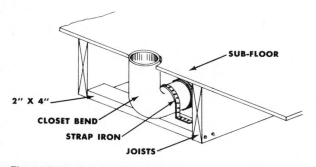

SUB-FLOOR

2" X 4"

CLOSET BEND

STRAP IRON

JOISTS

Figure 5-88. Bracing a closet bend in wood frame construction. (Cast Iron Soil Pipe Research Assn.)

Support of Closet Bends

Closet bends joined to a cast iron soil pipe stack by means of a rubber gasket joint or a no-hub coupling, or to a plastic stack with a solvent weld joint, must be adequately supported both horizontally and vertically to prevent movement in either direction. Figures 5-87 and 5-88 illustrate the methods of supporting closet bends that will prevent them from moving.

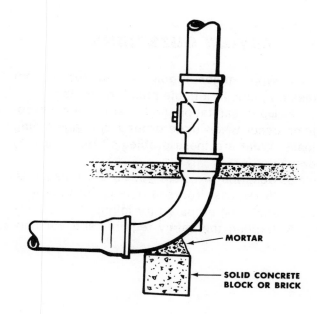

MORTAR

SOLID CONCRETE BLOCK OR BRICK

Figure 5-89. Underground stack base support.

Support of Stack Bases

Stacks must be adequately supported at their bases. To the plumber, this means placing a brick or concrete support under the fitting at the base of the stack when the stack base is below ground, as shown in Figure 5-89. If the stack base is above ground, a hanger would be placed on the base fitting or as close to it as possible (Figure 5-90).

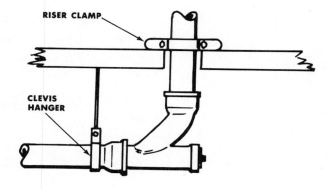

RISER CLAMP

CLEVIS HANGER

Figure 5-90. Support of above-ground stack base fitting using a riser clamp or a clevis hanger.

⬤ REVIEW QUESTIONS

1. What major problems may result from leaking joints in a home plumbing system.

2. Both sanitation and maintenance problems occur when underground drainage piping leaks. What are these problems? How are they prevented?

3. Describe two main types of soil pipe joints.

4. When cast iron soil pipe is cut to length, how must it be prepared for joining?

5. Explain the safety requirements for handling lead for caulking joints.

6. What special characteristic do all pipe threads have?

7. Why must a pipe be reamed before it is threaded?

8. Name and describe the three kinds of joints used with copper tubing.

9. Name three ways to join plastic pipe.

10. Name three problems that result from faulty support of piping.

CHAPTER 6

Sanitary Drainage, Vent, and Storm Drainage Piping

In this chapter we will discuss the principles that govern the installation of sanitary drainage (soil and waste) piping, vent piping, and storm drainage piping. These principles are important to a plumber because he will need a thorough understanding of them to be able to design plumbing systems.

The ability to design plumbing systems is an essential part of a plumber's knowledge for two reasons:

1. On many of the smaller jobs (especially new housing and remodeling work), there will be no blueprints of the piping to work from and the plumber will have to design the piping for the entire job.

2. On larger jobs, even if the plumbing systems are presented in detailed blueprints, the plumber is responsible for seeing that these systems are designed and installed within the requirements of the local plumbing code.

A plumber can design the sanitary drainage, vent, and storm water drainage systems of any

building if he understands the principles of this chapter and has the following information about the building:

1. The number and type of plumbing fixtures and appliances, floor drains, and roof drains on the job.

2. The locations of these items.

3. Where the pipes that supply the water to the fixtures and appliances and remove the waste can be run. (Has space been provided?)

4. The piping materials to be used.

5. Areas of conflict with structural features of the building and with other trades whose work must also be installed in the building.

 SANITARY DRAINAGE PIPING

Soil and Waste Pipes. The principles of sanitary drainage presented in this chapter apply to two types of pipes: soil pipes and waste pipes.

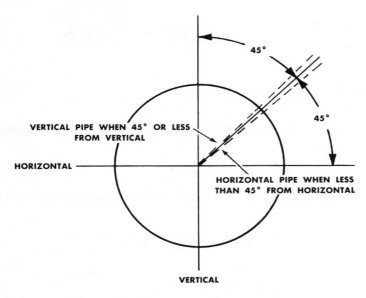

Figure 6-1: Definition of horizontal and vertical pipe.

A *soil pipe* is a pipe that conveys the discharge of water closets or similar fixtures containing fecal matter, with or without the discharge of other fixtures to the building drain or building sewer.

A *waste pipe* is a pipe that conveys only liquid waste free of fecal material.

These terms are important to the apprentice only in that a waste pipe is generally of a smaller size than a soil pipe.

Horizontal and Vertical Pipes. Horizontal and vertical are two terms that will be used frequently in this chapter. Vertical and horizontal pipe angles are shown in Figure 6-1.

A *horizontal pipe* is any pipe or fitting that makes an angle of less than 45 degrees with the horizontal.

A *vertical pipe* is any pipe or fitting that makes an angle of 45 degrees or less with the vertical.

Drainage Fixture Unit System

Before any sanitary drainage piping can be installed on a job it must be sized. Sanitary drainage piping is sized according to the *drainage fixture unit system.*

The drainage fixture unit system was formulated in tests conducted by the Uniform Plumbing Code Committee, a group consisting of representatives of management, labor, and agencies of the U.S. government. Standard plumbing fixtures were individually tested and the amount of liquid waste that could be discharged through their waste outlets in a given time interval was carefully measured. It was found that a lavatory, which is one of the smaller types of plumbing fixtures, would discharge through its waste outlet approximately $7\frac{1}{2}$ gallons of water in a 1-minute interval. This volume was so close to 1 cubic foot of water that it was established as the basis of the drainage fixture unit system:

1 drainage fixture unit (dfu) = $7\frac{1}{2}$ gallons of water (discharged in 1 minute).

Based on the above value of 1 drainage fixture unit being equal to $7\frac{1}{2}$ gallons of water per minute of waste discharge and tests of various plumbing fixtures, waste water discharge rates have been established. Table 6-1, Drainage Fixture Unit Values for Various Plumbing Fixtures, shows these rates.* In addition to listing the individual plumbing fixtures and their drainage fixture unit values, Table 6-1 also lists the minimum size trap and drain for each fixture. This is the smallest size pipe that this fixture may drain into. This size must be used even if one of the sizing tables in this chapter should indicate that a smaller size of pipe would convey an equal number of drainage fixture units of waste.

Sizing the Building Drain and the Building Sewer

Tests conducted on horizontal drainage pipes have indicated that given sizes of pipe would discharge up to a certain number of drainage fixture units of waste without subjecting the plumbing system to plus or minus pressure. These tests were made on various installations of standard design. Changes in direction, materials, grades, and other factors were carefully considered, and the discharge capacities for the different sizes of horizontal drainage pipes have

*Note that all of the tables presented in this chapter were taken from the Minnesota Plumbing Code and apply *only* to the State of Minnesota. The apprentice must consult the local plumbing code for his area for its equivalent tables before sizing any pipe on the job in another area, as there may be some differences.

TYPE OF FIXTURE	DRAINAGE FIXTURE UNIT VALUE	MINIMUM FIXTURE TRAP AND DRAIN SIZE
Clothes washer (domestic use)	2	1½
Clothes washer (public use in groups of three or more)	6 each	
Bathtub with or without shower	2	1½
Bidet	2	1½
Dental unit or cuspidor	1	1¼
Drinking fountain	1	1¼
Dishwasher, domestic	2	1½
Dishwasher, commercial	4	2
Floor drain with 2-inch waste	2	2
Floor drain with 3-inch waste	3	3
Floor drain with 4-inch waste	4	4
Lavatory	1	1¼
Laundry tray (one or two compartment)	2	1½
Shower stall, domestic	2	1½
Shower (gang) per head	1	
SINKS:		
Combination, sink and tray (with disposal unit)	3	1½
Combination, sink and tray (with one trap)	2	1½
Domestic	2	1½
Domestic, with disposal unit	2	1½
Surgeons	3	1½
Laboratory	1	1½
Flushrim or bedpan washer	6	3
Service	3	2
Pot or scullery	4	2
Soda fountain	2	1½
Commercial, flat rim, bar or counter	3	1½
Wash, circular or multiple (per set of faucets)	2	1½
URINAL Pedestal or Wall-Hung, with 3-inch trap (blowout and syphon jet)	6	3
Wall-hung with 2-inch trap	4	2
Wall-hung with 1½-inch trap	2	1½
Trough (per 6-foot section)	2	1½
Stall	3	2
WATER CLOSET	6	3
Unlisted Fixture or Trap Size		
1¼ inch	1	
1½ inch	2	
2 inch	3	
2½ inch	4	
3 inch	5	
4 inch	6	

Minnesota Plumbing Code

DIAMETER OF DRAIN (INCHES)	SLOPE			
	1/16 in/ft. (dfu)	1/8 in/ft. (dfu)	1/4 in/ft. (dfu)	1/2 in/ft. (dfu)
1 1/4	—	—	—	—
1 1/2	—	—	—	—
2	—	—	21	26
2 1/2	—	—	24	31
3*	—	36**	42**	50**
4	—	180	216	250
5	—	390	480	575
6	—	700	840	1,000
8	1,400	1,600	1,920	2,300
10	2,500	2,900	3,500	4,200
12	3,900	4,600	5,600	6,700
15	7,000	8,300	10,000	12,000

*No water closet shall discharge into a drain less than 3 inches in diameter.

**Not over two water closets.

***Every building drain that receives the discharge of three or more water closets shall not be less than 4 inches in diameter.

Minnesota Plumbing Code

been compiled into tables. Table 6-2 represents such a table for sizing building sewers, building drains, and building drain branches from stacks. Figure 6-2 illustrates these terms to help clarify them.

With the information contained in these tables, the plumber can establish the total discharge of all the fixtures in a building in drainage fixture units and can select a size of drain to serve the demand.

Horizontal Drainage Pipe Sizing EXAMPLE. To familiarize you with the drainage fixture unit system and its application to an installation of plumbing, the following example is offered.

Suppose a plumbing installation consisted of thirty water closets, twenty-eight lavatories, four drinking fountains, three wall-hung urinals with a 2-inch trap, four 2-inch floor drains, and two service sinks. The sum of the drainage fixture unit values of all the fixtures would be as given in the following summary:

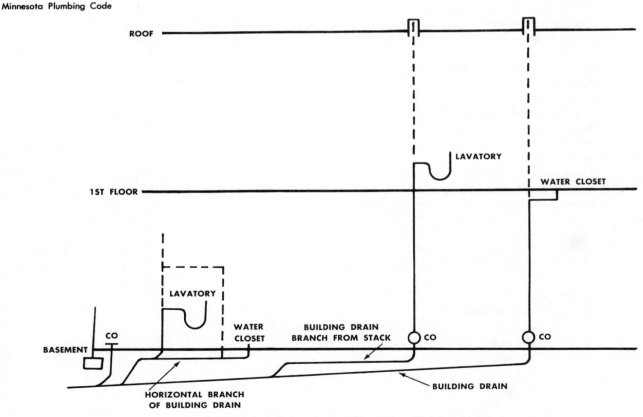

Figure 6-2: Building drain and branches. (Ralph R. Lichliter)

DRAINAGE FIXTURE UNIT (dfu) SUMMARY

No.			dfu		Total
30	Water closets	×	6	=	180
28	Lavatories	×	1	=	28
4	Drinking fountains	×	1	=	4
3	Urinals	×	4	=	12
4	2" Floor drains	×	2	=	8
2	Service sinks	×	3	=	6
	Total drainage fixture units				238

According to Table 6-2 (extending the diameter with a slope or pitch of $1/4$ inch to each foot) a 5-inch building drain and building sewer would be required.

Many plumbers believe that if they make a horizontal drainage pipe a size larger than is necessary they will increase its efficiency. However, this is not the case. Scouring action is lost by increasing the size of the drain. With a larger pipe, the solids are carried along the bottom of the pipe and, because the water flow within the larger pipe is shallow and slow, they become separated from the water and remain in the drain. This may result in stoppage of the drain or branch, and often the entire building drain is affected.

A horizontal drain of the proper size has a flow of about $1/3$ of the pipe diameter. (That is, the pipe is $1/3$ full of sewage.) This amount of sewage assures a scouring action. On the other hand, a drain too small in size is overtaxed by flow and is apt to produce syphonage, back pressure, and basement flooding.

In addition to the total number of drainage fixture units that drain into a horizontal drain pipe, other factors must be taken into consideration when using Table 6-2. If the apprentice will study the notes at the bottom of that table he or she will notice three important items:

1. No water closet is permitted to discharge its waste into a horizontal drain pipe less than 3 inches in size.

2. No more than two water closets may drain into a 3-inch horizontal drain pipe.

3. Every building drain that receives the discharge of three or more water closets shall not be less than 4 inches in diameter.

TABLE 6-3.
MAXIMUM LOADS FOR
HORIZONTAL BRANCH DRAINS*
IN DRAINAGE FIXTURE UNITS (dfu)

DIAMETER OF DRAIN (INCHES)	HORIZONTAL BRANCH DRAIN*— $1/4$ in/ft. (dfu)
$1^{1/4}$	1
$1^{1/2}$	3
2	6
$2^{1/2}$	12
3**	32***
4	160
5	360
6	620
8	——
10	——
12	——
15	——

*Includes horizontal branches of the building drain.
**No water closet shall discharge into a drain less than 3 inches.
***Not over two water closets.

Minnesota Plumbing Code

One other item required by most plumbing codes is that all building sewers must be at least 4 inches in diameter.

Sizing Horizontal Branch Drains

Another type of horizontal drainage piping that the plumber will install is the horizontal branch drain. A *horizontal branch drain* is a drain pipe extending horizontally from a soil or waste stack or building drain with or without vertical sections or branches. It receives the discharge from one or more fixture drains on the same floor as the horizontal branch and conducts it to the soil or waste stack or to the building drain. Figure 6-2 illustrates a horizontal branch drain in addition to several other key terms. Table 6-3 is a typical sizing table for horizontal branch drain pipes.

Horizontal Branch Drain Sizing EXAMPLE.
To familiarize you with the use of Table 6-3, the following example is offered.

What size horizontal branch drain would be required to serve 2 water closets, 2 lavatories, 2 bathtubs, and 2 domestic kitchen sinks?

By referring back to Table 6-1, we find the following information.

DRAINAGE FIXTURE UNIT (dfu) SUMMARY

No.			dfu		Total
2	Water closets	×	6	=	12
2	Lavatories	×	1	=	2
2	Bathtubs	×	2	=	4
2	Kitchen sinks	×	2	=	4
	Total drainage fixture units				22 dfu

Consulting Table 6-3, we find that a 3-inch horizontal branch drain would be required.

As was the case with Table 6-2, the apprentice should carefully read the three notes at the bottom of Table 6-3, which give restrictions on the use of the sizing table.

Another item of importance to the apprentice is that most plumbing codes require all underground drainage pipes to be at least 2 inches in diameter.

Sizing Soil and Waste Stacks

After the horizontal building sewer and building drain piping is sized, the plumber must size the vertical pipes or stacks that empty into these horizontal drainage pipes. Remember, *stack* is a general term for any vertical line of soil, waste, or vent piping extending through one or more stories.

Tests have been conducted on various stack installations just as they have on horizontal drainage piping. These tests indicate that, in spite of what the apprentice might think, water does not flow down a stack in slugs of waste water separated by pockets of air. Rather, the waste water flows down the stack in a sheet around the inside walls of the pipe. Another factor these tests have shown is that the waste does not continue to accelerate as it falls down a tall stack, but that it reaches its maximum velocity in about two normal stories or floor levels.

The results of the tests on stack installations have led to the formulation of a stack sizing table.

TABLE 6-4.
MAXIMUM LOADS FOR SOIL AND WASTE STACKS IN DRAINAGE FIXTURE UNITS (dfu)

DIAMETER OF STACK (INCHES)	STACKS OF NOT MORE THAN THREE STORIES OR BRANCH INTERVALS (BI)	STACKS OF MORE THAN THREE STORIES OR BRANCH INTERVALS (BI)	TOTAL AT ONE STORY OR BRANCH INTERVAL (BI)
1¼*	2	2	1
1½*	4	4	2
2*	9	18	6
2½*	20	42	9
3	36***	72***	24**
4	240	500	90
5	540	1,100	200
6	960	1,900	350
8	——	3,600	600
10	——	5,600	1,000
12	——	8,400	1,500

*No water closets permitted on a stack less than 3 inches in diameter.
**Not over two water closets permitted.
***Not over six water closets permitted, and not over six branch intervals on a 3-inch soil stack.

Minnesota Plumbing Code

based on the same drainage fixture unit method as the previously mentioned horizontal drainage pipes. Table 6-4 represents a typical stack sizing table.

In Table 6-4, notice the term *branch interval*. A *branch interval* is a vertical length of stack at least eight feet high (it corresponds *in general* to a story height), within which the horizontal branches from one story or floor of the building are connected to the stack. (That is, one branch interval equals one floor of plumbing fixture drains.) Figure 6-3 illustrates several different stack installations and how the branch intervals are counted on each stack. Note that some branch intervals run more than one story of building height.

Some additional explanation of Table 6-4 is required. Table 6-4 is actually three tables in one. It lists the *maximum number* of drainage fixture units which may empty into:

1. Stacks of *not more* than three stories, with a maximum of three branch intervals.
2. Stacks of *more* than three stories or three branch intervals.
3. A stack on any one story or branch interval.

Notice that in the 2-inch and larger sizes, the taller stacks (those over three stories or branch intervals) have a larger drainage fixture unit capacity than the shorter stacks.

There are also three important notes at the bottom of Table 6-4:

1. No water closets are permitted to drain into a stack less than 3 inches in diameter.

2. No more than two water closets may drain into a 3-inch stack in any one story or branch interval.

3. No more than six water closets may drain into a 3-inch stack, and 3-inch stacks may not have more than six branch intervals of waste emptying into them.

There are several other regulations that apply to the sizing of plumbing stacks within buildings:

1. No soil or waste stack is permitted to be smaller than the largest horizontal branch connected to it.

2. Any building in which plumbing is installed

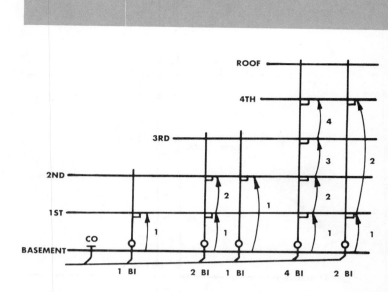

Figure 6-3: Branch intervals (BI). (Minnesota Plumbing Code)

must have at least one stack vent or vent stack extended full size through the roof using not less than 3-inch pipe. This full-size stack should be the stack that is most remote from the location where the building drain leaves the building.

Soil and Waste Stack Sizing EXAMPLES. The following examples are offered to help you understand Table 6-4. Figures 6-4 and 6-5 both illustrate stacks with two domestic kitchen sinks draining into them. These sinks are rated at 2 drainage fixture units each, and each has a 1½-inch trap and drain (Table 6-1). If the sinks empty into the stack on two different stories (or branch intervals) as illustrated in Figure 6-4, Table 6-4 indicates that a 1½-inch stack would serve the installation.

However, if both sinks are installed on the same story as illustrated in Figure 6-5, a 2-inch stack is required because the far right column of Table 6-4 indicates that only 2 drainage fixture units may drain into a 1½-inch pipe on any one story or branch interval.

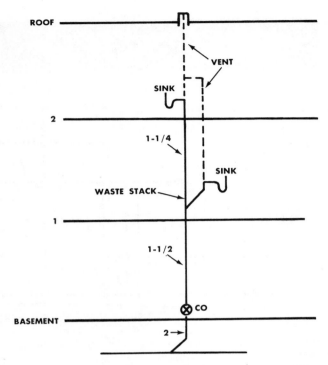

Figure 6-4: Sizing a stack with kitchen sinks on two different floors.

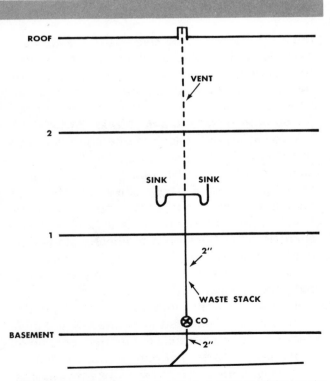

Figure 6-5: Sizing a stack with two kitchen sinks on the same floor.

Offset Stacks. There will be occasions in which the plumber must install stacks that cannot continue vertically for their entire length for a variety of reasons. If a stack cannot continue vertically, it must be offset. An *offset* is a combination of elbows or bends that brings one section of the pipe out of line but into a line parallel with the other section. Figures 6-6, 6-7, and 6-8 illustrate offset stacks.

Stacks that are offset 45 degrees or less from the vertical, as shown in Figure 6-6, are sized as though they were straight vertical stacks. (Use Table 6-4.)

However, if the stack must be offset *more* than 45 degrees from the vertical, as shown in Figure 6-7, the stack is sized using the following procedure.

1. The portion of the stack above the offset (section A in Figure 6-7) is sized the same as for a straight stack based on the total number of drainage fixture units above the offset. (See Table 6-4.)

2. The offset portion of the stack (B) is sized as though it were a horizontal building drain branch. (See Table 6-2.)

3. The portion of the stack below the offset (C) is sized at least as large as the offset.

4. In buildings of five or more stories with offset stacks, no fixtures should be set on the floor in which the offset occurs if there are fixtures that discharge their waste into the stack four or more stories above the offset. This applies as well to the horizontal building drain branch from the stack. Fixtures that drain into the horizontal portion of offset stacks in these tall buildings under the above condition are subject to *trap seal loss* from back pressure (a point that will be fully explained in Chapter 8). This difficulty can be avoided if the drains from the fixtures near the base of the stack or stack offset connect to the horizontal pipe at least eight feet away from the offset measured vertically or horizontally. The drain may also connect back into the vertical portion of the stack two feet below the offset. Figure 6-8 illustrates where these connections might be made.

The above rules do not affect stacks that are offset above the *highest* fixture drain connected to the stack. Offsets in this vertical portion of the

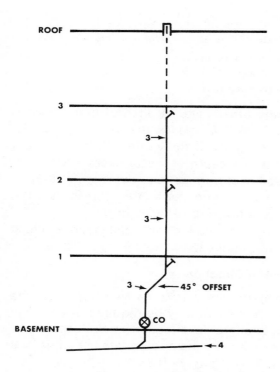

Figure 6-6: A 45° offset stack may be considered as a straight stack in sizing.

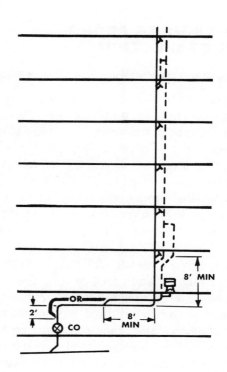

Figure 6-8: No fixture should connect within 8 feet of the base of the offset measured either vertically or horizontally. (Ralph R. Lichliter)

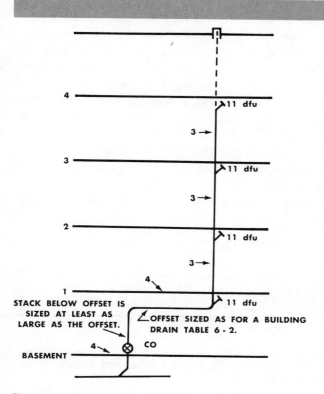

Figure 6-7: Sizing a stack offset more than 45°.

stack are sized the same as any stack vent, as will be explained later in this chapter.

Offset Stack Sizing EXAMPLE. To clarify the above example, suppose that in Figure 6-7 there are 11 drainage fixture units of waste that empty into the stack on each floor, or a total of 44 drainage fixture units (4 × 11 = 44) on the entire stack. A 3-inch stack would be large enough for the section of the stack above the offset, according to Table 6-4, since this is a stack of more than three stories.

The offset section of the stack is sized in accord with Table 6-2. If we assume there is a pitch of ¼ inch to the foot, the offset must be 4-inch pipe. The section of the stack below the offset must be at least as large as the offset, so it too is 4-inch pipe.

DRAINAGE PIPING INSTALLATION

Grade or Pitch of Horizontal Drainage Piping

After the size of any horizontal drainage piping has been determined, other factors in installing the pipe must be considered. The grade at which the pipe is to be installed is probably next in importance. Grade or *pitch* is the fall (slope) of a line of pipe in reference to a horizontal plane. As applied to plumbing drainage, pitch is usually expressed as the fall in a fraction of an inch per foot length of pipe (for example, ¼ inch per foot).

It is advisable to grade all horizontal drainage piping ¼ inch to each foot of run. Tests verify the soundness of this pitch. Horizontal pipes have been found to produce the necessary velocity and discharge capacity at this pitch so that they can scour themselves properly and function without producing plus or minus pressures in the plumbing system. (It may be necessary, however, because of the depth of the basement floor and inadequate depth of the sanitary sewer main, to pitch the building drain or building sewer at less than ¼ inch per foot.)

Also, an unusually long sewer would require less pitch per foot because the accumulated or total pitch would result in an exceptionally deep building drain outlet. If the grade is slight, the plumber should use an instrument, such as the builder's level, to grade the sewer so that it slopes evenly throughout its entire length and is free from sags (which will result in sections of the pipe being full of water).

The following are the minimum slopes for the stated sizes of pipe:

Size of Pipe	Minimum Slope Per Foot
Less than 3 inches	¼"
3 to 6 inches	⅛"
8 inches and over	¹/₁₆"

On the other hand, a pitch of more than ¼ inch per foot *increases* the velocity and discharge capacity of the waste, but it also might decrease the depth of waste necessary to provide self-scouring action (see Figure 6-9). It might also account for a minus pressure if the drain were taxed to capacity flow.

Further information on calculating pitch is given in Chapter 12.

Change in Direction

Change in direction is the term applied to the various turns that may be required in drainage pipes. Changes in direction in drainage pipes must be made by the appropriate use of fittings such as Ys, long or short sweep ¼ bends, ⅙, ⅛, or ¹/₁₆ bends, or by a combination of these fittings.

Care must be taken in selecting the proper fittings for changes in direction in drainage piping to prevent waste stoppages. Long-radius fittings lessen the probability of waste stoppage. Short-turn ¼ bends may be used for changes of direction from the horizontal to the vertical. Long-sweep ¼ bends, two ⅛ bends, or a combination Y and ⅛ bend may be used where the change of direction is from the vertical to the horizontal or from horizontal to horizontal. Figure 6-10 and its accompanying information illustrate the proper use of ¼ bends and elbows for changes in direction. Fittings that are not illustrated in Figure 6-10, but have the equivalent

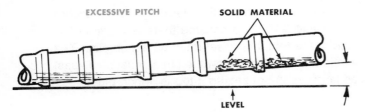

ABOVE: DEPOSIT OF SUSPENDED BODIES OCCURS WHEN THE DRAIN IS GIVEN EXCESSIVE PITCH.

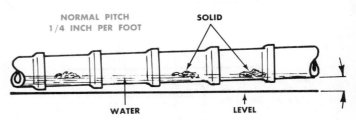

ABOVE: ADEQUATE DEPTH OF FLOW THROUGHOUT THE LENGTH OF THE DRAIN ASSURES THOROUGH DISCHARGE OF WASTE MATERIALS.

Figure 6-9: Proper and improper pitch on horizontal drainage pipes.

1. VERTICAL TO HORIZONTAL	LESS THAN 3 INCH SIZE	LONG SWEEP SOIL 1/4 BEND OR EXTRA LONG TURN 90° DRAINAGE ELL
2. VERTICAL TO HORIZONTAL	3 INCH AND LARGER SIZE	SHORT SWEEP SOIL 1/4 BEND OR LONG TURN 90° DRAINAGE ELL
3. HORIZONTAL TO VERTICAL	ALL SIZES	SOIL 1/4 BEND OR LONG TURN 90° DRAINAGE ELL
4. HORIZONTAL TO HORIZONTAL	ALL SIZES	LONG OR SHORT SWEEP SOIL 1/4 BEND OR EXTRA LONG TURN 90° DRAINAGE ELL

*COPPER, CAST BRASS, AND PLASTIC DRAINAGE FITTINGS SHALL HAVE THE EQUIVALENT SWEEP OF THE LISTED DURHAM DRAINAGE FITTINGS.

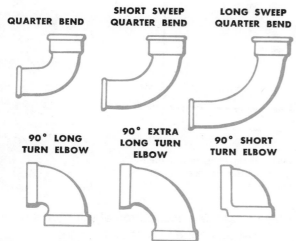

QUARTER BEND   SHORT SWEEP QUARTER BEND   LONG SWEEP QUARTER BEND

90° LONG TURN ELBOW   90° EXTRA LONG TURN ELBOW   90° SHORT TURN ELBOW

Figure 6-10: Changes in direction in drainage piping. (Ralph R. Lichliter)

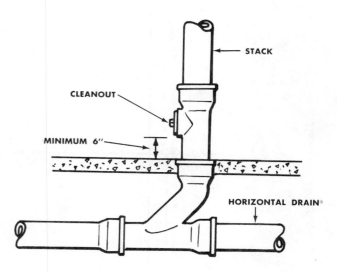

Figure 6-11: Stack base cleanout.

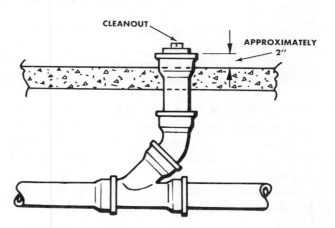

Figure 6-12: Building drain cleanout.

sweep of the illustrated fittings, can be used for the same type of changes in direction.

Cleanouts

The building sanitary drainage system must be equipped with an adequate number of cleanouts (symbol: CO) to make the entire system accessible without breaking up the basement floor or breaking into finished walls and ceilings should a stoppage occur. The plumber must use good judgment in the location of cleanouts.

The plumber provides cleanouts in drainage pipe lines by installing T or Y fittings in these lines and plugging the unused opening in the fitting with the appropriate, removable plug. Figure 6-11 illustrates a cleanout tee for use in a stack. Figure 6-12 illustrates the use of a Y and 1/8 bend for a cleanout on a horizontal building drain. Figure 6-13 illustrates a cleanout at a 90 degree change in direction.

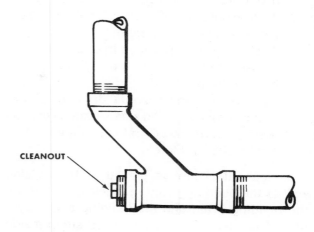

Figure 6-13: Cleanout at a 90° change in direction.

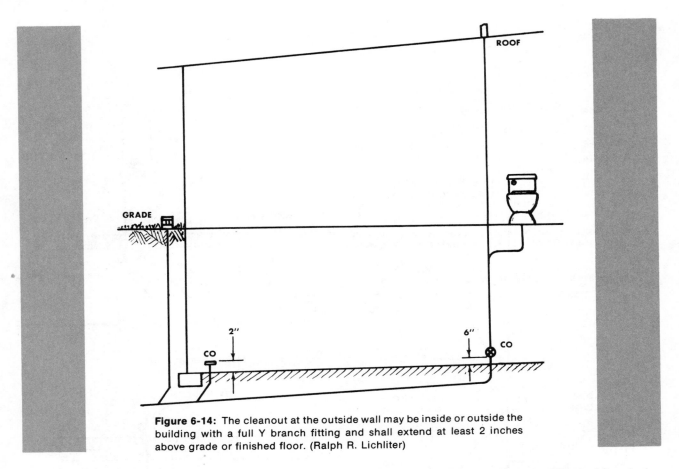

Figure 6-14: The cleanout at the outside wall may be inside or outside the building with a full Y branch fitting and shall extend at least 2 inches above grade or finished floor. (Ralph R. Lichliter)

In general, cleanouts are of the same size (up to 4 inches) as the drainage pipe they serve. However, a 4-inch cleanout is usually the largest size cleanout installed regardless of the size of the drain line it serves.

Cleanouts should be provided in all of the following locations:

1. At the outside wall of the building where the building sewer and building drain connect (front main cleanout).
2. At the base of all vertical soil or waste stacks (stack base cleanout).
3. At all 90 degree changes in direction.
4. At the upper terminal of all horizontal branch drains.
5. Every 50 feet on horizontal drainage piping that is 3-inch size and smaller and every 100 feet on horizontal drainage pipe that is 4-inch size and larger.

A cleanout is always placed at the outside wall of the building (front main cleanout) where the building drain and the building sewer connect. This cleanout may be placed either inside or outside of the building. It should be a full Y fitting placed in the direction of flow of the drain. Also, it should extend at least 2 inches above the finished floor or grade level so that the cover will not be removed and the cleanout opening used for a drain. (The cleanout opening would, of course, be placed flush with the floor if it is in a traffic area.) Figure 6-14 illustrates the location of this cleanout at the outside wall of the building.

Stack base cleanouts should be located at least 6 inches above the floor for easy access and to prevent their use as floor drains. A cleanout at the base of a stack is also shown in Figure 6-14.

The cleanouts at the upper terminals of horizontal branch drains can be eliminated if there is a plumbing fixture trap or a plumbing fixture with an integral trap that can be easily removed and used for a cleanout.

Remember: Cleanouts must be accessible. The plumber must see that access panels are left in ceilings and walls for access to the drainage pipe cleanouts. Cleanouts on buried piping must also be watched when the ditch is backfilled, for these are sometimes broken off and/or buried.

VENTING THE BUILDING SANITARY DRAINAGE SYSTEM

Vent piping is presented in the same chapter with sanitary drainage piping because, as will be shown, vent pipes are extensions of drain pipes, and they are installed at the same time as the drain pipes.

A *vent pipe* is any pipe provided to ventilate a building drainage system and to prevent trap syphonage and back pressure. This simply means that because every plumbing fixture has a water seal trap, every trap should have a vent to protect its water seal.

A properly installed system of vent pipes provides a free circulation of air within the sanitary drainage system and assures that at no time will the trap seals be subject to a pressure of more than 1 inch of water column. This free circulation of air within the drainage system allows the liquid waste to drain freely through the pipes. The vent pipe system also removes objectionable gases from the sanitary drainage system.

To vent a building drainage system properly, the plumber must have some knowledge of the earth's atmosphere.

Surrounding the earth's surface is a blanket of gases (called the atmosphere) about 100 miles thick. This blanket of gases contains approximately 21% oxygen, 78% nitrogen, and 1% other gases. About one-half the total weight of the atmosphere is *below* 18,000 feet. A column of air 1 inch square and as high as the atmosphere exerts a pressure on the earth's surface of 14.7 pounds. That is, at sea level, all objects are under a pressure of 14.7 pounds per square inch. Atmospheric pressure increases below sea level and decreases above sea level because the total volume of air is reduced as one goes higher in the atmosphere.

One of the common properties of the gases that compose the atmosphere is *compressibility.* Air can be compressed and, in this condition, develops a pressure that is greater than its normal sea-level pressure of 14.7 pounds per square inch. This basic scientific fact is illustrated by the automobile tire under pressure. Air

also can be withdrawn from a space or container, and this condition is termed a *vacuum* or *partial vacuum,* depending on the volume of air removed. A partial vacuum, therefore, would indicate a pressure less than that of the atmospheric pressure at sea level.

A pressure of *less* than one atmosphere (14.7 pounds per square inch) will be referred to as a *minus pressure.* A pressure *greater* than one atmosphere will be called a *plus pressure.*

Trap Seal Loss. One of the most common difficulties occurring in a drainage system is trap seal loss. This failure can be attributed directly to inadequate venting of the trap and the subsequent minus and plus pressures that occur. (How improperly vented traps can lose their seal will be discussed in Chapter 8.)

Retarded Flow in the Drainage System. Retarded flow may be the result of improper atmospheric conditions, insufficient venting, or incorrect installation of fittings.

Air, because of the elasticity of the gases it contains, may be compressed into pressures far in excess of atmospheric pressure. The flow of water in a soil pipe tends to compress the volume of air ahead of it, and pressures greater than atmospheric are bound to prevail unless the system is properly vented. Increased pressure causes retarded flow in a vertical stack and, therefore, affects the discharge capacity in its branches. Under these circumstances, the drainage system becomes unbalanced. It contains a pressure greater than atmospheric pressure and so it cannot function properly.

There is also the possibility that a partial vacuum may be developed in the drainage system, which would affect its discharge capacity. This is the result of atmospheric pressure on the flow side of the waste, creating resistance to the volume of water because a minus pressure develops on the opposite side. This condition indicates a lack of proper relief ventilation or partial closure of the vent pipe terminal. It may also be the result of an excessively long vent pipe.

Removal of Objectionable Gases. In addition to being foul-smelling, sewer gases contain chemical elements which, when they combine with the moist air, create acids that corrode vent pipes. Hydrogen, which is an important element

TABLE 6-5.
SIZE AND LENGTH OF VENTS—
INDIVIDUAL, BRANCH, CIRCUIT, AND STACK VENTS

FIXTURE UNITS CONNECTED (dfu)	DIAMETER OF VENT (INCHES)							
	1 1/4	1 1/2 *	2	2 1/2	3	4	5	6
	MAXIMUM DEVELOPED LENGTH OF VENT IN FEET							
2	50	ul						
4	40	200	ul					
8	np	150	250					
10		100	200	ul				
24		50	150	400	ul			
42		30	100	300	500			
72		np	50	80	400			
240			np	50	200	ul		
500				np	180	700	ul	
1100					50	200	700	

ul—Unlimited length.
np—Not permitted.   *Except 6 fixture unit fixtures.

Minnesota Plumbing Code

in all acids, is found in large quantities in the sanitary drainage system. It may be found in the free state, but it is usually combined with other elements. Hydrogen sulfide is one of these combinations. This compound is particularly objectionable because it picks up additional oxygen from the moisture in the sanitary drainage system and becomes sulfuric acid. Sulfuric acid is an extremely corrosive acid.

As was stated earlier in the chapter, horizontal drainage pipes are designed to flow about 1/3 full. In addition to providing a scouring action, pipes that flow 1/3 full have an air space at the top of the pipe for venting away objectionable sewer gases. This is especially important in the building sewer and building drain pipes. This air space allows the free passage of the sewer gases along the top of the pipe, up the stack, and out into the atmosphere.

Venting Methods

Many forms of venting can be applied to the plumbing installation. Choice is determined largely by the manner in which the plumbing fixtures are to be located and grouped. As a rule,

the completed vent pipe system is a combination of several methods.

At one time, all fixture traps were individually vented. But through experimentation and the introduction of different forms of soil and waste pipe relief vents, it was found that individual trap venting was costly and not entirely necessary. It is true that in a plumbing installation in which the vent pipe system is of the proper size, and where every trap is vented with an individual vent, the danger of trap seal loss is practically eliminated. However, as long as practical tests have indicated that these conditions are not *absolutely essential* to a *safe system* of drainage, various methods of venting may be adapted for the sake of economy.

There are several kinds of vent pipe systems and each has a definite function in the completed plumbing system. The various types may be grouped under two principal classifications:

1. The vent pipes used to ventilate the soil and waste pipes are the stack vents and vent stacks. The stack vents, vent stacks, and various other forms of relief vents are classified according to the purpose they serve, and are referred to as

relief and *yoke* vents. These methods of venting serve the fixture trap only in an indirect way. Their primary purpose is to maintain atmospheric pressure in the waste pipe system.

2. The venting methods whose primary purpose is to protect trap seals against back pressure and syphonage are *individual* or *back vents*, *common vents*, and *wet vents*. These methods are covered later.

Vent Grades. The vent pipe must be graded slightly back toward a soil or waste pipe so that water cannot accumulate in it. No definite amount of pitch is required.

Sizing Vent Pipes

Vent pipes are also sized according to the drainage fixture unit (dfu) method. Tests on vent pipe installations led to the formulation of vent sizing tables such as Table 6-5 for sizing stack vents, individual, branch, and circuit vents; and Table 6-6 for sizing vent stacks.

TABLE 6-6. SIZE AND LENGTHS OF VENT STACKS

SIZE OF SOIL OR WASTE STACK (INCHES)	FIXTURE UNITS CONNECTED (dfu)	DIAMETER OF VENT (INCHES)										
		1 1/4	1 1/2	2	2 1/2	3	4	5	6	8	10	12
		MAXIMUM DEVELOPED LENGTH OF VENT IN FEET										
1 1/4	2	50										
1 1/2	4	40	200									
2	9		100	200								
2	18		50	150								
2 1/2	42		30	100	300							
3	72			50	80	400						
4	240			40	70	250						
4	500				50	180	700					
5	540					150	600					
5	1100					50	200	700				
6	1900						50	200	700			
8	2200							150	500			
8	3600							60	250	800		
10	3800								200	600		
10	5600								60	250	800	
12	6000									200	600	
12	8400									100	300	900
15	10500									50	200	600
15	50000										75	180

Minnesota Plumbing Code

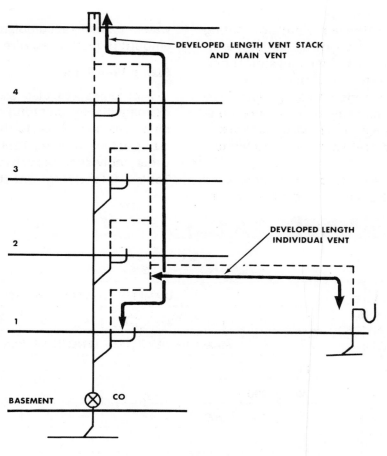

Figure 6-15: Measurement of developed length or vent stacks, main vents, and individual vents. (Ralph R. Lichliter)

Before explaining the use of these tables, it might be well to mention two things:

1. The apprentice will notice that there is no pipe size on either of these two tables smaller than 1¼ inches in diameter, just as there was no waste pipe smaller than the 1¼-inch size. A pipe of less than 1¼-inch size is unsuitable for venting. Although the flow of air may be sufficient, the vent can easily be stopped up. A stopped-up vent is, of course, worthless.

2. The term *developed length* appears in both tables. The developed length of a pipe is the length measured along the centerline of the pipe and fittings. Figure 6-15 illustrates how developed length is measured. Developed length is important in vent piping because friction between the air in motion within the vent pipes and the interior surface of the pipe tends to reduce both the flow and the volume of air moving through the vent pipe.

Stack Vents. A *stack vent* is the extension of a soil or waste stack above the highest horizontal drain connected to the stack. The stack vent admits air to the plumbing system. It also serves as a means of eliminating objectionable odors. The stack vent usually is the terminal for other vent pipes such as individual and group fixture vents and vent stacks. Figure 6-16 indicates the portion of the soil stack defined as the stack vent and shows tees that serve as the terminal for other vent pipes.

Stack vents are sized from Table 6-5 based on the total number of drainage fixture units that drain into the waste portion of the stack.

Stack Vent Sizing EXAMPLE. To familiarize you with the use of Table 6-5 the following example is offered.

What size of stack vent would be required for a group of fixtures consisting of six water closets, four lavatories, three urinals (2-inch trap), and two showers installed on the second floor of a 100-foot-high building. To determine the unit value of this group of fixtures, use the following procedure (the same as in sizing waste pipe):

Fixtures	dfu
6 Water closets	36
4 Lavatories	4
3 Urinals	12
2 Showers	4
Total	56

Assuming that each floor is 10 feet high, the stack vent would have a developed length of about 90 feet from the ceiling of the first floor (which is where the horizontal branch drain for fixtures set on the second floor would connect to the stack) to the top of the stack (100−10=90). Referring to Table 6-5, we find that up to 72 fixture units may be vented with either a 2-inch pipe with a developed length of 50 feet, a 2½-inch pipe with a developed length of 80 feet, or a 3-inch pipe with a developed length of 400 feet. Since the developed length of the vent in the example is 90 feet, a 3-inch stack vent would be required.

This example is given merely to acquaint the apprentice with the method of using Table 6-5 to size a stack vent. It is not an actual representation of what might occur on the job.

Vent Stacks and Main Vents. The *vent stack* is a vertical pipe installed to provide circulation of air to and from the drainage system. Vent stacks are also main vents. A main vent is the principal artery of the venting system to which vent branches may be connected.

Vent stacks or main vents are required in buildings that require individual vents, relief vents, or branch vents on three or more branch intervals.

Vent stacks or main vents are also terminals for smaller individual and group fixture vents. The vent stack or main vent may be considered a vent collecting line.

The vent stack usually is located within a few feet of the soil pipe stack. This is not a set policy, however, and its location depends a great deal on the building construction. It is installed at the same time as the soil pipe stack. Openings are

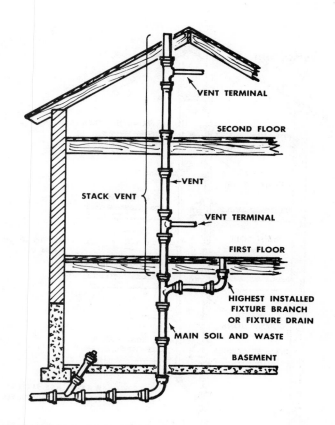

Figure 6-16: A typical stack vent.

left at the correct height and proper floors to accommodate the fixture trap vents.

The vent stack or main vent begins at the base of the soil stack. Its function is to relieve any back pressure that might occur at this location. Its connection to the waste stack should be made with a Y and ⅛ bend. The vent stack usually ties into the stack vent (the uppermost portion of the soil stack). However, if it is not run within a few feet of the soil stack, it may continue on through the roof separately.

Figure 6-17 illustrates a main vent installed close to the soil pipe stack. It indicates how the stack vent joins the soil pipe above the highest fixture branch, and shows its connection into the base of the stack. The connections shown on the various floors are fixture trap vent terminals. (See A, B, C, and D on Figure 6-17.)

In multistory buildings of more than five branch intervals, the waste stack and the vent

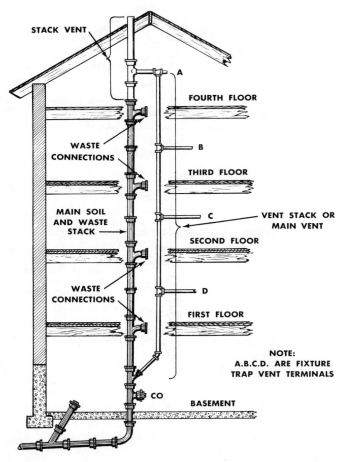

Figure 6-17: A picture of a typical vent stack or main vent installation.

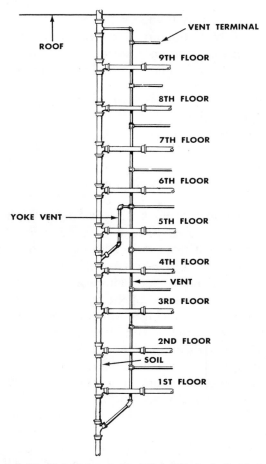

Figure 6-18: Pictorial representation of a typical yoke vent installation.

stack must be reconnected with a yoke vent every five branch intervals, counting from the top interval down. A yoke vent is a pipe that connects upward from a soil or waste stack to a vent stack and prevents pressure differences in the two stacks. Yoke vents are a type of relief vent. A relief vent can have either of two primary functions. It can provide additional circulation of air between drainage and vent systems, or act as an auxiliary vent on a specially designed system.

The yoke vent connection to the soil stack should be made with a Y and $\frac{1}{8}$ bend. It should connect to the waste stack below the horizontal branch drain for the fixtures on that floor and to the vent stack at least 3 feet above the floor level.

The size of the yoke vent connection is the same size as the vent stack to which it connects.

A yoke or relief vent connection is illustrated in Figure 6-18. Care must be taken to allow a sufficient amount of space between the soil stack and the vent stack so this connection can be made. The fittings usually are of large diameter and require space to install.

Vent stacks and main vents are sized from Table 6-6 based on the following information:

1. The size of the soil or waste stack.
2. The number of drainage fixture units connected to that stack.
3. The developed length of the vent.

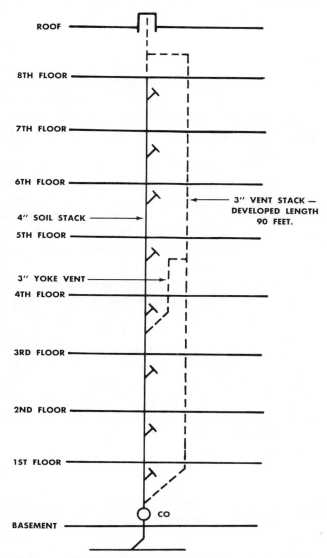

Figure 6-19: Sizing a vent stack in an eight-story apartment building with 88 drainage fixture units.

In the diagram: ROOF, 8TH FLOOR, 7TH FLOOR, 6TH FLOOR, 3" VENT STACK — DEVELOPED LENGTH 90 FEET., 4" SOIL STACK, 5TH FLOOR, 3" YOKE VENT, 4TH FLOOR, 3RD FLOOR, 2ND FLOOR, 1ST FLOOR, CO, BASEMENT

Vent Stack and Main Vent Sizing EXAMPLE.

The following example illustrates how to use Table 6-6 for determining the size of a vent stack:

What size vent stack is required in an eight-story apartment building when a 4-inch size soil stack is used to serve eighty-eight fixture units? (Assume 10 feet to a floor.)

This stack is illustrated in Figure 6-19. The developed length as shown is about 90 feet.

Referring to Table 6-6, we find that a 4-inch stack with 240 fixture units or less connected to it requires a vent stack of 3-inch size. (The apprentice should note that if the developed length had been 70 feet or less, a 2½-inch vent stack would be sufficient.)

Stack Terminals.

Stack vent and vent stack terminals are extended through the roof to vent the sewer to the outside air. The vent terminals usually extend at least 1 foot through the roof to prevent rainwater on the roof from draining into them and to prevent any objects on the roof from falling into the pipe and blocking the vent opening.

Because the vented sewer gases are foul-smelling, a vent terminal should not be located near any window, door, or other ventilating opening. Keep a vent terminal at least 10 feet away from any such openings, or if this is not possible, extend the pipe at least 2 feet *above* the opening. If a vent terminal passes through a roof that serves another purpose, such as a sun deck, it should rise at least 7 feet *above* that roof.

Roof terminals for vents must be sealed to the roof surface to prevent rainwater around the pipe from leaking into the building. Figure 6-20 illustrates a flat roof vent flange and Figure 6-21, a slanted roof vent flange. These vent flanges or roof jackets are installed around the vent pipe

Figure 6-20: Flat roof vent flange.

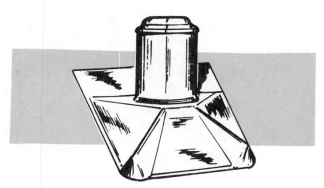

Figure 6-21: Slanted roof vent flange, adjustable for different pitches of roofs. See Figure 6-22 for a cross section of this roof flange.

terminals to make the roof watertight. Both of the illustrated roof jackets are adjustable for variations in stack height caused by expansion and contraction of the stack or settling of the building. In addition, the roof jacket illustrated in Figure 6-21 is adjustable for different pitches of slanted roofs. (Front corners may be bent to accommodate different roof pitches.)

In the colder climates, frost closure of the vent terminal is a common problem and may be responsible for trap seal loss. The air within the plumbing is usually close to the moisture saturation point. When this humid air is emitted through the stack terminal, condensation occurs. This condensation freezes rapidly. Precautionary methods are usually taken to prevent frost closure of the vent terminal. One method is to increase the size of the vent terminal before it passes through the roof. In cold climates, vent pipes that pass through the roof should be at least 2-inch pipe. Another method is to use a roof jacket constructed to allow 1 inch of free air space around the vent terminal. The roof jackets

illustrated in Figures 6-20 and 6-21 are both of this frost-proof type. Figure 6-22 illustrates a cross section through a frost-proof roof jacket. Figure 6-23 is a section through the standard style roof jacket.

Individual Vents. An individual vent is a pipe installed to vent a single fixture trap. It may either connect with the vent system above the fixture served or terminate into the open air. Individual vents are also called *back vents* because they are commonly installed directly in back of the fixtures they serve.

The individual vent is by far the most practical method of venting a fixture trap. Danger of trap seal loss when this type of vent is employed is negligible. When plumbing fixtures are all individually vented, the plumbing system is relieved of minus and plus pressure at every fixture trap. The most common type of individual vent is the continuous vent, shown in Figures 6-24 and 6-25. A continuous vent is simply a vertical vent that is a continuation of the drain it is connected to.

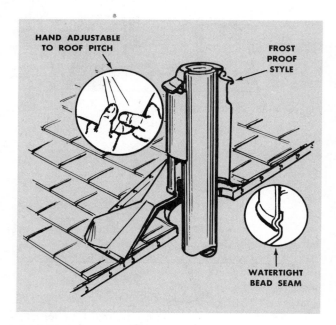

Figure 6-22: Section through a frost-proof vent flange. (F. J. Moore Mfg. Co.)

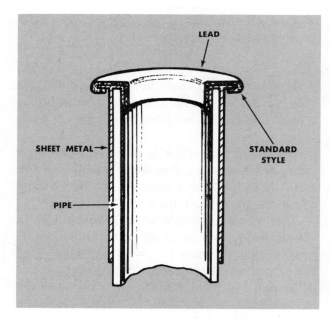

Figure 6-23: Section through a standard-style vent flange. (F. J. Moore Mfg. Co.)

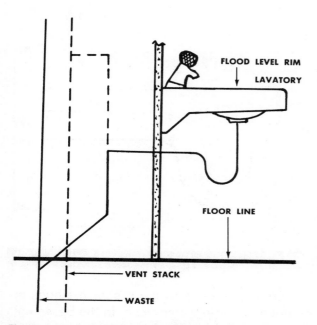

Figure 6-24: Individual venting (continuous vent) of wall-hung and cabinet-set fixtures.

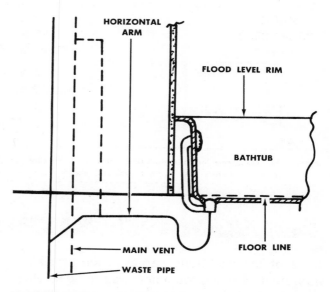

Figure 6-25: Individual venting (continuous vent) of floor-set fixtures.

Individual vents (and all other fixture trap vents as well) should be connected as close to the trap as possible. This connection, as shown in Figures 6-24 and 6-25, is usually made to the fixture drain pipe directly underneath and in back of the fixture.

If the individual vent is to be reconnected into the vent stack, it is reconnected at least 6 inches above the overflow line of the fixture which the vent serves. This fixture overflow line, which is also called the *flood level rim,* is pictured in Figures 6-24 and 6-25.

Sizing the Individual Vent. Individual vents are sized based on: (1) the drainage fixture unit value for each individual fixture; and (2) the developed length of the individual fixture vent to the vent stack, stack vent, or roof terminal. In no case should an individual fixture vent be smaller than $\frac{1}{2}$ the diameter of the fixture drain.

For example, the following are some common individual plumbing fixture vent sizes (as taken from Table 6-5, assuming the developed length of the vent pipe is 50 feet or less).

Type of Fixture	Minimum Size of Vent	dfu
Lavatory	$1\frac{1}{4}''$	1
Drinking fountain	$1\frac{1}{4}''$	1
Domestic sink	$1\frac{1}{4}''$	2
Shower stalls, domestic	$1\frac{1}{4}''$	2
Bathtub	$1\frac{1}{4}''$	2
Laundry tray	$1\frac{1}{4}''$	2
Service sink	$1\frac{1}{2}''$	3
Water closet	$2''$	6

The apprentice will notice that the vent for a water closet is larger than the table values indicate. However, remember the note at the bottom of Table 6-5, which states that 6 fixture unit fixtures (water closets) may not vent into a $1\frac{1}{2}$-inch vent pipe. As a general rule, most plumbing codes require a 2-inch vent for all water closets.

Individual Venting of Plumbing Fixtures

Figure 6-24 illustrates the recommended method of individual venting of plumbing fix-

TABLE 6-7.   DISTANCE OF FIXTURE TRAP FROM VENT

SIZE OF FIXTURE DRAIN (INCHES)	DISTANCE— TRAP TO VENT
1¼	2 ft. 6 in.
1½	3 ft. 6 in.
2	5 ft.
3	6 ft.
4	10 ft.

Note—The developed length between the trap of the water closet or similar fixture and its vent shall not exceed four (4) feet.

Minnesota Plumbing Code

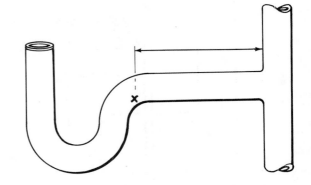

tures of the wall-hung variety, such as sinks, lavatories, drinking fountains, wall-hung service sinks, and laundry trays. It is also applicable to kitchen sinks mounted in a cabinet or lavatories mounted in a vanity. As may be noted, the trap discharges into the side opening of a drainage tee. The maximum distance between the trap and its vent is as shown in Table 6-7.

The top opening of the drainage tee is used as a connection for the individual vent, which must be extended at least 6 inches above the overflow line (flood level rim) of the fixture so if the fixture waste pipe is stopped up, the vent will not serve as a waste for the fixture. There is a decided advantage in connecting the vent pipe to the tee, as indicated in Figure 6-24. Accumulation of dust, rust, or foreign material in the vent automatically drops into the waste line and is carried away by the discharge of the fixture, thus maintaining a clear vent.

Figure 6-25 illustrates the method by which fixtures such as bathtubs are vented. This same method can be used to vent urinals, shower baths, mop basins, and other fixtures that are set on the floor. The fixture trap discharges into the top opening of a drainage tee fitting, the side opening of which is used as a connection for the individual vent. This form of back vent, which is called a *line vent*, is not as satisfactory as the method previously described. The waste discharge from the fixtures into the fixture drain may tax its capacity. The waste materials are liable to back into the vent pipe.

Figure 6-26 illustrates the proper methods of venting a floor-set water closet in the basement of a home, as well as the improper flat vent. Flat venting of plumbing fixtures should be avoided because stoppage of the fixture drain would result in waste backing up into the vent and plugging it.

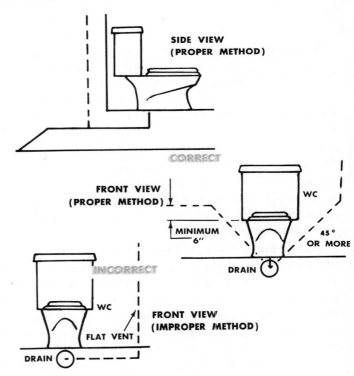

Figure 6-26: Top: Side view of properly vented basement water closet. Middle: Front view showing proper methods of venting this water closet. Bottom: Front view showing improper method (flat vent) of venting this water closet.

Whenever the installation may be designed to permit the venting of floor-type fixtures by the method shown in Figure 6-27, it is best to do so.

Common Vents. The common vent is a practical form of venting that may be defined as a vent connecting at the junction of two fixture drains and serving as a vent for both fixture drains. The two fixtures discharge the waste into a short pattern cross. Common vents are also called *unit vents.* This method of venting is used on two fixtures of similar design installed on opposite sides of a partition. The design of it is practically identical with back-venting, and the principles involved are the same in both instances. This form of venting is commonly applied to fixture traps serving apartment and hotel toilet rooms. Because of the economy obtained, bathroom facilities usually are located in back-to-back rooms so that the waste, vent, and water pipes may be run in the common wall. The common vent and waste can be applied when fixtures have the same vertical drain heights.

Sizing the Common Vent. The diameter of pipe required for a common vent serving two fixture traps is determined in the same manner as for an individual vent except that the size of the pipe must be sufficient to vent the sum of the two drainage fixture unit values. For example, suppose two lavatory traps are to have a common vent. Each lavatory is rated at 1 drainage fixture unit, and the developed length of the vent is 60 feet. The sum of their drainage fixture unit values would be 2. Since 50 feet is the maximum developed length of a 1¼-inch vent pipe that vents 2 drainage fixture units, a 1½-inch vent would be required for this installation. The same method is employed for other fixtures and the size of the vent naturally increases as the drainage fixture unit value of each fixture becomes greater.

Figure 6-28 illustrates the common venting of two wall-hung lavatories. This method can also be applied to other wall-hung or cabinet-mounted fixtures. The fixture traps discharge

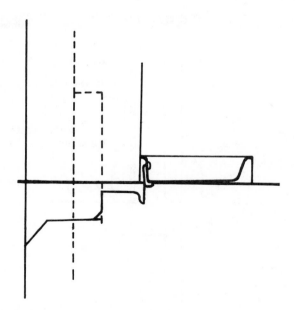

Figure 6-27: The preferred method of individual venting of floor-set plumbing fixtures.

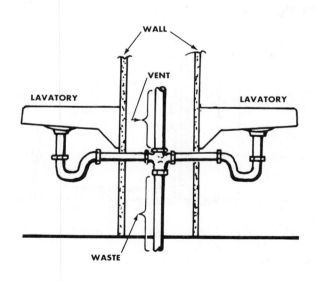

Figure 6-28: Common vent method of venting wall-hung plumbing fixtures.

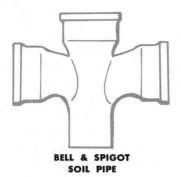

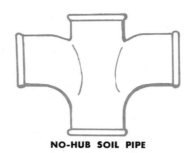

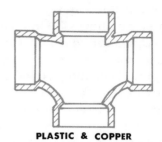

BELL & SPIGOT
SOIL PIPE

NO-HUB SOIL PIPE

PLASTIC & COPPER

Figure 6-29: Short pattern crosses.

into a sanitary drainage cross. The top opening of the cross is used for the common vent connection, which is completed in the same manner as an individual vent. *Short pattern crosses,* like those illustrated in Figure 6-29, should be used on common vent installations.

Bathtubs, showers, and other floor-set fixtures can be vented by the common vent method. However, it is sometimes rather inconvenient to use this method of venting fixture traps because of space limitations. Figure 6-30 illustrates the common venting of two bathtubs.

Water closets and other fixtures of similar design may also be common vented, as shown in Figure 6-31. When water closets are the highest

fixtures on a stack, they are sometimes common vented to that stack. This is a common installation in homes with back-to-back bathrooms. In this installation, shown in Figure 6-32, the stack vent also serves as the common vent.

Branch Vents. A branch vent is a vent connecting one or more individual vents with a vent stack or a stack vent. Branch vents are commonly run horizontally behind a group of fixtures. When a vent pipe is run horizontally to tie several individual fixture vents together, it must be run at least 6 inches above the flood level rim of the highest fixture connected to the vent.

A branch vent is sized according to the number of drainage fixture units connected to it and

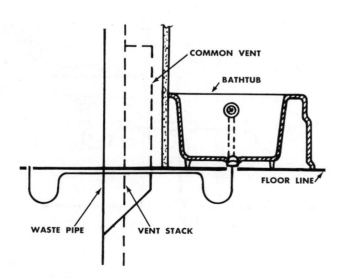

Figure 6-30: Vent used in bathtub installation.

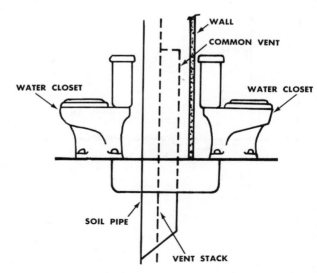

Figure 6-31: Vent used in water closet installation.

the developed length of the branch vent from its vent stack or stack vent to the farthest fixture drain served by the branch vent.

Branch Vent Sizing EXAMPLE. What size branch vent is required for a toilet room containing six water closets, four urinals (wall-hung, with 2-inch trap), four lavatories, and one service sink with a developed length of 76 feet to the nearest vent stack?

Fixture			dfu
6 Water closets	× 6	=	36
4 Urinals	× 4	=	16
4 Lavatories	× 1	=	4
1 Service sink	× 3	=	3
		Total	59

Using Table 6-5, we can see that a 2½-inch branch vent is required.

Wet-Venting. A wet vent is a method of venting used rather extensively for small groups of residential bathroom fixtures. A wet vent is that portion of the vent pipe system through which liquid wastes flow. The practice of wet-venting plumbing fixtures, one that is discussed often, has both advocates and opponents. Some plumbing authorities believe wet-venting is an efficient method of maintaining adequate atmos-

pheric pressure to prevent loss of fixture trap seal. The strongest contention of those supporting wet-venting is the fact that a fixture placed on a dry vent line tends to scour the line at each discharge of the fixture. Laboratory tests on new and clean fixture wastes have proved wet-venting to be adequate.

Those who oppose the above contend that the portion of the vent pipe used as a waste for another fixture becomes fouled very rapidly and that its diameter is greatly reduced. Complete stoppage often occurs and trap seal loss can result.

Both arguments have merit. An improperly designed wet vent does present stoppage problems and can cause a reduced diameter of the vent line. Situations of this kind are common. When the waste and vent installation is clear, there appears to be adequate ventilation. Many states have permitted wet-venting for years and apparently have experienced no great amount of trouble with it. This method has a place in the venting of small fixture groups.

Sizing Wet Vents. The simplest type of wet vent is the common vent with two fixtures whose horizontal drain openings are at different heights installed back to back, as shown in Figures 6-33 and 6-34. This type of vent is sized as follows: The vertical drain is sized one pipe size larger than the upper fixture drain, but in no case smaller than the lower fixture drain.

For example, in Figure 6-33 when the *lavatory waste* is above the kitchen sink waste, the wet vent portion of the vertical drain pipe would be 1½-inch pipe instead of the 1¼-inch pipe that would normally be required for a lavatory drain. However, in the installation in which the *kitchen sink drain* is above the lavatory drain (Figure 6-34), the wet vent portion is sized as 2-inch pipe instead of 1½-inch pipe, which is the normal size for a kitchen sink drain.

As a rule, not more than 1 drainage fixture unit may drain into a 1½-inch wet vent or more than 4 drainage fixture units into a 2-inch wet vent.

Figure 6-35 illustrates a typical wet vent installation of a domestic bathroom. In this installation, the vertical portion of the lavatory waste is the wet vent for the bathtub drain. In this case, the lavatory waste is increased to 1½-inch pipe,

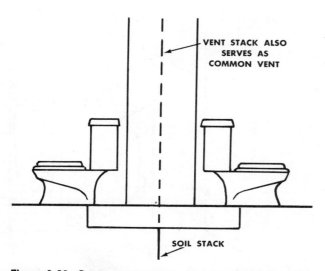

Figure 6-32: Common vent used when two water closets are the highest fixture drain connected to a stack.

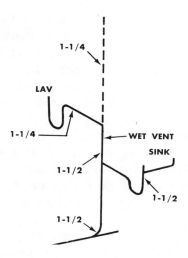

Figure 6-33: Sizing of a wet vent when the lavatory waste is the wet vent for a kitchen sink.

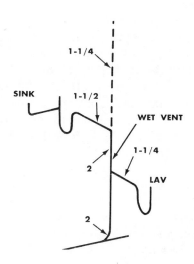

Figure 6-34: Sizing of a wet vent when the kitchen sink waste is the wet vent for a lavatory.

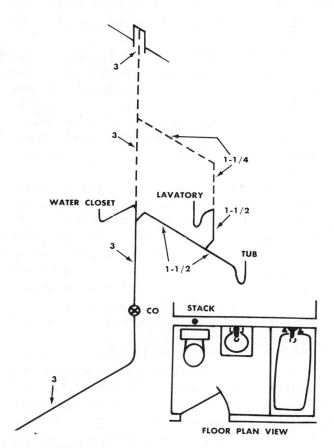

Figure 6-35: Sizing of a wet vent installation—domestic bathroom. (Ralph R. Lichliter)

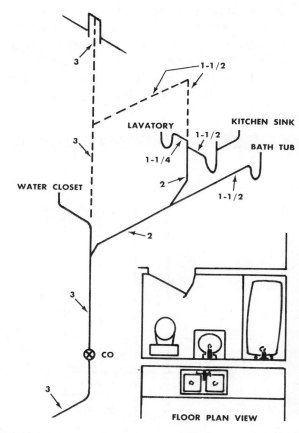

Figure 6-36: Sizing of a wet vent installation—domestic bathroom and kitchen sink. (Ralph R. Lichliter)

but the lavatory vent, which vents 3 drainage fixture units, remains 1¼-inch pipe.

Figure 6-36 illustrates an installation in which the kitchen sink and lavatory are back to back and the waste for these two fixtures serves for the wet vent for the bathtub. In this installation, the wet vent portion of the vertical drain must be 2-inch pipe because it drains 3 drainage fixture units, and the branch vent to the stack is 1½-inch pipe because it vents 5 drainage fixture units. (Lavatory 1 dfu + kitchen sink 2 dfu + bathtub 2 dfu = 5 dfu)

Wet vents are quite often used in basement toilet room installations in homes. Figure 6-37 illustrates an installation in which the lavatory drain is the wet vent for the water closet. In this instance, the wet vent portion of the pipe is not increased in size because the 2-inch pipe required for a water closet vent is more than one

pipe size larger than the normally installed 1¼-inch lavatory waste.

The last type of wet vent illustrated (Figure 6-38) is called a *stack group*. A stack group is a group of fixtures located next to the stack so that by means of proper fittings, vents may be reduced to a minimum. This installation is commonly used in homes of one story or when the bathroom is located on the top floor. In an installation of this type, the individual fixture drains are piped their normal size, but the stack vent is the same size as the soil stack. However, if the individual fixture trap arms exceed the maximum distance of Table 6-7, the fixtures must be re-vented.

Further illustrations of the use of the principles presented in the previous sections of this chapter on sanitary waste and vent piping will be given in Chapter 7, "Sizing of Sanitary Waste and Vent Piping Installations."

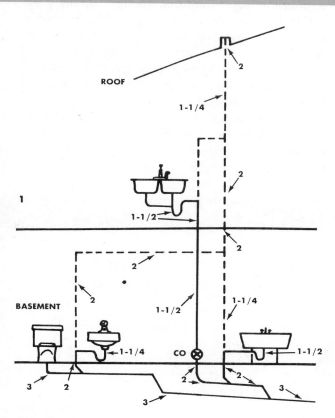

Figure 6-37: Sizing of a wet vent basement toilet room installation. (Ralph R. Lichliter)

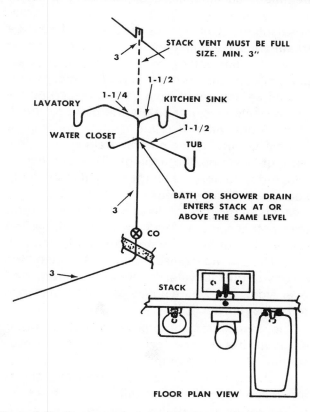

Figure 6-38: Stack group venting of a bathroom and kitchen sink.

STORM WATER DRAINAGE PRINCIPLES

The storm water drainage pipes are installed at the same time as the sanitary drainage and vent pipes. In fact, the storm water and sanitary drainage pipes are often piped in the same ditch and are run parallel to one another in pipe shafts in tall buildings.

The storm water drainage system is that part of the plumbing system that conveys the water from precipitation to the storm sewer or other place of disposal. The disposal of surface and storm water, up to a few years ago, was accomplished by discharging it into the building drain, which served the plumbing fixtures of the building. This practice was considered satisfactory, because the public sewer was drained into a river, lake, or natural drainage basin, and the discharge of a large volume of clear water waste did not create a serious problem. However, as cities grew in population, the need for sewage treatment became more apparent. At present, virtually all municipalities maintain sewage disposal plants. One of the processes of sewage treatment is the liquefying of suspended organic materials. Rainwater passing through a disposal plant affects this process; hence the clear water should be separated from the water-carried organic waste before it enters the plant for treatment. Storm water is relatively pure and can be discharged into some natural drainage terminal without affecting the ecology. This practice is adapted to factory buildings located in outlying, sparsely populated areas.

Most municipalities have constructed storm sewers that serve privately owned buildings. The connection of storm drains to them is compulsory. This practice is wise because it eliminates the discharge of rainwater leaders into gutters and over sidewalks where it may become a nuisance to pedestrians.

The installation of piping for storm water drainage has several features in common with sanitary drainage piping:

1. Changes in direction in storm water drains are made with the same fittings as those required for sanitary drainage piping.

2. Storm water drain pipes are required to have cleanouts in the same locations as sanitary drain pipes.

3. Horizontal storm drains are graded at least $1/4$ inch to the foot. However, excessive pitch is not a problem in storm drainage because storm water is relatively free from solids.

Installation of the Building Storm Drain

Building storm drains can be classified as *inside* or *outside drains.* The inside building storm drain can be located either under the basement floor, or it may be suspended from the basement ceiling when the storm sewer main in the street is not deep enough to permit gravity drainage of an under-the-floor drain.

When the building storm drain is installed beneath the basement floor, the plumber must take care to plan the installation of the below-ground storm drain so it will not interfere with the installation of the sanitary drainage piping. The greatest difficulty encountered is in crossing over the main runs of the building sanitary drain with branches of the building storm drain.

Careful planning of an overhead storm drain is also necessary. The drain should not conflict with heating mains, ventilation ducts, windows, or beams. It must be installed so that sufficient headroom is provided and the appearance of basement areas is not marred.

The outside storm drain is located around the outer foundation wall of the building and is placed in the ground below the frost level, as shown in Figure 6-39.

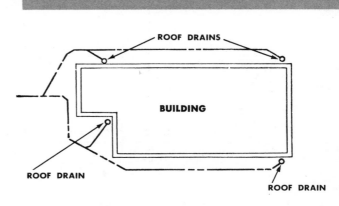

Figure 6-39: Outside storm drain.

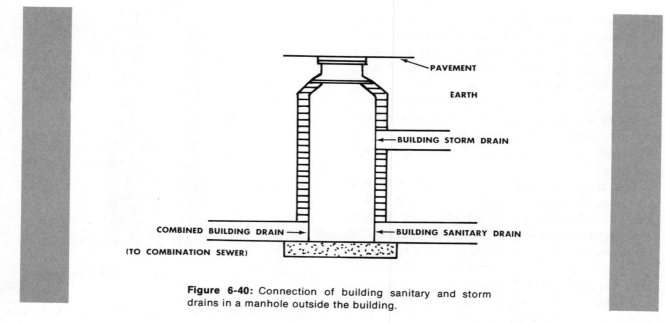

Figure 6-40: Connection of building sanitary and storm drains in a manhole outside the building.

Both inside and outside building storm drains must be piped to the storm sewer main in the street, if there is one available, or to a natural drainage basin, such as a stream, or a lake. However, the only place of disposal available in some older municipalities might be a combination sewer main in the street. (A *combination sewer* is one that conveys both sewage and storm water.) In this case, both the building storm drain and the building storm sewer should be piped out of the building separately and then tied together in a manhole either outside the building or at the curb line so that when a storm sewer main is installed in the street the building storm sewer can be connected to it. Figure 6-40 illustrates a typical manhole connection of the building storm drain and the building sanitary drain.

Rainwater Leaders. A rainwater leader is often referred to as a *conductor* or *downspout*. It is a pipe that conveys water from a roof drain to the building storm drain or other point of disposal. Rainwater leaders are usually vertical pipes.

A rainwater leader must never be used as a soil or waste pipe for a plumbing fixture.

Location of Rainwater Leaders. There are two varieties of rainwater leaders, each designated according to its location in the building. The *outside rainwater leader* is located on the

outside wall of the building and the *inside rainwater leader* is installed within the building walls.

Outside Rainwater Leaders. The outside rainwater leader generally is constructed of sheet metal or plastic and is normally installed by the tinner or sheet metal man. The plumber is responsible for the base fitting installation and its connection to the building storm drain. The base fitting illustrated in Figure 6-41 is usually cast iron soil pipe, although any material approved by the local plumbing code for underground building storm drains can be used. The

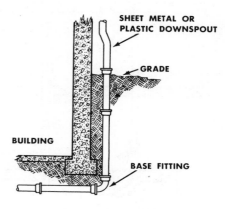

Figure 6-41: Base fitting connection for outside rainwater leaders.

Sanitary Drainage, Vent, and Storm Drainage Piping   **147**

hub end of the pipe, which connects to the downspout, should be left at least 6 inches above grade. Care should be taken to plumb the vertical portion of a base fitting so that its appearance will be neat, as this pipe will be exposed to view on the outside of the building.

Inside Rainwater Leaders. Rainwater leaders piped inside the building are usually run up alongside a column or are placed in vertical pipe shafts.

The rainwater leaders are usually extended vertically through the floors to a point just below the roof and then offset horizontally to the roof drain body to reduce the danger of breakage. A leaky roof drain connection can also result because of expansion and contraction of the rainwater leader piping. Expansion and contraction are important in rainwater leader piping because of the wide variation in temperature of the water flowing through the pipe. In the colder climates, this water temperature can vary from an ice-water mixture of melting snow in the winter months to very warm water from a summer rainfall on the hot roof.

The change in direction at the high point of the rainwater leader should be made by means of a *double-ell offset,* as shown in Figure 6-42. Fit-

Figure 6-43: Expansion joint. (Josam Mfg. Co.)

tings installed in this manner form a *swing joint* to compensate for movement that occurs with expansion and contraction of the rainwater leader pipe. If for some reason a rainwater leader must be extended straight up and through the roof, it should be provided with an expansion joint of the type illustrated in Figure 6-43.

Roof Drains. There are many types of manufactured roof drain bodies for rainwater leaders. Figures 6-44 and 6-45 illustrate two common roof drain bodies. The drain body illustrated in Figure 6-44 is of the *bottom outlet type* and is available with either a hub connection or threaded connection. The drain in Figure 6-45 is of the *side outlet type* and is available only with a threaded connection.

Both of these drains are provided with clamping rings so that a piece of sheet lead flashing can be installed around the drain body and sealed to the roof to make a watertight connection. The installation of this lead flashing is the plumber's work, while the sealing of the lead to the bonded roof surface is roofer's work.

A cast iron strainer basket that extends several inches above the roof is attached to the drain to prevent stones, leaves, and other materials from entering the rainwater piping. If the roof is used for a sun deck or in some other capacity, a flat drain cover is installed.

Traps or Storm Drains. Traps are not required on storm drains that connect to storm

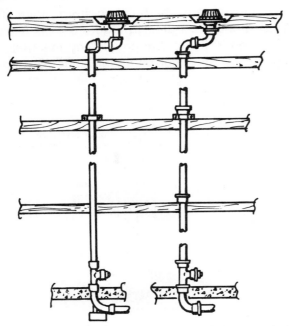

Figure 6-42: Double-ell offset (swing joint) on an inside rainwater leader.

Figure 6-44: A typical bottom outlet roof drain body. (Josam Mfg. Co.)

Figure 6-45: A typical side outlet roof drain body. (Josam Mfg. Co.)

sewer mains in the street. But, if they connect to a combination sewer in the street, there will be objectionable sewer gases from this sewer and traps will be required on some installations.

Two installations that require traps on storm drains are illustrated in Figure 6-46. These installations are:

1. A roof drain body that is located within 10 feet of a door, window, or any other opening to a building.

2. An outside sheet metal conductor that is used and located within 10 feet of a door, window, or any other opening to a building. This trap should be located inside the building and provided with an accessible cleanout, as shown in Figure 6-46.

Sizing Building Storm Drains and Rainwater Leaders. Both building storm drains and rainwater leaders are sized based on the projected roof area. The projected roof area is the area in

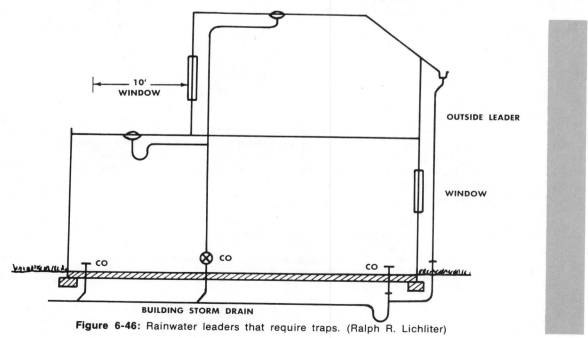

Figure 6-46: Rainwater leaders that require traps. (Ralph R. Lichliter)

TABLE 6-8. SIZE OF HORIZONTAL STORM DRAINS

DIAMETER OF DRAIN	MAXIMUM PROJECTED ROOF AREA FOR DRAINS OF VARIOUS SLOPES (SQUARE FEET)		
(INCHES)	$^1/_8$ -in. Slope	$^1/_4$ -in. Slope	$^1/_2$ -in. Slope
3	822	1,160	1,644
4	1,880	2,650	3,760
5	3,340	4,720	6,680
6	5,350	7,550	10,700
8	11,500	16,300	23,000
10	20,700	29,200	41,400
12	33,300	47,000	66,600
15	59,500	84,000	119,000

Table 6-8 is based on a maximum rate of rainfall of 4 inches per hour. If in any locality, the maximum rate of rainfall is more or less than 4 inches per hour, then the figures for the roof area must be adjusted proportionately by multiplying the figure by 4 and dividing by the maximum rate of rainfall in inches per hour.

Minnesota Plumbing Code

TABLE 6-9. SIZE OF VERTICAL RAINWATER LEADERS

SIZE OF LEADER OR CONDUCTOR* (INCHES)	MAXIMUM PROJECTED ROOF AREA (SQUARE FEET)
2	720
2$^1/_2$	1,300
3	2,200
4	4,600
5	8,650
6	13,500
8	29,000

Table 6-9 is based on a maximum rate of rainfall of 4 inches per hour. If in any locality, the maximum rate of rainfall is more or less than 4 inches per hour, then the figures for roof area must be adjusted proportionately by multiplying the figure by 4 and dividing by the maximum rate of rainfall in inches per hour.

*Note that all of the tables presented in this chapter were taken from the Minnesota Plumbing Code and apply *only* to the State of Minnesota. The apprentice must consult the local plumbing code for his area for its equivalent tables before sizing any pipe on the job in another area, as there may be some differences.

square feet of that portion of the roof drained by a particular pipe. Table 6-8 is used for sizing horizontal storm drains for building storm sewers, building storm drains, and horizontal rainwater pipes. Table 6-9 is used for sizing vertical rainwater leaders. The apprentice should note that Tables 6-8 and 6-9 are based on a maximum rainfall of 4 inches per hour and can be adjusted for greater or lesser amounts of rainfall to suit local weather conditions, as explained in the notes below the tables.

Building Storm Drain and Rainwater Leader Sizing EXAMPLE. To familiarize the apprentice with the use of Table 6-8 and 6-9 the following example is offered:

Size the rainwater leaders and building storm drain pipes illustrated in Figure 6-47.

Assume the following:

1. The building roof which is 60 feet × 100 feet or 6000 square feet is divided into 4 equal (1500 square foot) areas by roof dividers.

2. The building storm drain pipes are pitched 1/8 of an inch to the foot.

From Table 6-8:

Pipe A, which drains 6000 square feet, is 8 inch.

Pipe B, which drains 6000−1500=4500 sq. ft., is 6 inch.

Pipe C, which drains 4500−1500=3000 sq. ft., is 5 inch.

Pipe D: All branch drain pipes D drain 1500 square feet and are 4 inch.

From Table 6-9:

All the vertical rainwater leaders drain 1500 square feet and are 3-inch pipes.

The above example of the vertical rainwater leader being a smaller size than the horizontal pipe is not unusual with a 1/8-inch pitch. The underground pipe is reduced on the vertical rise, as illustrated in Figure 6-48.

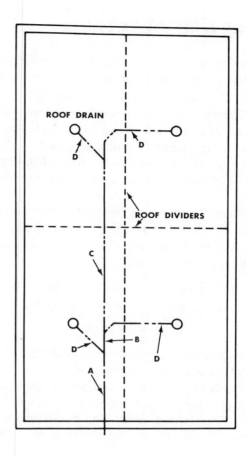

Figure 6-47: Sizing storm drainage pipes within a building. (Ralph R. Lichliter)

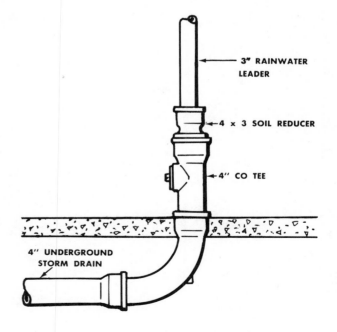

Figure 6-48: Reducing the size of a 4-inch horizontal storm drain pipe to a 3-inch rainwater leader.

REVIEW QUESTIONS

1. In general terms, what is the plumber's responsibility for correct and safe installation of sanitary drainage facilities in a building?

2. Define the term "drainage fixture unit." How was this unit determined as a standard for calculating the sizes of drainage pipe?

3. What is the minimum size of building sewer pipe permitted by most state plumbing codes?

4. From the tables in the text, determine the number of drainage fixture units (dfu) and the size of horizontal branch drain pipe required to serve this list of fixtures:

3 Water closets	4 Lavatories
1 Bathtub	2 Kitchen sinks
	2 Laundry trays

5. Define the term "stack." Distinguish between a soil stack and a stack vent.

6. At what point does a stack change from being a soil stack and become a stack vent?

7. Define "stack vent" and "vent stack" in your own words to the satisfaction of your instructor.

8. List as many kinds of approved fittings or combinations of fittings as you can for making the following changes of direction in a drainage pipe line:

Horizontal turns	Horizontal to vertical turns
Vertical to horizontal turns	Combination turns with cleanouts

9. What is the minimum size a soil stack and building drain may be in a residence? What is the minimum size of a drain branch under a concrete floor (basement floor)?

10. Give two important purposes served by the vent piping of the plumbing system.

11. What is a "stack group"?

12. What is a "trap arm"?

13. Define the term "continuous waste and vent."

14. What is the purpose of a yoke vent?

15. Why do some plumbing codes require that a vent stack be enlarged to at least 2-inch size before it passes through the roof of a building?

16. On what basis is the proper size of the storm water drainage system calculated?

CHAPTER 7

Sizing of Sanitary Drainage and Vent Piping

In the preceding chapter, sanitary drainage and vent pipe installation practices were discussed as separate units of the drainage system. It may be difficult for the apprentice to correlate these units into a complete plumbing system because of the many complex problems that arise.

The drainage system, as a whole, consists of many forms of waste and vent piping correlated into one specific unit. Problems in graduating the sizes of the waste and vent pipes become more difficult, and many important details can be overlooked, unless the apprentice becomes familiar with the complete plumbing installation and sees how to apply the information presented.

So that you may acquire a better understanding of the principles involved, several plumbing layouts are presented in this chapter. Each of these layouts serves to explain basic principles of sanitary drainage and vent piping installations, as these principles are commonly applied.

To simplify the drawings in this chapter, the following abbreviations and symbols will be used:

– – –	Vent pipe
⎯⎯	Waste pipe
(8)	Size of pipe (number in circle)
BT	Bathtub
CO	Cleanout
FD	Floor drain
KS	Kitchen sink
Lav	Lavatory
LT	Laundry tray
SP	Stand pipe (for automatic clothes washer drain)
U	Urinal
WC	Water closet

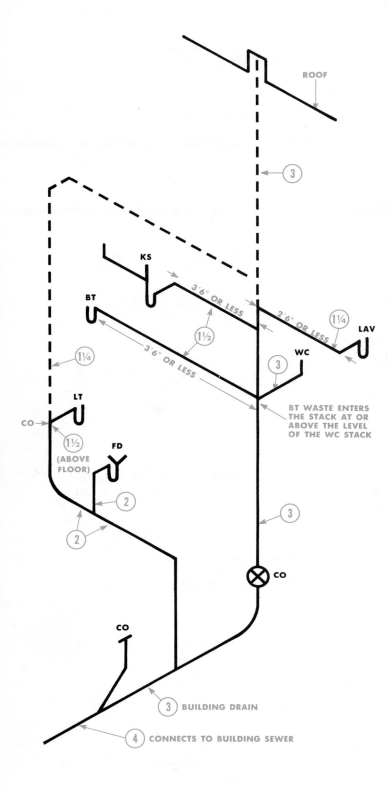

ROOF

KS

BT

3'6" OR LESS

2'6" OR LESS

1¼

LAV

1½

WC

3'6" OR LESS

3

1¼

BT WASTE ENTERS
THE STACK AT OR
ABOVE THE LEVEL
OF THE WC STACK

LT

CO

1½
(ABOVE
FLOOR)

FD

2

3

2

CO

CO

3 BUILDING DRAIN

4 CONNECTS TO BUILDING SEWER

Figure 7-1. Sanitary waste and vent piping layout for a one-story home.

Single Family Home— Minimum Venting Requirements

Figure 7-1 illustrates the sanitary drainage and vent pipe arrangement of a single family home that is permitted in some areas. The installation consists of a water closet, lavatory, bathtub, and kitchen sink on the first floor, and a laundry tray and a floor drain in the basement.

The building drain would be constructed of 3-inch pipe. The 3-inch building drain would be increased to 4-inch pipe for the building sewer, and the building sewer would be extended to a septic tank or the sanitary sewer main in the street. A cleanout should be provided on the inside of the basement wall (front main cleanout) and at the end of each horizontal run. The front main cleanout should be extended at least 2 inches above the floor to discourage the homeowner from removing the cleanout plug and using the cleanout as a floor drain, or from setting a water closet on it.

The branch of the building drain that serves the laundry tray and the basement floor drain is 2-inch pipe. This pipe reduces to 1½-inch pipe above the basement floor and a cross fitting is installed to provide an opening for the laundry tray waste and a cleanout. At the top of the cross the pipe is again reduced to 1¼-inch size to vent the laundry tray trap. This 1¼-inch vent pipe connects back into the 3-inch stack at a point 6 inches above the spill line of the kitchen sink (3 feet 6 inches above the first-floor line).

The piping layout shown in the bathroom area of Figure 7-1 is a *stack group.* It represents the minimum requirement for a waste and vent pipe installation. An arrangement of this kind might not be permitted in areas with strict plumbing codes. In this installation, the bathtub, lavatory, and kitchen sink are not individually vented but rather their drains are just *arms* from the stack. (Note that the bathtub drain enters the stack at or above the water closet opening.)

Tests have proved that the separate re-venting of these fixtures is not necessary, provided that the distance between the trap and the stack, which in this instance is the point of vent, does not exceed 2 feet 6 inches for the 1¼-inch lavatory trap, or 3 feet 6 inches for the 1½-inch bathtub and kitchen sink traps, as indicated in Figure 7-1.

Single Family Home—
All Fixtures Individually Vented

Figure 7-2 presents the layout of a sanitary drainage and vent pipe installation for the home illustrated in Figure 7-1. This installation is designed to satisfy the maximum requirements of nearly all localities. This layout is recommended over that pictured in Figure 7-1. In this installation, all fixtures are trapped, and each trap is vented except the floor drain.

The building drain from the stack is 4-inch pipe with a 4-inch cleanout at the inside of the basement wall.

The branch serving the floor drain, laundry tray, and stand pipe for the automatic clothes washer is 2-inch pipe, as in Figure 7-1.

All of the fixtures draining into the stack have a continuous waste and vent. The horizontal branch drain to the right of the stack, which serves the lavatory, is constructed of 1¼-inch pipe. The branch to the left of the stack, which serves the kitchen sink and bathtub wastes, is 2-inch pipe for the portion that serves both of these drains. It then reduces to 1½-inch pipe for the bathtub drain.

The horizontal vent connecting to the stack on the right side is 1¼-inch size to serve the lavatory drain. On the left side, the vent is 2-inch pipe to the first tee, which is the future vent. A future vent is provided so that if at some future time the homeowner wishes to install a basement bathroom, there will be a proper size vent opening for these fixtures. After the future vent opening, the vent reduces to 1½-inch pipe to receive the kitchen sink and bathtub vents. After the bathtub vent, it again reduces to 1¼-inch size for the basement laundry tray and stand pipe vents.

Cleanouts have been indicated at the outside wall on the building drain (front main cleanout), at the base of the 3-inch soil stack (stack base cleanout), and at the ends of the horizontal branches draining into the stack. A cleanout is also shown on the vent pipe above the laundry tray and stand pipe to serve this horizontal branch drain.

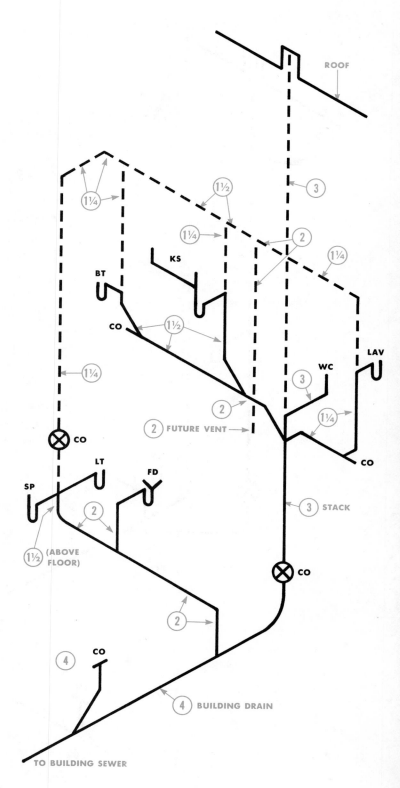

Figure 7-2. Sanitary waste and vent piping layout for a one-story home showing all fixtures individually vented.

Sizing of Sanitary Drainage and Vent Piping **155**

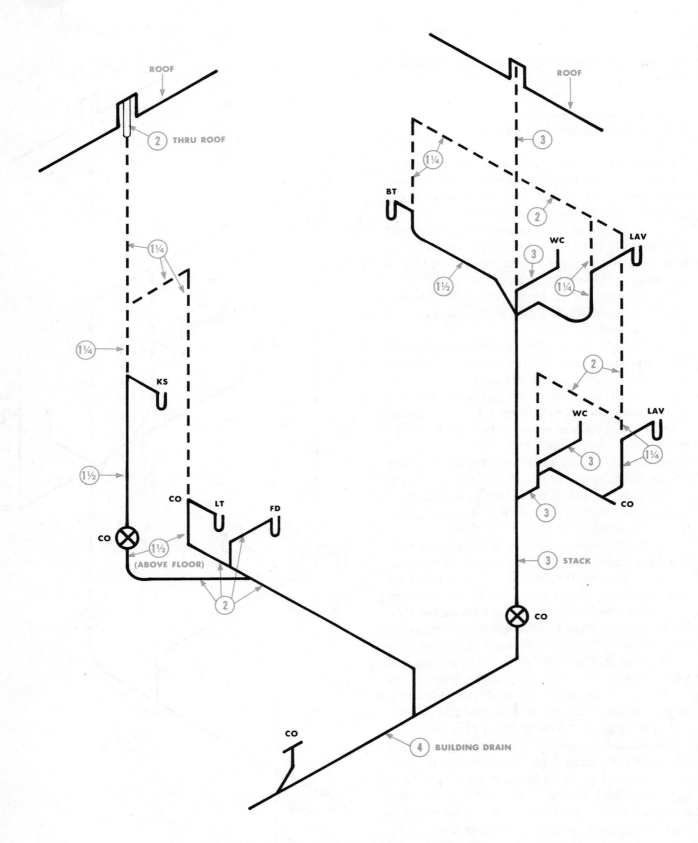

Figure 7-3. Plumbing layout for a two-story home showing individual venting of fixtures.

Two-Story Single Family Home— All Fixtures Individually Vented

Figure 7-3 is a layout of plumbing consisting of a water closet, lavatory, and kitchen sink on the first floor, and a water closet, lavatory, and bathtub on the second floor of a residence. A laundry tray and floor drain are located in the basement. This figure serves to illustrate the principles of individual fixture venting.

The building drain and stack installation that serve the fixtures in Figure 7-3 are similar to those of Figures 7-1 and 7-2 except that a provision for the second-floor bathroom must be provided. The kitchen sink in this figure is located so that it is not practical to drain it into the stack but rather, it drains into the building drain branch serving the basement laundry tray and floor drain.

The building drain branch, which receives the wastes from the laundry tray and floor drain in the basement and the kitchen sink on the first floor, is sized at 2 inches below the basement floor. Above the floor, this pipe reduces in size to 1 1/2 inches for both the laundry tray and kitchen sink wastes. The vent for the laundry tray is 1 1/4-inch pipe, and this vent continues up to the first floor where it ties into the 1 1/4-inch kitchen sink vent at a point 6 inches above the kitchen sink spill line (or about 3 feet 6 inches above the kitchen floor). This 1 1/4-inch vent pipe then continues up to a point below the roof where it is increased to 2-inch pipe before passing through the roof.

The first-floor water closet is set off to the side of the 3-inch stack, and drains into a 3-inch sanitary tee and then a wye in the stack. There are also specially designed fittings for offsetting a water closet on a lower floor from the stack so that the water closet may be properly vented. These fittings are called *vented closet tees* (and *crosses*). They are available with side openings to accept the wastes of other fixtures on the same floor. Figure 7-4 illustrates vented closet tees in both hub and no-hub type soil pipe.

Waste from the lavatory on the first floor is drained by 1 1/4-inch pipe into either a wye below the closet bend or into the side opening of the vented closet tee.

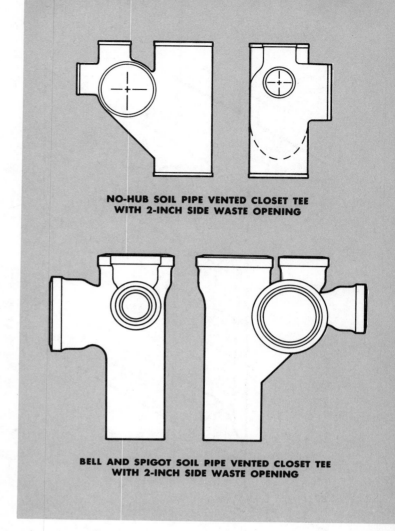

NO-HUB SOIL PIPE VENTED CLOSET TEE WITH 2-INCH SIDE WASTE OPENING

BELL AND SPIGOT SOIL PIPE VENTED CLOSET TEE WITH 2-INCH SIDE WASTE OPENING

Figure 7-4. Vented closet tees. (Tyler Pipe)

The first-floor 2-inch water closet vent receives the 1 1/4-inch lavatory vent and continues up to the second floor. There, it picks up the 1 1/4-inch vent from the second-floor lavatory and ties back into the stack at a point 6 inches above the spill line of the lavatory (about 3 feet above the second floor).

The waste from the second-floor lavatory discharges into a double wye placed in the stack below the closet bend. The 1 1/2-inch bathtub waste discharges into the other side of this double wye. The vent from the bathtub is 1 1/4-inch pipe. It also ties back into the stack.

Cleanouts have been indicated at the outside wall on the building drain, at the base of the 1 1/2-inch kitchen sink waste stack, on the laundry tray vent stack, and on the horizontal branch

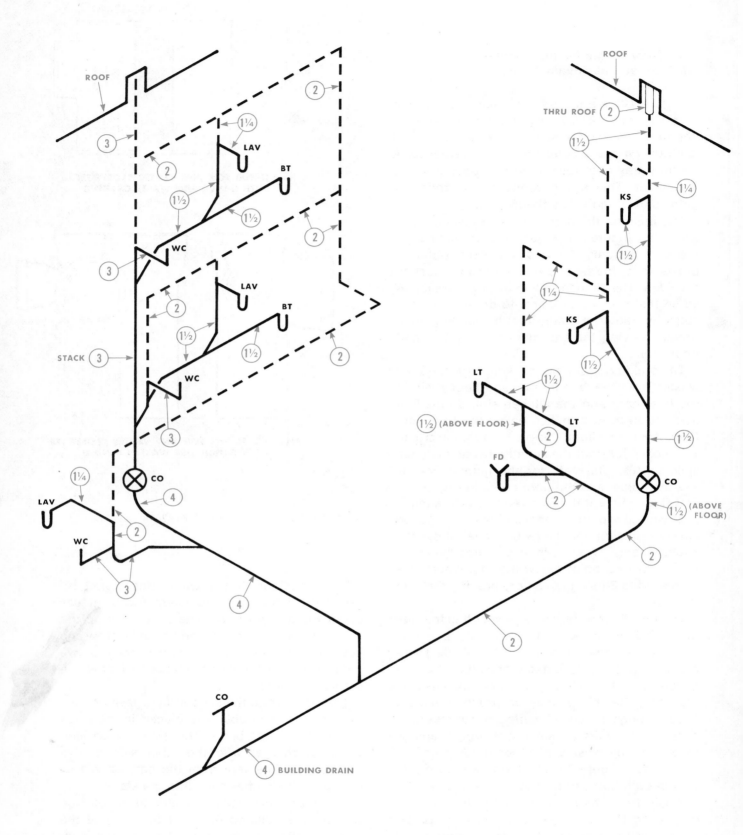

Figure 7-5. Plumbing layout for a duplex residence.

serving the first-floor lavatory waste. However, they have not been shown on the horizontal branches serving the bathtub and lavatory on the second floor because proper access to them would not usually be provided in a residence.

Duplex Residence

Figure 7-5 represents the plumbing installation for a duplex residence. It consists of a water closet, lavatory, and two laundry trays in the basement, as well as bathrooms on the first and second floors, each of which has a water closet, lavatory, and bathtub. A sink is provided in each kitchen.

The building drain, with the exception of variations in stack and waste locations, is 4-inch size, as in the preceding illustrations. The stack is 3-inch pipe and has bathroom branch connections on both the first and second floors. The stack is installed in the partition directly in back of the water closet.

The basement water closet is vented with a 2-inch wet vent into which the 1¼-inch basement lavatory waste has been connected. The 2-inch vent is extended from the basement toilet room to the first floor, where it ties into the vent from the first-floor fixtures.

The bathtub waste of the first-floor bathroom is made of 1½-inch pipe. This waste pipe also receives the lavatory waste, and then it discharges into the stack at or below the closet bend. The bathtub is wet-vented through the lavatory waste, which has been increased to 1½-inch pipe. The 1¼-inch lavatory and bathtub vent connects to the 2-inch vent for the first-floor water closet, and this vent extends to the second floor, as shown.

The second-floor bathroom waste pipe is identical with that of the first floor. On the second-floor level, the 2-inch vent from the basement and first-floor fixtures ties back into the stack.

The laundry trays are connected to a common waste of 1½-inch pipe and are vented by the common vent method—with a 1¼-inch pipe that connects into the vent portion of the kitchen sink stacks.

Both kitchen sinks discharge into the same 1½-inch waste pipe. Each kitchen sink is individually vented with 1¼-inch pipe. At the tee, where the laundry tray vent ties into the first-floor kitchen sink vent, the vent increases to 1½-inch pipe. This 1½-inch pipe continues to the second floor to pick up the 1¼-inch vent from that sink. The vent then continues as 1½-inch to a point below the roof, where it increases to 2-inch pipe before passing through the roof.

The underground pipe serving the kitchen sink and the basement laundry trays and floor drain is 2-inch pipe because it is considered to be a building drain branch from a stack. As a branch from a stack, it is sized from Table 6-2, and it may drain 21 drainage fixture units (dfu) at ¼-inch pitch. If this pipe was considered a horizontal fixture branch, it would be sized from Table 6-3, and it could only drain 6 dfu in the 2-inch size. (The apprentice should recognize that the portion of the underground pipe serving the basement laundry trays is a horizontal fixture branch.)

The installation shown in Figure 7-5 may be individually vented if necessary to conform with the local plumbing code. To do this, the basement lavatory and the bathtubs on the first and second floors would require individual vents.

Apartment Building Bathroom Stack— Minimum Venting Requirements

Figure 7-6 is a typical installation of a bathroom stack in a two-story apartment building with no apartments in the basement. The bathrooms are back-to-back installations. This design is typical of many hotel and apartment building layouts.

The waste portion of the stack may be 3-inch pipe since not more than two water closets are entering the stack at any one-story or branch interval. The vent portion of this soil stack (the stack vent) need not continue at the full 3-inch size through the roof if there is another full-size stack in the building.

The second-floor water closets would empty into a 3-inch sanitary cross in the stack. The first-floor water closets could drain either into a 3-inch sanitary cross, which would then waste into a 3-inch wye placed in the stack, or they could waste into a vented closet cross fitting similar to the vented closet tees illustrated in Figure 7-4.

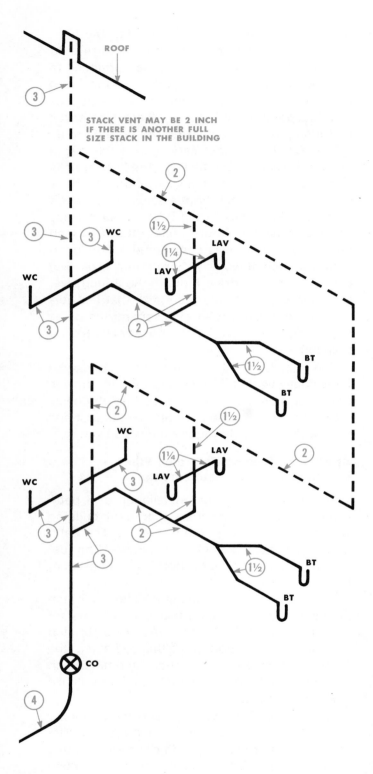

ROOF

STACK VENT MAY BE 2 INCH
IF THERE IS ANOTHER FULL
SIZE STACK IN THE BUILDING

Figure 7-6. Typical bathroom stack in a two-story apartment building with back-to-back bathrooms. The bathtubs are wet-vented.

The two 1½-inch bathtub wastes on each floor drain into a common 2-inch waste pipe. This waste drains either into a wye placed below the water closet waste fitting or into the side opening of a vented closet cross. Since 2 dfu drain into the common lavatory waste on each floor, and this lavatory waste is also the wet vent for the bathtubs, the pipe is 2 inches in size. (Remember what was said in the last chapter: no more than 1 dfu may drain into a 1½-inch wet vent and no more than 4 dfu may drain into a 2-inch wet vent.)

The 2-inch first-floor water closet vent pipe picks up the 1½-inch vent serving the lavatories and bathtubs. This 2-inch vent continues up to the second floor to pick up the 1½-inch lavatory and the 2-inch water closet vent there, and then it ties back into the stack.

Apartment Building Bathroom Stack— All Fixtures Individually Vented

Figure 7-7 illustrates how the apartment stack shown in Figure 7-6 would be piped and sized with every fixture vented.

In this installation, the back-to-back lavatory waste is reduced to 1½-inch pipe with a 1¼-inch common vent for the two lavatories. The two bathtubs are vented with a 1¼-inch common vent pipe. The rest of the waste and vent pipe sizing is the same as Figure 7-6.

Apartment Building Kitchen Sink Waste Stack

Figure 7-8 illustrates the sizing of a kitchen sink waste stack of the type that might be found in apartment buildings that use a bathroom stack like the one shown in Figures 7-6 and 7-7.

The lowest portion of this stack is 2-inch pipe because it receives 8 dfu. The portion that receives only the waste of the two kitchen sinks on the first floor is also sized as 2-inch pipe, as is the portion that receives only the waste from the second-floor kitchen sinks. The reason is that the stack sizing table (Table 6-4) allows only 2 dfu to drain into a 1½-inch waste stack on any one floor or branch interval.

The vent from the first floor sinks is 1¼-inch pipe. This pipe is extended to the second floor, where it ties back into the stack vent above the second-floor sinks. The 1¼-inch vent from the

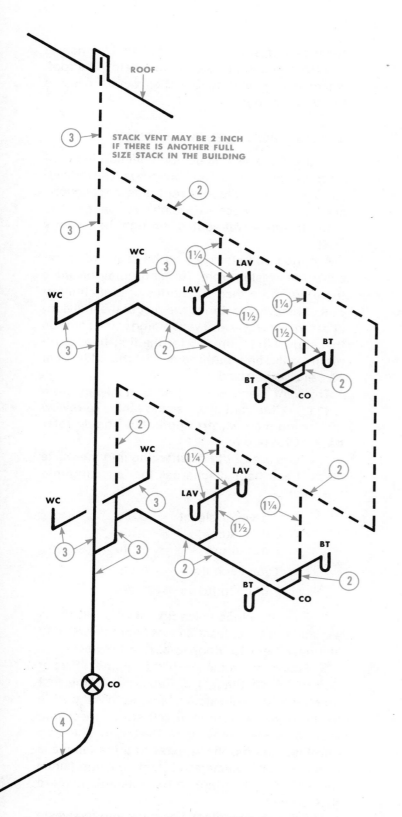

Figure 7-7. Typical bathroom stack in a two-story apartment building with back-to-back bathrooms. All fixtures are common vented.

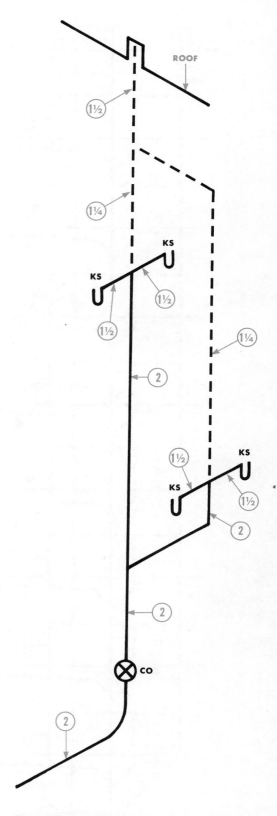

Figure 7-8. Typical kitchen sink stack in a two-story apartment building with back-to-back kitchens.

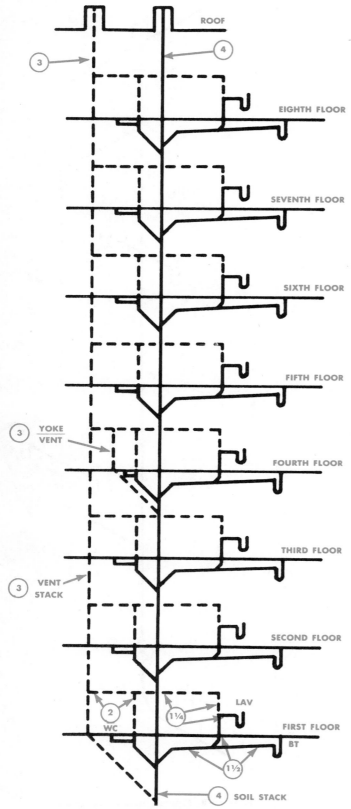

Figure 7-9. Multistory bathroom stack, showing bathtubs wet-vented through lavatory waste.

second-floor sinks is increased to 1½-inch at this tee and continues to a point below the roof, where it increases to 2 inches before passing through the roof.

Multistory Building Bathroom Stack

Figure 7-9 is an installation common to multistory apartment and hotel buildings. The individual bathroom waste and vent pipe connections are typical of those in a small residence. The bathtub trap is wet-vented through the lavatory waste.

A 4-inch soil stack would be required to receive the wastes from the bathrooms because there are more than six stories or branch intervals of bathrooms. Since the rest of the individual bathroom waste and vent piping is the same as that shown in Figure 7-5 for the first-floor bathroom, only one bathroom has the individual pieces of pipe sized.

The vent stack is sized as 3-inch pipe from Table 6-6, the vent stack sizing table. To review this sizing method, the apprentice should refer back to Table 6-6:

1. There are 72 dfu discharging into the waste stack. Each floor discharges the following into the stack:

1 water closet	=	6 dfu
1 lavatory	=	1 dfu
1 bathtub	=	2 dfu
Total each floor		9 dfu
9 dfu × 8 floors	=	72 dfu

2. The developed length of the stack is assumed to be at least 80 feet (eight floors × 10 feet each floor for floor-to-ceiling height).

3. Follow down the left-hand column ("Size of Soil or Waste Stack") of Table 6-6 to the first number 4 (4" waste stack). Move across the table to the right—as less than 240 dfu connect into the stack, keep moving to the right. The only developed length that applies to this example is in the 3-inch "Diameter of Vent" column (since the length of the stack in this example is more than 70 feet).

The 3-inch vent stack ties back into the waste stack below the first-floor bathroom, as shown. It also has a 3-inch yoke or relief vent, which is

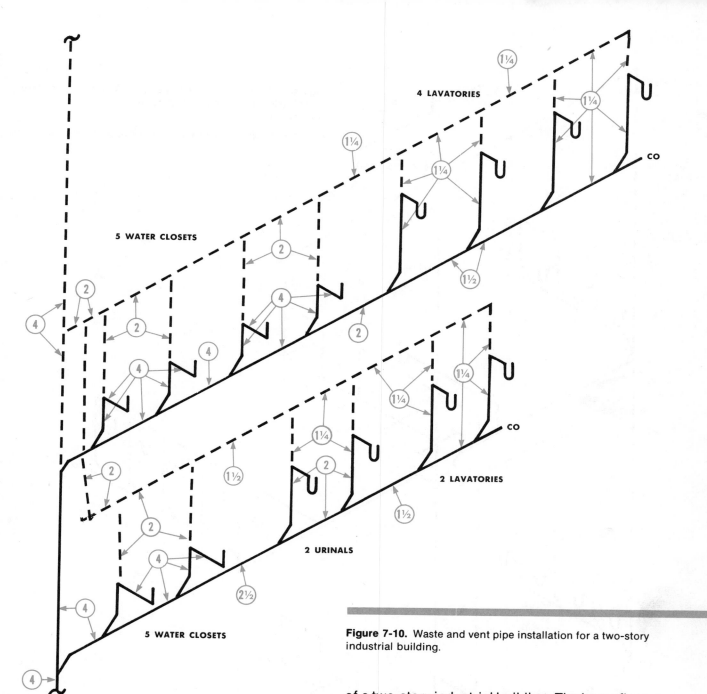

5 WATER CLOSETS

4 LAVATORIES

CO

5 WATER CLOSETS

2 LAVATORIES

2 URINALS

CO

Figure 7-10. Waste and vent pipe installation for a two-story industrial building.

installed between the third and fourth floors. The vent stack extends up to the roof where it either continues on through the roof or reconnects into the 4-inch soil stack.

Two-Story Industrial Building Toilet Room Piping

Figure 7-10 represents the waste and vent pipe installation for a small toilet room on each floor

of a two-story industrial building. The lower floor men's toilet room contains two water closets, two wall-hung urinals (2-inch trap), and two lavatories, all of which are located on the same wall. On the upper floor is a women's toilet room containing four water closets and four lavatories. The fixtures in this toilet room are also located on the same wall.

The first-floor water closets are all connected to the same 4-inch horizontal branch pipe, which connects into the 4-inch stack. This horizontal branch pipe is sized from Table 6-3. The end of this branch extends 2½-inch pipe to receive the

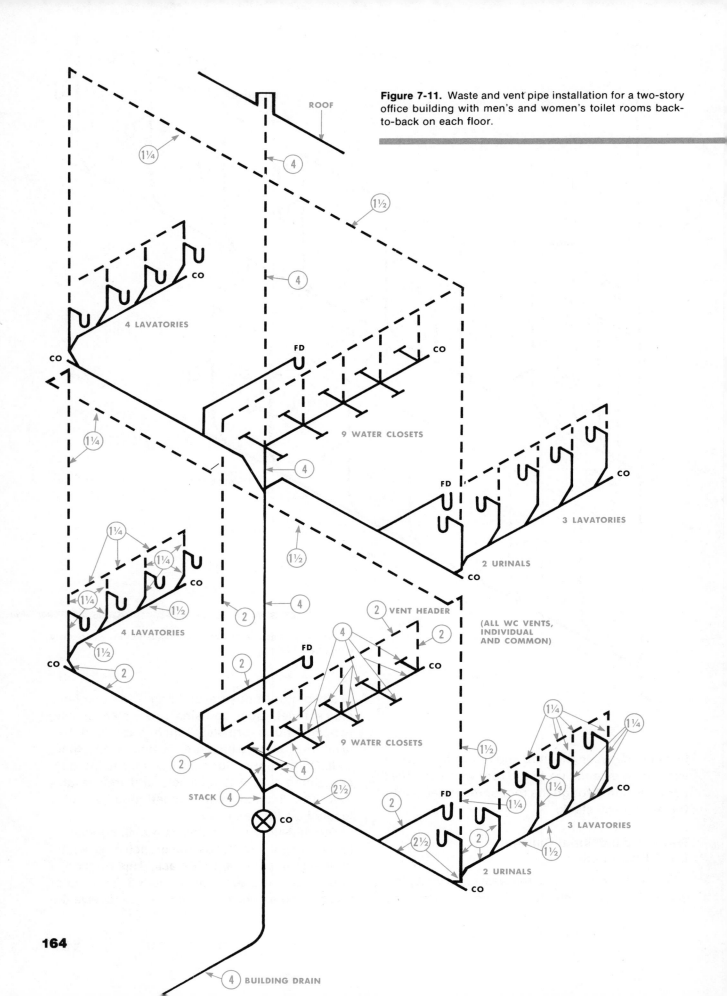

Figure 7-11. Waste and vent pipe installation for a two-story office building with men's and women's toilet rooms back-to-back on each floor.

ROOF

CO

4 LAVATORIES

CO

FD

9 WATER CLOSETS

CO

FD

3 LAVATORIES

2 URINALS

CO

VENT HEADER

(ALL WC VENTS, INDIVIDUAL AND COMMON)

4 LAVATORIES

CO

FD

9 WATER CLOSETS

CO

3 LAVATORIES

CO

FD

2 URINALS

CO

STACK

CO

2 WATER CLOSETS

BUILDING DRAIN

wastes from the two urinals and two lavatories located on the first floor:

2 wall-hung urinals   4 dfu each = 8 dfu
   plus 2 lavatories @ 1 dfu each = 2 dfu

            Total   10 dfu

At the first urinal, the horizontal waste pipe reduces to a 2-inch size. After the second urinal, this pipe reduces to 1½-inch size for the portion that receives the waste from both lavatories, and then reduces again to 1¼-inch pipe for the last lavatory waste.

The urinals and lavatories are individually vented with 1¼-inch pipe. These 1¼-inch individual fixture vents tie together into a 1¼-inch horizontal vent pipe, which increases to 1½-inch pipe at the second urinal vent. The horizontal vent pipe receives the 2-inch individual vents from the water closets and extends 2-inch pipe up to the second floor, where it ties together with the second-floor horizontal vent pipe.

The second-floor toilet room water closets empty into a 4-inch horizontal branch pipe. The end of this branch extends 2-inch pipe to receive the wastes from the four lavatories. After the first lavatory waste, this pipe reduces to 1½-inch pipe for the next two lavatory wastes and then reduces again to 1¼-inch pipe for the last lavatory waste.

The four lavatories on the second floor are all individually vented with 1¼-inch pipe. These four 1¼-inch lavatory vents connect to a 1¼-inch horizontal vent pipe. This horizontal vent pipe increases to 2-inch pipe at the point where the first 2-inch individual water closet vent ties into it. The vent then continues 2-inch size picking up the other three 2-inch water closet vents. Another 2-inch tee is provided for the connection for the 2-inch vent from the first-floor toilet room, before the vent connects 2-inch size into the stack vent.

Two-Story Office Building
Toilet Room Piping

Figure 7-11 illustrates the waste and vent piping for back-to-back men's and women's toilet rooms on each floor of a two-story office building. The men's toilet rooms contain four wall-hung water closets, two urinals (with 2-inch traps), three lavatories, and a 2-inch floor drain set between the urinals. In the women's toilet rooms, there are five wall-hung water closets, four lavatories, and a 2-inch floor drain set between the second and third water closet stalls. In this installation, the wall-hung water closets in both toilet rooms are back-to-back on the common wall, while the other fixtures are on the opposite wall of the toilet rooms.

The water closet waste openings indicated in Figure 7-11 are different than others in this chapter because the water closets here are wall-hung. The symbol shown represents a wall-hung water closet supporting chair carrier fitting. Wall-hung water closet supporting chair carrier fittings are available in many styles to accommodate a large variety of different installations. One of these fittings is pictured in Figure 7-12.

The water closets on both floors are connected to a 4-inch horizontal branch constructed of water closet carrier fittings. The water closets that are back-to-back are common vented with 2-inch pipe, and the single water closet in the women's toilet room also has a 2-inch vent. The vents from all the water closets on the first floor tie together into a 2-inch horizontal vent that is piped up to the ceiling where it receives the 1¼-inch vent from the lavatories in the women's toilet room and the 1½-inch vent from the urinals and lavatories in the men's toilet room. This 2-inch vent then continues up to the second floor, where it ties back into the stack vent.

The four lavatories in the first-floor women's toilet room are piped with individual 1¼-inch wastes. The horizontal waste, which receives the individual 1¼-inch lavatory wastes, increases to 1½-inch pipe at the second lavatory waste and then to 2-inch pipe at the fourth lavatory waste. This 2-inch waste pipe then turns back toward the stack where it receives the 2-inch waste from the floor drain in the women's toilet room before it discharges into a double wye in the stack.

The horizontal waste pipe for the first-floor men's toilet room is sized as 1¼-inch pipe for the first lavatory waste and 1½-inch pipe for the second and third lavatory wastes. This pipe then increases to 2-inch size to receive the first urinal waste and finally to 2½-inch pipe for the second

Figure 7-12. No-hub vertical closet carrier. (Wade Division, Tyler Pipe)

urinal waste. The 2½-inch waste pipe turns back toward the stack, and while on the way, it receives the waste from the 2-inch floor drain under the urinals. It then discharges into the double wye in the stack.

The vent pipe from the lavatories and urinals in the first-floor men's toilet room is 1¼-inch pipe for the three lavatory vents. It is increased to 1½-inch pipe at the first urinal and continues as this size up to the ceiling, where it ties into the 2-inch vent from the water closets.

The vent from the four lavatories in the first-floor women's toilet room is all 1¼-inch pipe.

The sizes of the waste and vent pipes in the second-floor toilet rooms are identical to those in the first-floor toilet rooms and for that reason were not sized in Figure 7-11.

The only difference in the piping of the toilet rooms between the first and second floors is that the 2-inch horizontal vent pipe for the water closets connects directly into the 4-inch stack vent. This is true also for the 1¼-inch vent from the lavatories in the women's toilet room and the 1½-inch vent from the lavatories and urinals in the men's toilet room.

Note: As a point of information, it should be mentioned that all the drainage and vent piping sized as 2½-inch pipe in Figures 7-9, 7-10 and 7-11

would most likely be piped in a 3-inch size on the job. The reason for this is that although the 2½-inch pipe size is shown in plumbing code sizing tables, 2½-inch pipe and fittings are not available in most piping materials.

 REVIEW QUESTIONS

1. Give the abbreviations or symbols for:

Waste pipe	Cleanout
Vent pipe	Floor drain
Size of pipe	Stand pipe
Laundry tray	Floor drain

2. Define a stack group.

3. Sketch the piping in your apartment or home bathroom and size the piping.

4. Sketch the entire sanitary drainage and vent piping system for a single family home you are familiar with and size it.

5. Sketch and size the sanitary drainage and vent piping system for the men's and women's toilet rooms in your school.

6. Explain why 2½-inch drainage piping is seldom installed.

CHAPTER 8

The Plumbing Trap

One basic principle of a plumbing installation is that every plumbing fixture must have a *trap*. A trap is a fitting or device that provides, when properly vented, a liquid seal to prevent the emission of sewer gases without materially affecting the flow of sewage or waste water through it.

Since the innovation of the first patented trap used on plumbing fixtures in the United States (in 1856), manufacturers have designed and offered to the plumbing industry hundreds of these devices, each one varying in construction. Testing of these traps demonstrated that some offered advantages over others under certain conditions, but most failed to come up to the standard expected of it when subjected to actual installation conditions.

The gases that occur in public sewage systems, caused by the decomposition of organic materials within the sewers, have been discussed briefly in Chapter 6. It is improbable that water-borne diseases, such as dysentery, typhoid, and cholera, can be transmitted through the gases of the public sewers. However, this fact does not lessen the importance of trap installations in a plumbing system. The properties, both physical and chemical, of the many gases found in sewage systems are known, and their effect on the human body is often serious. No individual could maintain his health if he is required to breathe large quantities of hydrogen, hydrogen sulfide, methane, or carbon dioxide; and even a small amount of carbon monoxide in the atmosphere within a building may be fatal. Many of these sewer gases are obnoxious, and, if not fatal when breathed by human beings, are nauseating and undesirable.

The basic function of a trap on a plumbing fixture is to prevent these objectionable sewer gases from entering the building. The trap is not designed to catch or retain any items that fall in the fixture.

Because of extreme conditions caused by simultaneous fixture use and overtaxed waste conditions, the plumbing system is subjected to minus and plus pressures, which affect the liquid content of the trap. (This will be discussed later.) Traps consisting of movable internal mechanisms that form their seals were produced to overcome these difficulties. In principle, these devices were to act as mechanical barriers against the passage of sewer gas. But these various types of traps failed because the dissolved and suspended materials contained in the waste corroded their movable parts and caused them to fail. Most plumbing codes *prohibit* their installation for these reasons.

Traps also have been designed with internal metal partitions, which were subject to acid conditions of the waste. The objective of this design was compactness and neat appearance. But these forms of traps also have failed and are *prohibited*.

Today, sanitary authorities depend on and design plumbing systems that use traps with a *water seal*. A well-designed vent system is essential to maintain a constant pressure of 1 atmosphere (14.7 pounds per square inch at sea level) to make the water seal trap effective.

 ## TYPES OF TRAPS

The most practical form of water seal trap is constructed in the form of the letter P. Hence its name, the P-trap. There are other forms of water seal traps: the *drum trap*; the *running trap*; the *S-trap* (shaped like the letter S); and the ³⁄₄ *S-trap*. These traps will be discussed in detail later in this chapter.

The P-Trap

A P-trap and its parts are illustrated in Figure 8-1. It must be installed as close to the fixture as is possible. This practice overcomes the tendency for the inlet side of the trap to become fouled. Each time that the fixture is discharged, a quantity of the liquid waste is arrested and retained in the dip of the trap. The liquid content is termed the *trap seal,* and may be defined as the column of water retained between the *crown weir* and the *top dip* of the trap. There are two forms of water-sealed P-traps. These are known as the *common seal* (Figure 8-2) and the *deep seal* (Figure 8-3). The common seal trap has a depth of 2 inches between the crown weir and the top dip. The deep seal retains twice this column of

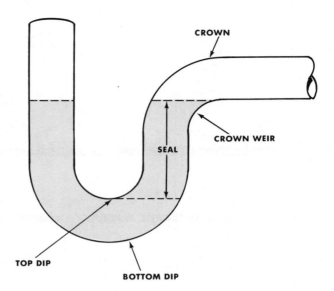

Figure 8-1. A P-trap.

Figure 8-2. Common seal P-trap.

Figure 8-3. Deep seal P-trap.

water, or 4 or more inches of liquid content.

The common seal trap is normally used on plumbing fixtures. The deep seal trap also may be used under normal conditions, but it is intended for abnormal situations, such as extreme heat, and for use where complete venting cannot be obtained. Deep seal traps are also used for floor drains and indirect waste receivers. (An *indirect waste* is a waste pipe that does not connect directly with the drainage system but conveys liquid wastes by discharging into a plumbing fixture, interceptor, or receptacle that is directly connected to the drainage system.)

There are two disadvantages to deep seal P-traps:

1. Because water is contained in the trap, there is increased resistance to the flow of waste water through the trap.

2. Because the trap seal is 2 inches deeper, the distance between the outlet and the bottom of the trap is 2 inches lower than a regular seal trap, and more space is required for installation.

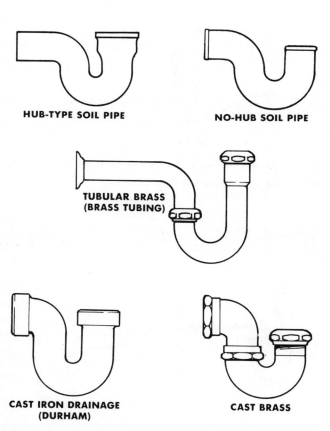

HUB-TYPE SOIL PIPE

NO-HUB SOIL PIPE

TUBULAR BRASS (BRASS TUBING)

CAST IRON DRAINAGE (DURHAM)

CAST BRASS

Figure 8-4. P-traps of various piping materials.

TABLE 8-1.   TRAP SIZES FOR VARIOUS PLUMBING FIXTURES.

TYPE OF FIXTURE	MINIMUM SIZE OF TRAP IN INCHES
Clothes washer	1 1/2
Bathtub with or without shower	1 1/2
Bidet	1 1/2
Dental unit or cuspidor	1 1/4
Drinking fountain	1 1/4
Dishwasher, domestic	1 1/2
Dishwasher, commercial	2
Floor drain	2, 3, or 4
Lavatory	1 1/4
Laundry tray (one or two compartment)	1 1/2
Shower stall, domestic	1 1/2
SINKS:	
Combination, sink and tray (with disposal unit)	1 1/2
Combination, sink and tray (with one trap)	1 1/2
Domestic with or without disposal unit	1 1/2
Surgeon's	1 1/2
Laboratory	1 1/2
Flushrim or bedpan washer	3
Service	2 or 3
Pot or scullery	2
Soda fountain	1 1/2
Commercial, flat rim, bar, or counter	1 1/2
Wash, circular or multiple	1 1/2
URINALS:	
Pedestal	3
Wall-hung	1 1/2 or 2
Trough (per 6-foot section)	1 1/2
Stall	2
Water closet	3

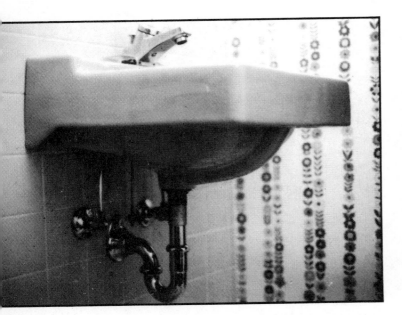

Figure 8-5. P-trap installed on a wall-hung lavatory.

The P-trap may be obtained in sizes from 1¼ to 6 inches in diameter, although, as is shown in Table 8-1, only floor drains, pedestal urinals, and water closets have traps larger than a 2-inch size, and these three types of fixtures are usually constructed with integral (built-in) traps. (Wall-hung service sinks may also have a trap larger than a 2-inch size, but this fixture is usually supplied with a P-trap standard.) The P-trap may be used for all of the fixtures listed on Table 8-1 with the exception of the water closet, which is manufactured with an integral S-trap.

The P-traps used for the fixtures listed on Table 8-1 that do not have an integral trap are constructed of one of the materials illustrated in Figure 8-4. In fact, most of the smaller plumbing fixtures (those which have 1¼- or 1½-inch traps) are at present installed with either a tubular brass or plastic P-trap, as illustrated in Figure 8-4.

Figure 8-5 shows how a P-trap may be used in connection with a wall-hung fixture. It must be installed as close to the fixture as is practical, and care should be taken so that the vertical inlet leg of the trap does not exceed 24 inches between the fixture outlet and the trap weir. This practice eliminates high velocities, which can cause trap seal loss, and lessens the possibility of fouling. The horizontal leg (outlet) connection to the waste system must also be short so that the trap will be properly vented.

Figure 8-6 illustrates the use of a P-trap on a fixture that is set on the floor. In this type of installation, the trap is concealed between the building joists.

Anti-syphon P-Traps. Anti-syphon P-traps have been designed to increase resealing quality. The additional reseal is obtained by enlarging the volume of water in the trap through building a bowl of increased diameter into the outlet leg of the trap. Figure 8-7 illustrates a typical anti-syphon P-trap. These traps are effective when installed under normal conditions and when the installation is comparatively new. However, continued discharge of greasy wastes tends to reduce the diameter of the bowl, and when this occurs, no benefits can be expected of the trap.

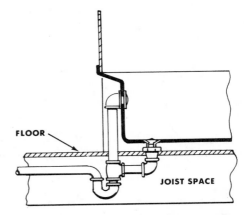

Figure 8-6 P-trap installed on a bathtub drain.

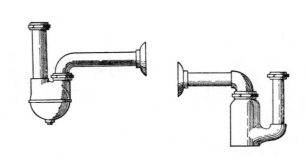

Figure 8-7. Anti-syphon P-traps.

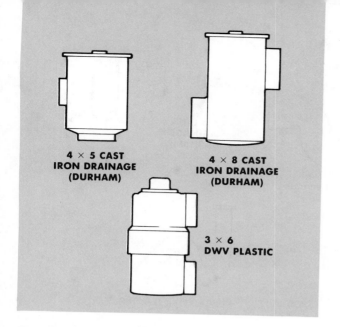

Figure 8-8 Drum traps of various piping materials.

4 × 5 CAST
IRON DRAINAGE
(DURHAM)

4 × 8 CAST
IRON DRAINAGE
(DURHAM)

3 × 6
DWV PLASTIC

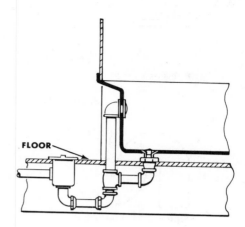

FLOOR

Figure 8-9. Drum trap on a bathtub drain.

The Drum Trap

The drum trap, also a water-sealed device, gets its name from its large diameter. These traps are designed in a variety of styles depending on the material from which they are made. (See Figure 8-8.)

A drum trap has the advantage of containing a larger volume of water than a P-trap. It was the opinion of some plumbing authorities that a greater volume of water could be passed through a drum trap than a P-trap in a given time interval. The resealing quality of a drum trap is greater than that of a P-trap, and because trap seal loss is more common in fixtures discharging greater volumes of water, the drum trap was considered more practical for this kind of installation.

The most common use of a drum trap in a home plumbing installation was on the bathtub, as illustrated in Figure 8-9. This installation prevailed because the bathtub is a fixture that usually contains a large volume of water.

The drum trap has some disadvantages, however. This type of trap is large and cumbersome, and because it retains a large volume of waste water, it tends to become a miniature cesspool. Drum traps are also objectionable because their cover or cleanout opening is above the water seal, and unless the plumber uses a lubricant and a washer in the joint between the cover and the body of the trap, it may leak sewer gas into

the room in which the trap is installed. For these reasons the use of drum traps is *not recommended* by modern plumbing codes.

The use of drum traps is usually restricted to laboratory tables, dental chairs, and other fixtures that are difficult to vent properly. In these types of installations, the additional volume of water contained within the drum trap helps to prevent the loss of the trap seal by syphonage. (This point will be discussed later in this chapter.)

The Running Trap

The running trap is another type of water seal trap that plumbers sometimes install. Running traps are usually used to provide trap seal protection on area drains, rainwater leaders, and downspouts, and also on the house trap in areas where the local plumbing code requires that the building drain be trapped. Several different types of running traps are illustrated in Figure 8-10.

S-Traps and ¾ S-Traps

The full S- and ¾ S-traps (Figures 8-11 and 8-12) normally should *not* be used in plumbing installations because they present obstacles to proper venting of the trap. These traps were commonly installed when the crown venting method was permitted. S- and ¾ S-traps form perfect syphons and are objectionable for that reason alone.

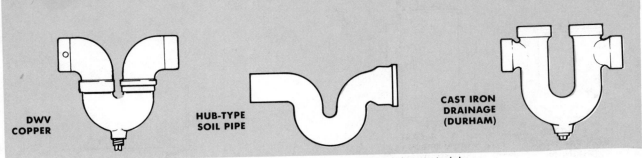

Figure 8-10. Running traps of various piping materials.

Figure 8-11. Full S-trap.

Figure 8-12. ¾ S-trap.

The plumber should avoid the installation of a P-trap into either long turn T-Y fittings or unvented vertical waste pipes, as shown in Figure 8-13, for in these installations a P-trap is made into an S-trap. The use of two fittings to offset the outlet of a P-trap downward, as shown in Figure 8-14, also creates an S-trap and should be avoided.

Plumbing codes *prohibit* the installation of S- and ¾ S-traps except in two cases: (1) in replacing a similar trap (for example, in repair work), and (2) when the S-trap is an integral part of a water closet. It is permitted on a water closet because the self-syphoning action of an S-trap is essential to the flushing function of a water closet.

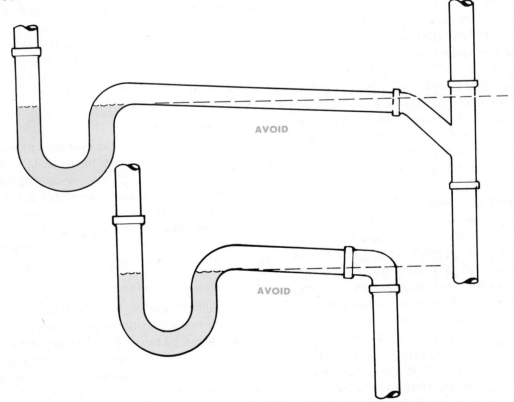

Figure 8-13. The improper use of fittings has made the above two P-trap installations into S-traps.

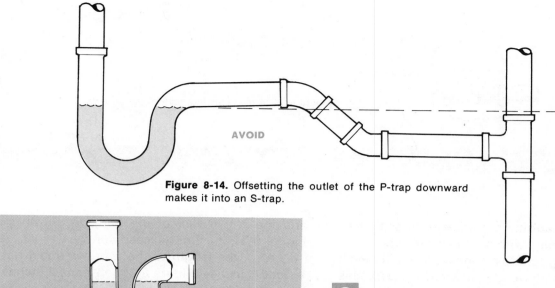

Figure 8-14. Offsetting the outlet of the P-trap downward makes it into an S-trap.

Figure 8-15. Cleanout opening on a P-trap.

Size of Traps

The size of the trap required for a specific plumbing fixture has been established by laboratory tests. In Chapter 6, Table 6-1 ("Fixture Unit Valves for Various Plumbing Fixtures") listed the minimum size trap and drain for various plumbing fixtures in addition to their individual fixture unit values. This portion of the table has been reproduced in Table 8-1, "Trap Sizes for Various Plumbing Fixtures."

Materials Used for Traps

Traps available at present are made from nearly all of the different types of piping materials discussed in Chapter 3. Figure 8-4 illustrates P-traps constructed from a variety of different materials, and Figure 8-8 illustrates drum traps available in different piping materials.

Cleanouts

Because of their construction all traps are subject to stoppage, and each must be provided with a cleanout or be so designed that it can be easily disassembled. Figure 8-15 illustrates a typical cleanout opening on a P-trap.

PROHIBITED TRAPS

Bell Traps. Bell traps (Figure 8-16) were intended for use with certain specialized plumbing fixtures. These traps were commonly installed for indirect or local wastes, such as old-fashioned ice boxes and similar fixtures. The purpose of a bell trap was to prevent the passage of sewer gases. The seal in a bell trap is formed by a raised metal rim that is cast into a depressed bowl. A furrowed bell or cap, placed over the rim, forms a small water seal.

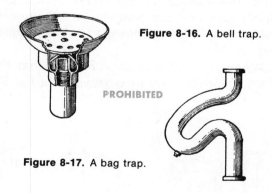

Figure 8-16. A bell trap.

PROHIBITED

Figure 8-17. A bag trap.

Bag Trap. The bag trap (Figure 8-17) is an extreme form of S-trap and one that is seldom found except in old plumbing installations. Where is is encountered in repair work, it should be replaced with a trap of approved design.

Other objectionable traps are traps that depend wholly or partially on the action of movable parts for their seal. Figure 8-18 shows a form of trap (now obsolete) that depended in part on an

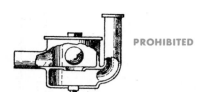

Figure 8-18. A mechanically sealed trap.

internal mechanism to complete its seal. This trap was provided with a hollow ball, which was supposed to drop into a concave seat after each flush of the fixture. Constructed of lead, this device was used mainly in connection with bathtubs and similar fixtures. Tests and experience proved it inadequate.

Figures 8-19 and 8-20 show traps that use internal partitions as the means of providing water seal. Although some traps of this type were permitted in certain areas, their efficiency is doubtful.

Traps Made from Fittings. From the previous illustrations of the different varieties of P-traps, drum traps, S-traps, and running traps, you can easily visualize how these traps could be made up on the job with the proper assortment of fittings. However, plumbing codes usually *prohibit* the use of traps that are made up with fittings. There are two reasons for this requirement:

1. Plumbing codes usually require that the trap seal depth (the distance between the top dip and the crown weir of the trap), as shown in Figure 8-21, must not exceed 4 inches.

2. Also, the length of the trap seal must not exceed 6 inches (Figure 8-21) on traps 2 inches in size or smaller.

Traps made up from fittings usually end up having a trap seal depth greater than 4 inches, and the horizontal length of the trap seal is usually more than 6 inches, because of the length and radius of the fittings used to make up the trap.

Perhaps the only combination of fittings for a P-trap (or running trap) that will meet the above requirements for trap seal depth and length is the use of the combination of a return bend and a 90° street elbow, as illustrated in Figure 8-22.

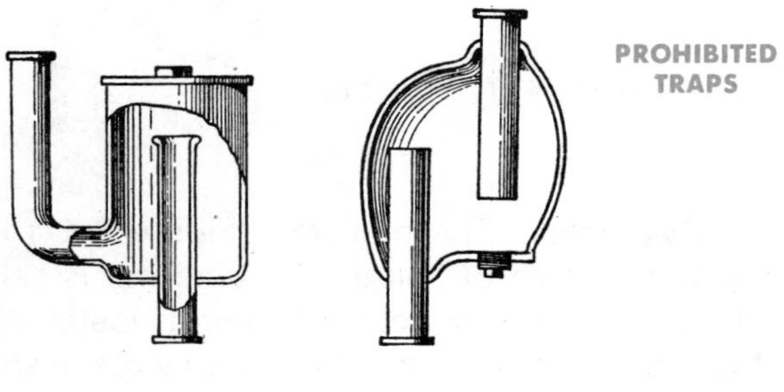

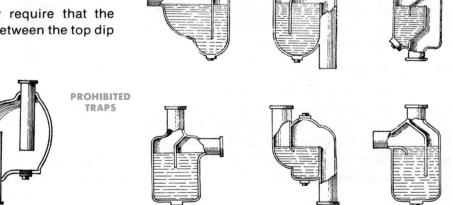

Figure 8-19. Internal partition traps.

Figure 8-20. Light metal partition traps.

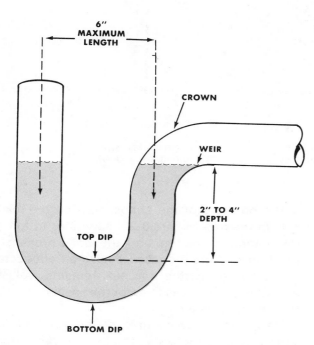

Figure 8-21. Recommended plumbing code trap requirements: (1.) Fixture traps shall have a water seal depth of not less than two inches and not more than four inches. (2.) The length of the seal of any fixture trap shall not exceed 4 inches where the waste pipe required is 2 inches or less. (Ralph Lichliter)

6" MAXIMUM LENGTH

CROWN

WEIR

2" TO 4" DEPTH

TOP DIP

BOTTOM DIP

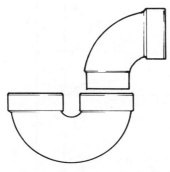

Figure 8-22. A P-trap made from a return bend and a street 90° elbow.

⊙ TRAP SEAL LOSS

As was stated earlier in Chapter 6, one of the most common and objectionable difficulties occurring in a drainage system is trap seal loss. This failure can be attributed directly to inadequate venting of the trap and the subsequent minus and plus pressures that occur. There are at least six ways in which a trap seal may be lost:

1. Syphonage
   a. Direct or self-syphonage
   b. Indirect or momentum syphonage
2. Back pressure
3. Evaporation
4. Capillary action
5. Leaks
6. Wind effect

Syphonage

Loss of the trap seal by syphonage is the result of a minus pressure in the drainage system. Through the phenomenon of syphonage, the seal content of the trap is forced into the waste piping of the drainage system by the atmospheric pressure on the fixture side of the trap seal.

Syphonage of the trap seal occurs in two forms: self-syphonage (or direct syphonage) and indirect syphonage (or syphonage by momentum).

Self-Syphonage. Self-syphonage is commonly found in unventilated traps that serve oval-bottomed fixtures, such as a lavatory or a small service sink. It is the result of unequal atmospheric pressures caused by the rapid flow of water through the trap. A plumbing fixture with a rounded oval bottom discharges its content abruptly and does not offer the small amount of trickling waste needed to reseal the trap. Figures 8-23, 8-24, and 8-25 give examples of fixtures of this type, with their connection to an unvented S-trapped waste.

The seal content of the trap in Figure 8-23 is in a neutral position. Both the inlet and outlet sides of the trap are exposed to the atmosphere and are under identical pressure.

Suppose the lavatory were filled with water and the waste plug suddenly would be removed, as shown in Figure 8-24. The water then would

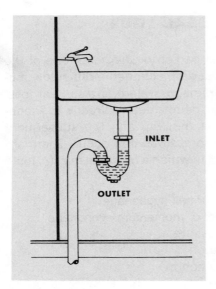

Figure 8-23. Syphonage: direct seal intact.

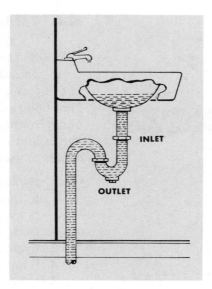

Figure 8-24. Syphonage: direct fixture discharging.

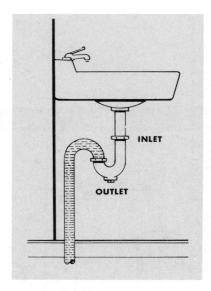

Figure 8-25. Syphonage: direct loss of seal.

rush through the trap and into the vertical waste pipe. It is a scientific fact that water is practically incompressible, and when it is forced into a vessel containing air, it compresses the extremely elastic atmospheric gases; or, if the vessel is open to the atmosphere, it replaces them entirely. This is precisely what occurs in the installation. The atmosphere in the vertical leg is replaced by the liquid discharge of the fixture, and a *minus pressure* in the waste results.

The pressure of the atmosphere on the inlet side of the trap (Figure 8-25) continues to force the water from the trap until the seal is broken. If the trap or fixture does not have an adequate resealing quality, the trap seal remains broken and allows sewer gas to enter the room in which the fixture is located.

Traps that serve fixtures of the flat bottom variety, such as sinks, bathtubs, shower baths, and laundry trays, are subject to the same diffi-

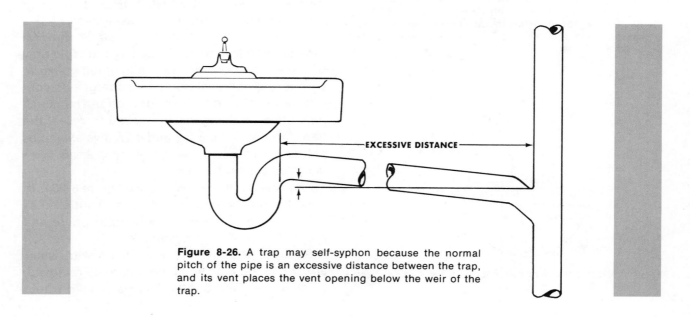

Figure 8-26. A trap may self-syphon because the normal pitch of the pipe is an excessive distance between the trap, and its vent places the vent opening below the weir of the trap.

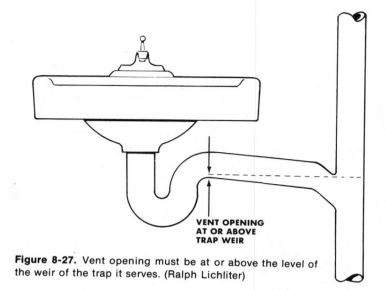

Figure 8-27. Vent opening must be at or above the level of the weir of the trap it serves. (Ralph Lichliter)

VENT OPENING
AT OR ABOVE
TRAP WEIR

culty, but the seal content is never entirely removed because the last trickle of water is sufficient to replenish the depleted trap seal.

Figure 8-26 illustrates an installation in which it is possible for the trap to lose its seal by self-syphonage. In this installation, the trap is vented. However, the distance from the trap to its vent is excessive, and because of the normal pitch on the drain pipe from the trap, the vent opening is below the weir of the trap. To prevent self-syphoning of traps in installations of this type, plumbing codes require that the vent opening be at or above the weir of the trap it serves. (See Figure 8-27.) The code will also specify a maximum distance from the trap to its vent to avoid the self-syphoning installations pictured in Figure 8-26. A typical code requirement of *maximum trap-to-vent distances for fixtures* is listed in Table 8-2 and illustrated in Figure 8-28.

From the above, the apprentice would be led to believe that the closer he can install a vent to the trap it serves the better the installation. This

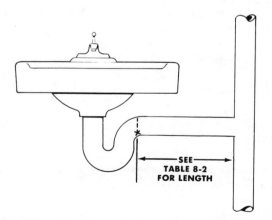

SEE
TABLE 8-2
FOR LENGTH

Figure 8-28. The vent must be located so that the developed length from the fixture trap to the vent does not exceed the distance given in Table 8-2.

TABLE 8-2.   MAXIMUM DISTANCES OF FIXTURE TRAP FROM VENT.*

SIZE OF FIXTURE DRAIN IN INCHES	DISTANCE— TRAP TO VENT
1¼	2 ft. 6 in.
1½	3 ft. 6 in.
2	5 ft.
3	6 ft.
4	10 ft.

*Note: The developed length between the trap of the water closet or similar fixture and its vent shall not exceed four (4) feet.

Minnesota Plumbing Code

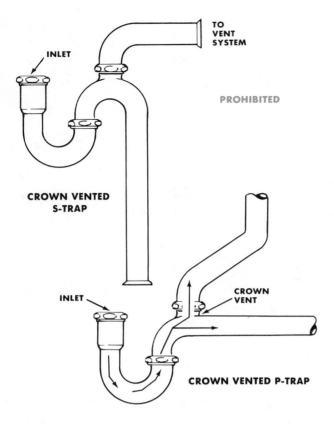

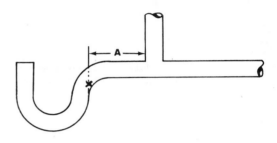

Figure 8-29. A crown-vented trap.

Figure 8-31 shows a lavatory on the first floor and a sink on the second floor of a building, both installations having a common waste. When the fixtures are not in use, the waste line and the room where the fixtures are located are under atmospheric pressure. Thus, the trap seals of both fixtures remain stationary because of the equality. When the sink (on the second floor) is discharged into the waste, however, the volume of water rushes past the trap outlet of the lavato-

Figure 8-30. A vent opening may not be installed closer than two pipe diameters (distance A) to the weir of the trap it serves.

is not the case. At one time, the traps were crown-vented, as illustrated in Figure 8-29. However, when the vent was installed in this location some of the waste drained from the fixture was carried into the vent pipe by momentum. This waste would eventually block the vent pipe. For this reason, crown-venting is specifically *prohibited* by plumbing codes.

A typical plumbing code also requires that no vent be installed within two pipe diameters of the trap weir as shown in Figure 8-30. For example: A $1\frac{1}{2}$-inch trap must be at least 3 inches ($2 \times 1\frac{1}{2}$ inches = 3 inches) from its vent; a 2-inch trap at least 4 inches; etc.

Syphonage by Momentum. Loss of the trap seal by syphonage caused indirectly or by the momentum of water, as it passes a fixture trap outlet, is a difficulty commonly experienced in plumbing systems. This form of trap seal loss is the result of a *minus pressure* in the waste piping caused by discharge of waste from a fixture into a drain line that also serves a fixture placed at a lower elevation. The discharge of the higher fixture syphons the trap of the lower fixture.

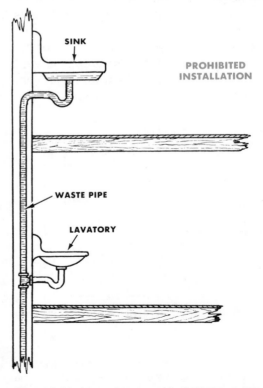

Figure 8-31. Loss of trap seal by indirect syphonage.

ry (on the first floor) and tends to lower the pressure on the trap leg. A minus pressure occurs in the trap outlet, and the seal is forced from the trap by the atmospheric pressure contained in the room. There is no possibility of resealing under these circumstances, and once the trap seal of the lavatory is lost, sewer gas may enter the room through the fixture trap. This type of installation (Figure 8-31) should never be used; Figure 8-31 is used merely to illustrate the condition described.

Back Pressure

Back pressure, which is caused by a *plus pressure,* is responsible for trap seal loss often experienced in large plumbing installations. It is a serious form of trap seal loss, for not only does it allow sewer gases to enter the building, but if a person is using the fixture at the time back pressure occurs he may be injured, or he may receive a thorough and unpleasant bathing. Back pressure actually blows the water out of the fixture into the room. Where sufficient pressure is produced, the content of the fixture trap often strikes the ceiling of the room. The fixtures in which this occurs are usually located at the base of soil stacks or where soil pipe changes its direction abruptly.

The flow of water in a soil pipe varies. A single fixture may produce only a small trickle, which spirals down the sides of the soil pipe. Large flows may completely fill the pipe, and the compressed atmospheric gases, offering resistance, are unable to slip past the flow of water and find an exhaust at the roof terminal. As the water falls the pressure grows as the area into which the air is squeezed is reduced. Soon the resistance of the trap seal is overcome and the trap seal *blows* from back pressure.

Figure 8-32 illustrates how back pressure can occur. The soil stack illustrated is connected to the building drain in the normal manner. A basement water closet is connected to the building drain close to the base of the soil stack. The soil stack serves toilet rooms on the upper floors of the building. Assume fixtures on the upper floors have been discharged. As the water falls down the stack and hits the horizontal building drain, there is a turbulent action as the water

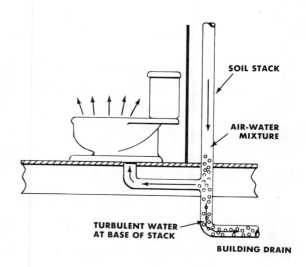

Figure 8-32. Loss of trap seal as a result of back pressure.

changes from a vertical to a horizontal direction. This turbulent water completely fills the building drain, as illustrated. The downward flow of waste water, which has completely filled the stack, compresses the trapped air. The back pressure thus created has blown the trap seal from the basement water closet into the room. This condition can be corrected by moving the water closet at least 8 feet away from the base of this stack (as was pointed out in Chapter 6 in the rules pertaining to the installation of stacks) and by the installation of a proper vent on this water closet.

Evaporation

Evaporation of the trap seal is one of the lesser forms of trap seal loss, and is a phenomenon of nature. The atmosphere absorbs moisture, and the amount absorbed increases with higher temperatures. A trap seal located in a room where the air is not saturated with water serves as a water source, and it gradually evaporates.

Under ordinary conditions, it would require many weeks to evaporate a trap seal, and frequent use of the fixture would eliminate this difficulty entirely. Venting of the trap is not a solution to this problem; neither does it affect the trap content as one might suppose, because

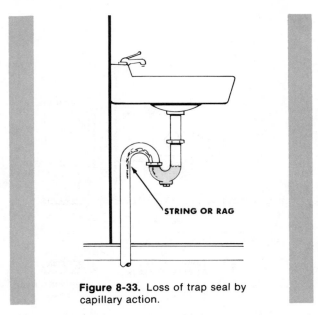

Figure 8-33. Loss of trap seal by capillary action.

the air circulating within the plumbing system usually is in a saturated condition.

The use of a deep seal trap is recommended as a means of prolonging the interval of total loss of the trap seal in the hope that the fixture will be used before complete evaporation of the trap's liquid content has occurred.

Capillary Action

Loss of a trap seal by capillary action sometimes occurs. It is caused by a foreign object (such as a rag, string, or lint) in the trap extending over the outlet arm. Figure 8-33 illustrates capillary action. The string in this instance forms an absorbing syphon. It soaks up water until it drips from the end reaching into the outlet arm of the trap. Once it reaches this stage, the water flows from it rather rapidly, and the seal of the trap is soon displaced.

Leaks

Obviously, any trap that leaks will lose its protective seal of water. All fixture traps must be tested after they are installed to see that they retain the proper amount of water to seal out the sewer gases without leakage. Traps should also be protected from freezing, because a trap that freezes and cracks or breaks will leak its water seal.

Wind Effects

Wind of high velocity passing over the top of the soil pipe roof terminal affects the trap seal. A downdraft occurring in the plumbing system tends to ripple the liquid content of the trap and spill a quantity of it over its outlet leg into the system. This is not a serious problem because it is quite improbable that the entire seal will be removed. Some precaution can be taken to terminate the soil stack away from valleys, gables, or abrupt roof projections where the wind may strike and be directed into or across the soil pipe roof terminal.

In conclusion, it should be emphasized that since every plumbing fixture has a trap (and only one trap, as the double trapping of plumbing fixtures is prohibited by plumbing codes), this trap must be protected. A trap is not a device designed to withstand pressure variations. To be effective and maintain its water seal, it must be vented properly.

◎ REVIEW QUESTIONS

1. Why must every plumbing fixture have a trap?

2. There are two major areas of concern in designing plumbing traps. One is mechanical; one is biological or related to health. What are these areas? Why are they design problems?

3. How is drainage system venting related to trap seals? What has atmospheric pressure to do with water seals in traps?

4. Explain what is meant by the term *reseal* in a plumbing trap.

5. What has length of the trap arm to do with the effectiveness of the trap seal and its vent?

6. Why should a P-trap *not* be installed close to a long turn T-Y drainage fitting? Or into a trap arm that has a vertical offset? Or into a nonvented vertical waste pipe?

7. Explain the meaning of these terms as they relate to trap seal loss:

Self-syphonage
Back pressure
Syphonage by momentum
Capillary action

CHAPTER 9

Sizing Water Supply Piping

The building water supply system must supply the plumbing fixtures and appliances with enough potable water to enable them to function properly.

There are five factors that determine the size of water supply piping:

1. Available pressure.
2. Demand.
3. Length of piping.
4. Height of the building.
5. Flow pressure needed at the top floor.

Available Pressure

The available pressure is the water pressure in the street water main or other source of supply. Normal street water main pressure is between 45 and 60 pounds per square inch (psi). In no case should the water pressure within the building exceed 80 psi. If the street water main pressure exceeds 80 psi, a pressure-reducing valve (Figure 9-1) must be installed on the water service pipe at the point where it enters the building to reduce the pressure to 80 psi or less.

When pressure in the street water main fluctuates widely throughout the day, the water supply system of the building must be designed on the basis of the minimum pressure available.

Demand

Each plumbing fixture served by the water supply system has a certain demand for water. This demand is called the *flow rate.* The flow rate for a given plumbing fixture is the volume of water it uses in a given amount of time. The flow rate is usually expressed in gallons per minute (gpm). Table 9-1 lists a variety of plumbing fixtures and each fixture's minimum flow rate.

Adding up the minimum flow rates for all the plumbing fixtures within a building would give the total demand on the water supply system in gallons per minute if all the fixtures were to be used at the same time. However, all the plumbing fixtures in any building are rarely used at the same time. A more reasonable estimate of the total demand on the water supply system would be based on the extent to which the various plumbing fixtures might actually be used simultaneously.

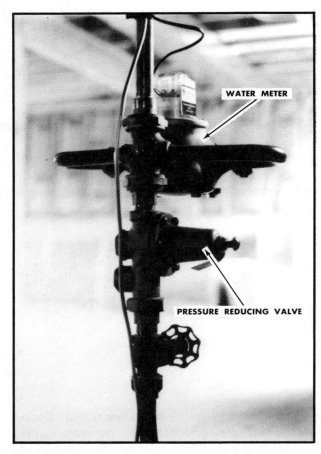

Figure 9-1. A pressure-reducing valve installed on a water service.

The *water supply fixture unit system* (wsfu) was devised to estimate the total demand on a water supply system, based on reasonable assumptions of the likelihood of simultaneous use of plumbing fixtures. Each plumbing fixture is assigned a wsfu value, which reflects:

1. The fixture's demand for water (its flow rate in gpm) when it is used.

2. The average time water is actually flowing when the fixture is being used.

3. The frequency with which the fixture is likely to be used.

Table 9-2 lists a variety of plumbing fixtures and the wsfu values assigned to each fixture when it is used in private or public use.

One plumbing code* defines private and public use of plumbing fixtures as follows:

1. *Private or Private Use.* In the classification of plumbing fixtures, *private* applies to fixtures in residences and apartments, and to fixtures in private bathrooms of hotels, as well as similar installations in other buildings where fixtures are intended for use of one family or an individual.

2. *Public or Public Use.* In the classification of plumbing fixtures, *public* applies to fixtures in general toilet rooms of schools, gymnasiums, hotels, railroad stations, bars, public comfort stations, and other installations (whether pay or free) where fixtures are installed so that their use is similarly unrestricted.

Length of Piping

The street water mains are usually under a pressure of 45 to 60 psi, which is enough to serve a plumbing installation of moderate size. A water supply pipe connected to the street water main would be under the same pressure as the main as long as the water within the pipe is at rest. However, should someone suddenly draw water from the pipe, a decided drop in pressure would occur at the outlet orifice. The variation in pressure between the street water main and the outlet end of the water supply pipe where the water is being used is referred to as *pressure loss by friction.*

It is impossible to use a pipe for the conveyance of water without encountering a loss in

TABLE 9-1. MINIMUM FLOW RATES FOR PLUMBING FIXTURES.

FIXTURE	FLOW RATE (gpm)
Ordinary basin faucet	2.0
Self-closing basin faucet	2.5
Sink faucet, $^3/_8$-inch	4.5
Sink faucet, $^1/_2$-inch	4.5
Bathtub faucet	6.0
Laundry tub cock, $^1/_2$-inch	5.0
Shower	5.0
Ball cock for water closet	3.0
Flushometer valve for water closet	15–35
Flushometer valve for urinal	15.0
Drinking fountain	0.75
Sill cock or wall hydrant	5.0

*Minnesota Plumbing Code.

TABLE 9-2.   WATER SUPPLY FIXTURE UNITS (wsfu).

FIXTURE	NUMBER OF FIXTURE UNITS	
	Private Use	Public Use
Bar sink.	1	2
Bathtub (with or without shower)	2	4
Dental unit or cuspidor	—	1
Drinking fountain (each head)	—	1
Hose bibb or sill cock (standard-type)	3	5
House trailer (each).	6	6
Laundry tray or clothes washer (each pair of faucets) . .	2	4
Service sink	—	4
Lavatory	1	2
Lavatory (dental)	1	1
Lawn sprinklers (standard-type, each head)	1	1
Shower (each head).	2	4
Sink (bar).	1	2
Sink or dishwasher	2	4
Sink (flushing rim, clinic)	—	10
Sink (washup, each set of faucets)	—	2
Sink (washup, circular spray)	—	4
Urinal (pedestal or similar type).	—	10
Urinal (stall)	—	5
Urinal (wall)	—	5
Urinal (flush tank)	—	3
Water closet (flush tank)	3	5
*Water closet (flushometer valve).	—	10

Water supply outlets for items not listed above shall be computed at their maximum demand, but in no case less than:

3/8 inch	1	2
1/2 inch	2	4
3/4 inch	3	6
1 inch	6	10

*Sizing for flushometer valves: Branches and mains serving water closet or similar flushometer valves may be sized from Table 9-2 when the following values are assigned to each flushometer valve, beginning with the most remote valve on each branch.

For the first flushometer valve . . . . 40 fixture units
For the second flushometer valve . . . . 30 fixture units
For the third flushometer valve. . . . . 20 fixture units
For the fourth flushometer valve. 15 fixture units
For the fifth flushometer valve (and all additional
flushometer valves) . . . . 10 fixture units

Minnesota Plumbing Code

TABLE 9-3. PRESSURE LOSS DUE TO FRICTION IN TYPE M COPPER TUBE.*

| FLOW (gpm) | PRESSURE LOSS PER 100 FEET OF TUBE (psi) | | | | | | | | | | | |
| | STANDARD TYPE M TUBE SIZE IN INCHES | | | | | | | | | | | |
	3/8	1/2	3/4	1	1 1/4	1 1/2	2	2 1/2	3	4	5	6
1	2.5	0.8	0.2									
2	8.5	2.8	0.5	0.2								
3	17.3	5.7	1.0	0.3	0.1							
4	28.6	9.4	1.8	0.5	0.2							
5	42.2	13.8	2.6	0.7	0.3	0.1						
10		**46.6**	8.6	2.5	0.9	0.4	0.1					
15			17.6	5.0	1.9	0.9	0.2					
20			**29.1**	8.4	3.2	1.4	0.4	0.1				
25				12.3	4.7	2.1	0.6	0.2				
30				**17.0**	6.5	2.9	0.8	0.3	0.1			
35					8.5	3.8	1.0	0.4	0.2			
40					11.0	4.9	1.3	0.5	0.2			
45					**13.6**	6.1	1.6	0.6	0.2			
50						7.3	2.0	0.7	0.3			
60						10.2	2.7	1.0	0.4			
70						**13.5**	3.6	1.2	0.5	0.1		
80							4.6	1.6	0.7	0.2		
90							5.7	2.0	0.9	0.2		
100							**7.5**	2.7	1.0	0.3	0.1	
200								**8.5**	3.6	1.0	0.3	0.1
300									**8.0**	2.0	0.7	0.3
400										**3.3**	1.2	0.5
500											1.7	0.7
750											**3.6**	1.5
1000												**2.5**

*Numbers in boldface correspond to flow velocities of just over 10 ft per sec.

Copper Development Association, Inc.

pressure resulting from friction within the pipe itself, as well as in the valves and fittings used in the construction of the building water supply system. Friction in the building water supply piping is the resistance produced by the contact of flowing water with the interior surface of the pipe. It is also the resistance between the molecules of water. The more pipe, fittings, valves, and other devices placed on the water supply system, the greater the pressure loss and, therefore, the lower the discharge capacity of the water supply pipes at the plumbing fixtures.

Table 9-3 indicates the pressure loss due to friction in type M copper tubing of different sizes for a variety of different flow rates, while Table 9-4 indicates the approximate loss of pressure due to friction in pipe fittings and valves of various design (expressed as equivalent length of tubing, in feet). From these tables, you can realize the necessity of giving the problem of friction some consideration.

Height of Building

Some of the available water pressure is also lost because of the height to which the water must flow. However, pressure loss resulting from height (head) is a relatively simple problem. A column of water loses 0.434 psi of pressure for every one foot of elevation or head. If a building is 50 feet high, the pressure loss because of head would be 0.434 × 50, or 21.7 pounds. This pressure loss would be deducted from the available pressure to determine whether or not there is sufficient pressure to raise the water to the required height.

Flow Pressure Needed at the Top Floor

Every plumbing fixture requires a minimum amount of working water pressure or *flow pressure* to function properly. (Flow pressure is the pressure in the water supply pipe near the faucet or water outlet while the faucet or outlet is wide open and flowing.) This flow pressure varies from 8 psi for most faucets and tank-type water closets to 25 psi for some types of flushometer valves. Table 9-5 lists a variety of plumbing fixtures and the minimum flow pressure required by each fixture.

To determine if there is sufficient flow pres-

TABLE 9-4. ALLOWANCE FOR FRICTION LOSS IN VALVES AND FITTINGS EXPRESSED AS EQUIVALENT LENGTH OF TUBE.*

FITTING SIZE IN INCHES	EQUIVALENT LENGTH OF TUBE IN FEET						
	Standard Ells		90° Tee		Coupling	Gate Valve	Globe Valve
	90°	45°	Side Branch	Straight Run			
3/8	0.5	0.3	0.75	0.15	0.15	0.1	4
1/2	1	0.6	1.5	0.3	0.3	0.2	7.5
3/4	1.25	0.75	2	0.4	0.4	0.25	10
1	1.5	1.0	2.5	0.45	0.45	0.3	12.5
1 1/4	2	1.2	3	0.6	0.6	0.4	18
1 1/2	2.5	1.5	3.5	0.8	0.8	0.5	23
2	3.5	2	5	1	1	0.7	28
2 1/2	4	2.5	6	1.3	1.3	0.8	33
3	5	3	7.5	1.5	1.5	1	40
3 1/2	6	3.5	9	1.8	1.8	1.2	50
4	7	4	10.5	2	2	1.4	63
5	9	5	13	2.5	2.5	1.7	70
6	10	6	15	3	3	2	84

*Allowances are for streamlined soldered fittings and recessed threaded fittings. For threaded fittings, double the allowances shown in the table.

Copper Development Association, Inc.

sure for the type of fixture being supplied with water, the plumber must subtract from the available pressure the pressure loss due to friction and the loss due to the height (or head) to which the water must rise from the main to the highest fixture.

TABLE 9-5. MINIMUM FLOW PRESSURES FOR PLUMBING FIXTURES.

FIXTURE	FLOW PRESSURE (psi)
Ordinary basin faucet	8
Self-closing basin faucet	8
Sink faucet, 3/8-inch	8
Sink faucet, 1/2-inch	8
Bathtub faucet	8
Laundry tub cock, 1/2-inch	8
Shower	8
Ball cock for water closet	8
Flushometer valve for water closet	15–25
Flushometer valve for urinal	15
Drinking fountain	15
Sill cock or wall hydrant	10

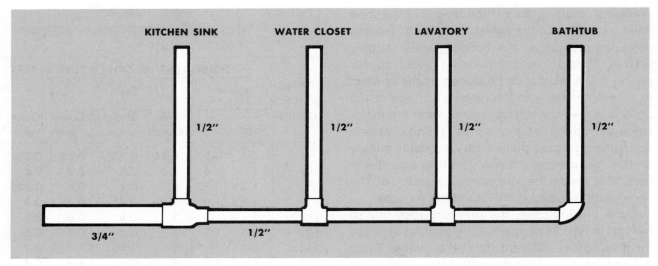

Figure 9-2. No more than three fixtures may be supplied by a ¹/₂-inch size pipe.

Of the above-described five factors that determine the size of water supply piping, the only factor over which the plumber has any control is the length of piping (or friction loss) factor. This factor can be changed by making the pipe sizes larger to reduce the amount of friction within the pipe and fittings.

Minimum Sizes of Water Pipe

In addition to the previously mentioned five factors which determine the size of water supply piping, the plumber is restricted by the following minimum sizes:

1. ³/₄-inch pipe is the minimum size water service for any building from the street to the water meter.

2. ³/₄-inch pipe is the minimum size of the building supply pipe (the first section of water distribution piping within the building).

3. ³/₄-inch pipe is the minimum size to a sill cock or lawn faucet.

4. ³/₄-inch pipe is the minimum size cold water supply to a water heater.

5. ³/₄-inch pipe is the minimum size for the first section of hot water pipe on the outlet side of a water heater.

6. No concealed water piping may be less than ¹/₂-inch size.

7. No more than three fixtures in the same bathroom of a house may be supplied by ¹/₂-inch size water pipe, as illustrated in Figure 9-2.

8. ¹/₂-inch pipe is the minimum size fixture branch pipe, except that no individual fixture branch piping may be smaller than the sizes listed in Table 9-6.

9. Individual fixture water supply piping may be ³/₈- or ¹/₄-inch size, if it is not concealed within the building partitions or over 30 inches in length.

TABLE 9-6. MINIMUM SIZES OF FIXTURE WATER BRANCH PIPE.

FIXTURE	NOMINAL PIPE SIZE IN INCHES
Automatic clothes washer (domestic)	¹/₂
Bathtub	¹/₂
Dishwasher (domestic)	¹/₂
Drinking fountain	¹/₂
Kitchen sink (domestic)	¹/₂
Kitchen sink (commercial)	³/₄
Lavatory	¹/₂
Laundry tray	¹/₂
Shower (single head)	¹/₂
Sill cock or wall hydrant	³/₄
Sinks (service and slop)	¹/₂
Sink (flushing rim)	³/₄
Urinal (flush tank)	¹/₂
Urinal (flushometer valve)	³/₄
Water closet (flush tank)	¹/₂
Water closet (flushometer valve)	1

◎ SIZING WATER SUPPLIES

In the following sections of this chapter, three building water supply systems will be sized using the previously-described five design factors and minimum sizing requirements. These buildings are a single-family house, a four-unit apartment building, and a public building.

To simplify the water piping drawings of these buildings, the following abbreviations and symbols will be used:

——————	Cold water pipe
- - - - - -	Hot water pipe
⑧	Size of pipe (number in circle)
[22]	wsfu load on pipe (number in square)
Auto Washer	Automatic clothes washer
BT	Bathtub
KS	Kitchen sink
Lav	Lavatory
LT	Laundry tray
SS	Service sink
U	Urinal
WC	Water closet
⋈	Gate valve
RV	Relief valve
SC	Sill cock
	Water meter

Sizing the Water Pipe in a Single-Family Home

Figure 9-3 illustrates the cold and hot water supply piping of a typical single-family home. In the basement of this home are located a laundry tray, an automatic clothes washer, and a water heater. On the first floor, there is a kitchen sink and a bathroom containing a water closet, a lavatory, and a bathtub. Two sill cocks are also indicated on the water piping, one at the front left side of the house and one at the rear.

The building water service is sized $3/4$-inch pipe, which is the size of the water service for most single family homes. A $3/4$-inch gate valve is provided on the water service. The $3/4$-inch water meter is then installed. Another $3/4$-inch gate valve is provided on the outlet side of the water meter.

From the second gate valve, the building supply pipe continues as $3/4$-inch size to a $3/4$-inch tee to provide cold water to the sill cock at the front of the house. The $3/4$-inch sill cock branch pipe is provided with a $3/4$-inch gate valve before it passes through the wall of the house to the sill cock.

From the sill cock branch tee, the cold water main pipe continues as $3/4$-inch size to another $3/4$-inch tee for the cold water supply to the water heater. This $3/4$-inch cold water supply to the water heater is provided with a $3/4$-inch gate valve before it connects to the water heater.

After the tee for the cold water supply to the water heater, the cold water main pipe continues as $3/4$-inch size to another $3/4$-inch tee to provide cold water to the first-floor bathroom fixtures and the kitchen sink and to the basement laundry tray and automatic clothes washer. From this tee, the cold water pipe continues as $3/4$-inch size to a gate valve and the sill cock at the rear of the house.

The $3/4$-inch tee provided for the cold water supply to the first floor and basement fixtures is reduced to $1/2$-inch size to supply cold water to the three bathroom fixtures (the water closet, the lavatory, and the bathtub), on the left side of the drawing.

The right side of the tee is also reduced to $1/2$-inch size to supply cold water to the first-floor kitchen sink and the basement laundry tray and automatic clothes washer.

The hot water piping begins as $3/4$-inch size at the outlet of the water heater. The hot water pipe continues as $3/4$-inch size toward the rear of the house to provide hot water to the first-floor and basement fixtures.

The hot water pipe is reduced to $1/2$-inch size on the left side to provide hot water to the

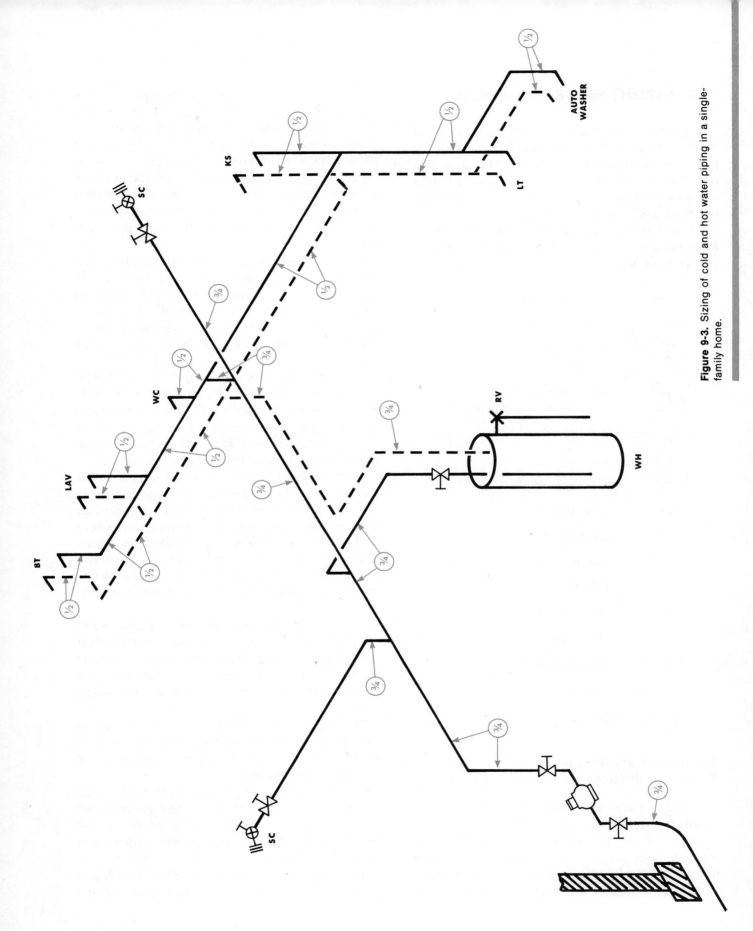

Figure 9-3. Sizing of cold and hot water piping in a single-family home.

lavatory and bathtub. It is also reduced to ½-inch size on the right side to provide hot water to the first-floor kitchen sink and the basement laundry tray and automatic clothes washer, as pictured in Figure 9-3.

An example of the sizing of the cold and hot water piping for a larger single-family home is illustrated in Figure 12-38 and explained in the text of Chapter 12.

Sizing Water Pipe in Larger Installations

The sizing of water supply and distribution piping for the single-family home described in the previous section is a relatively simple process. It does not really require knowledge of the details of the five water piping design factors because all of the water pipe sizing needs of a single-family home can be provided for by the minimum piping requirements discussed previously. However, when sizing water piping in larger installations, these five design factors play a very important role.

The method for sizing water pipe in larger installations presented in this chapter is based on the five water piping design factors taken from the Minnesota Plumbing Code. This sizing method is adequate for sizing the water piping for buildings that require a water service of 2-inch size or smaller and in which the building supply and distribution piping does not exceed 2½-inch size. The sizing of water pipe installations that require larger pipe sizes should be left to mechanical engineers. (On any water piping installation, if a drawing of the water piping was prepared by a mechanical engineer, the plumber should install the water pipe as sized on this drawing.)

The water pipe sizing method presented here for larger installations uses two tables: Table 9-2, "Water Supply Fixture Units"; and Table 9-7, "Fixture Unit Table for Determining Water Pipe and Meter Sizes for Water Supply Systems."

Table 9-7 is actually three tables for water supply pressures in three different ranges: 30 to 45 psi; 46 to 60 psi; and over 60 psi.

To size a water pipe installation using Tables 9-2 and 9-7, the twelve steps outlined below are followed.

STEP 1A. For buildings without flushometer valve water closets, total the *cold water supply fixture unit* (cwsfu) demand for the entire building in water supply fixture units from Table 9-2.

STEP 1B. For buildings with flushometer valve water closets, total the cold water supply fixture unit (cwsfu) demand for the entire building in water supply fixture units (wsfu) from Table 9-2, but assign the following values to each flushometer valve beginning with the most remote flushometer valve on each branch.

For the first flushometer valve . . . . . 40 wsfu
For the second flushometer valve . . . 30 wsfu
For the third flushometer valve . . . . . 20 wsfu
For the fourth flushometer valve . . . . 15 wsfu
For the fifth flushometer valve (and all
    additional flushometer valves . . . . 10 wsfu

The piping supply for a flushometer valve may never be smaller in size than the size of the valve inlet.

STEP 1C. Total the *hot water supply fixture unit* (hwsfu) demand for the entire building in water supply fixture units from Table 9-2 and assign this value to the water heater.

STEP 2. Determine the longest developed length of piping from the water meter to the most remote outlet, or from the water main to the most remote outlet, if the water pressure at the meter is not known.

STEP 3. Determine the difference in elevation (head) between the water meter (or other source of supply) and the highest water supply outlet, and multiply this figure by 0.5 psi.

STEP 4. Deduct the figure calculated in Step 3 from the available water pressure.

STEP 5. Select the pressure range from Table 9-7 that includes the pressure figure arrived at in the calculations of Step 4.

STEP 6. Select the maximum allowable length column that equals or exceeds the total length determined in Step 2. Once this column has been selected, *the water service, the water meter, and every piece of water supply pipe within the building will be sized from this column.*

STEP 7. Follow across the table to select the water service and water meter size that equals or exceeds the cold water fixture unit demand calculated in Step 1A or 1B.

TABLE 9-7. FIXTURE UNIT TABLE FOR DETERMINING WATER PIPE AND METER SIZES FOR WATER SUPPLY SYSTEMS.

PRESSURE RANGE—30 TO 45 psi

Meter and Water Service (in inches)	Building Supply and Branches (in inches)	MAXIMUM ALLOWABLE LENGTH IN FEET									
		40	60	80	100	150	200	250	300	400	500
3/4	1/2	6	5	4	4	3	2	——	——	——	——
3/4	3/4	18	16	14	12	9	6	——	——	——	——
3/4	1	29	25	23	21	17	15	13	12	10	9
1	1	36	31	27	25	20	17	15	13	12	10
1	1 1/4	54	47	42	38	32	28	25	23	19	17
1 1/2	1 1/4	90	68	57	48	38	32	28	25	21	19
1 1/2	1 1/2	151	124	105	91	70	57	49	45	36	31
2	1 1/2	210	162	132	110	80	64	53	46	38	32
1 1/2	2	220	205	190	176	155	138	127	120	105	96
2	2	372	329	292	265	217	185	164	147	124	107
2	2 1/2	445	418	390	370	330	300	280	265	240	220

PRESSURE RANGE—46 TO 60 psi

3/4	1/2	9	8	7	6	5	4	3	2	——	——
3/4	3/4	27	23	19	17	14	11	9	8	6	5
3/4	1	44	40	36	33	28	23	21	19	17	14
1	1	60	47	41	36	30	25	23	20	18	15
1	1 1/4	102	87	76	67	52	44	39	36	30	27
1 1/2	1 1/4	168	130	106	89	66	52	44	39	33	29
1 1/2	1 1/2	270	225	193	167	128	105	90	68	62	52
2	1 1/2	360	290	242	204	150	117	98	84	67	55
1 1/2	2	380	360	340	318	272	240	220	198	170	146
2	2	570	510	470	430	368	318	280	250	205	173
2	2 1/2	680	640	610	580	535	500	470	440	400	365

PRESSURE RANGE—OVER 60 psi

3/4	1/2	11	9	8	7	6	5	4	3	2	——
3/4	3/4	34	28	24	22	17	13	11	10	8	——
3/4	1	63	53	47	42	35	30	27	24	21	18
1	1	87	66	55	48	38	32	29	26	22	19
1	1 1/4	140	126	108	96	74	62	53	47	39	34
1 1/2	1 1/4	237	183	150	127	93	74	62	54	43	37
1 1/2	1 1/2	366	311	273	240	186	154	130	113	88	73
2	1 1/2	490	395	333	275	220	170	142	122	98	82
1 1/2	2	*380	*380	*380	*380	370	335	305	282	244	212
2	2	*690	670	610	560	478	420	375	340	288	245
2	2 1/2	*690	*690	*690	*690	*690	650	610	570	510	460

*Maximum allowable load on meter.

Minnesota Plumbing Code

STEP 8. Next, size the first piece of water supply pipe out of the meter (called the building supply pipe). Once this pipe is sized, no piece of pipe in the system need be larger.

STEP 9. Now go to the most remote fixture from the water meter and size each piece of pipe back toward the meter.

STEP 10. At the point where the cold water supply to the water heater is provided, size the cold water supply to the heater and the hot water supply out of the heater using the hot water fixture unit demand calculated in Step 1C.

STEP 11. At the point where the cold water supply to the water heater left the cold water main, add the hot water fixture unit demand to the cold water main, but if the total water supply fixture unit demand requires the size of the cold water main to exceed the size of the building supply pipe, this pipe *need not* be increased in size past the size originally selected.

STEP 12. Size the hot water supply piping by using the same column of Table 9-7, starting at the hot water supply opening that is most remote from the water heater and size back toward the heater.

The following two sections of this chapter will be used to illustrate the use of this water pipe sizing method.

Sizing the Water Pipe in a Four-Unit Apartment Building

Figures 9-4 and 9-5 illustrate the cold and hot water supply piping for a four-unit apartment building. Each apartment in the building has a kitchen sink, a flush tank water closet, a lavatory, and a bathtub. The water heater and a laundry room with one laundry tray and two automatic clothes washers are located in the basement of the apartment building.

For the purpose of this example, the following conditions will be assumed:

1. The water pressure at the meter is 60 psi.
2. The developed length of the cold water piping from the water meter to the most remote cold water outlet is 60 feet.
3. There is an elevation (head) difference of 12 feet between the water meter and the highest water supply outlet.

To size the water pipe in this apartment building according to the sizing method previously described, the following steps are followed.

STEP 1A. The cold water supply fixture unit (cwsfu) demand for the entire building is calculated as follows from the "Private Use" column of Table 9-2. (The numbers in the squares at each fixture and/or section of pipe indicate the water supply fixture unit demand at that location.)

Each apartment has:

1 kitchen sink	2 cwsfu
1 water closet	3 cwsfu
1 lavatory	1 cwsfu
1 bathtub	2 cwsfu
Total cwsfu each apartment	8 cwsfu
×	4 apts.
	32 cwsfu

plus: 2 sill cocks @ 3 cwsfu each	6 cwsfu
1 laundry tray	2 cwsfu
2 automatic clothes washers @ 2 cwsfu/each	4 cwsfu
Total cwsfu demand of building	44 cwsfu

STEP 1C. The hot water supply fixture unit demand for the entire building is calculated as follows from the "Private Use" column of Table 9-2.

Each apartment has:

1 kitchen sink	2 hwsfu
1 lavatory	1 hwsfu
1 bathtub	2 hwsfu
Total hwsfu each apartment	5 hwsfu
×	4 apts.
	20 hwsfu

Plus: 1 laundry tray	2 hwsfu
2 automatic clothes washers @ 2 hwsfu each	4 hwsfu
Total hwsfu demand of building	26 hwsfu

This demand of 26 hwsfu is assigned to the water heater in Figures 9-4 and 9-5.

STEP 2. The developed length is given as 60 feet.

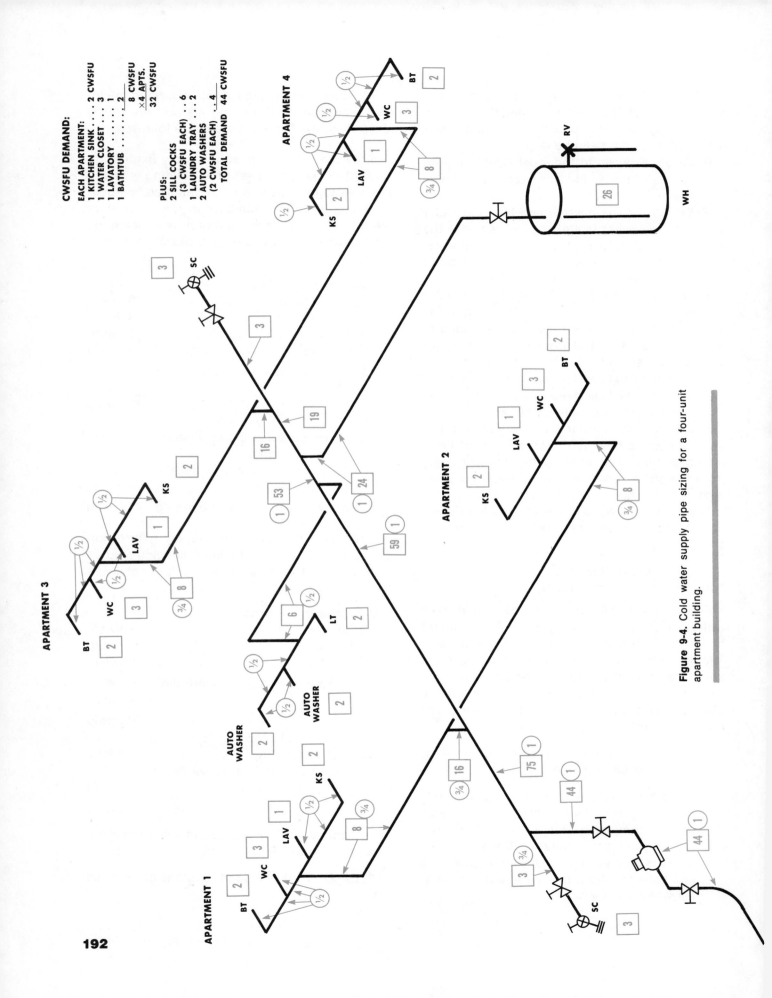

Figure 9-4. Cold water supply pipe sizing for a four-unit apartment building.

CWSFU DEMAND:

EACH APARTMENT:
1 KITCHEN SINK . . . . . 2 CWSFU
1 WATER CLOSET 3
1 LAVATORY 1
1 BATHTUB 2
                          8 CWSFU
                        ×4 APTS.
                        32 CWSFU

PLUS:
2 SILL COCKS
  (3 CWSFU EACH) . . 6
1 LAUNDRY TRAY . . . 2
2 AUTO WASHERS
  (2 CWSFU EACH) . . 4
TOTAL DEMAND 44 CWSFU

192

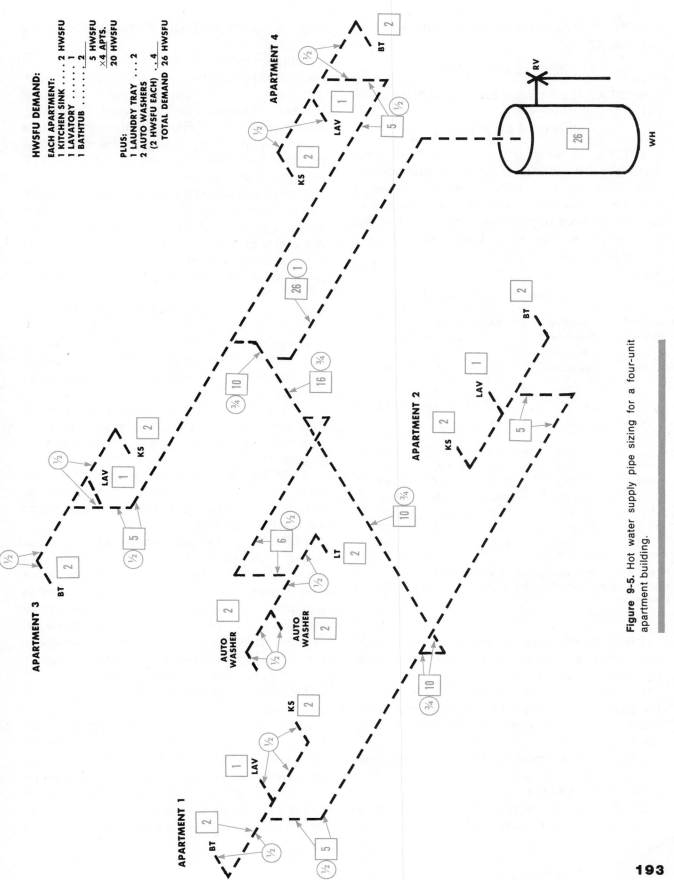

Figure 9-5. Hot water supply pipe sizing for a four-unit apartment building.

HWSFU DEMAND:

EACH APARTMENT:
1 KITCHEN SINK .... 2 HWSFU
1 LAVATORY 1
1 BATHTUB 2
        5 HWSFU
       ×4 APTS.
     20 HWSFU

PLUS:
1 LAUNDRY TRAY 2
2 AUTO WASHERS
(2 HWSFU EACH) ... 4
  TOTAL DEMAND 26 HWSFU

STEP 3. The elevation (head difference is given as 12 feet:

$$12 \times .5 \text{ psi} = 6 \text{ psi.}$$

STEP 4. The available pressure is given as 60 psi:

$$60 \text{ psi} - 6 \text{ psi} = 54 \text{ psi.}$$

STEP 5. A building with 54 psi of working water pressure is sized from the second portion of Table 9-7—the 46 to 60 psi pressure range.

STEP 6. The water pipe for this entire building will be sized from the 60 foot column of the 46 to 60 psi pressure range portion of Table 9-7, which is reproduced in Figure 9-6.

PRESSURE RANGE—46 TO 60 psi

Meter and Water Service (in inches)	Building Supply and Branches (in inches)	MAXIMUM ALLOWABLE LENGTH IN FEET		
		40	60	80
$3/4$	$1/2$	9	8	7
$3/4$	$3/4$	27	23	19
$3/4$	1	44	40	36
1	1	60	47	41
1	$1^1/4$	102	87	76
$1^1/2$	$1^1/4$	168	130	106
$1^1/2$	$1^1/2$	270	225	193
2	$1^1/2$	360	290	242
$1^1/2$	2	380	360	340
2	2	570	510	470
2	$2^1/2$	680	640	610

Figure 9-6. A reproduction of the portion of Table 9-7 that applies to the sizing of the water pipe illustrated in Figures 9-4 and 9-5.

STEP 7. The size of the water meter and water service is determined to be 1-inch pipe because the cold water supply fixture unit value calculated in Step 1A (44 cwsfu) falls between the 40 and 47 fixture unit value lines on the chart, and the higher 47 fixture unit value line must be used.

STEP 8. The size of the building supply pipe is also determined to be 1-inch pipe from this same line of the chart.

STEP 9. Since the fixtures in apartment 3 are most remote from the water meter, the sizing will begin in apartment 3 and proceed back toward the water meter.

Although there are only 8 cwsfu of demand on the pipe supplying the cold water to apartment 3, it is sized as $3/4$-inch pipe because it serves four fixtures. However, the individual fixtures in apartment 3 are all supplied with $1/2$-inch fixture branch piping. (The sizing of the cold water piping of the other three individual apartments is identical to that of apartment 3.)

At the point where the apartment 3 cold water supply joins the apartment 4 supply, the pipe serves a demand of 16 cwsfu. However, it does not increase in size, since a $3/4$-inch pipe will supply 23 cwsfu. (This sizing also applies to the point where the cold water supply for apartments 1 and 2 join.)

Since a $3/4$-inch pipe will serve 23 cwsfu, the cold water main continues as $3/4$-inch size past the point of supply to the sill cock at the rear of the building. This sill cock, however, must be supplied with a $3/4$-inch branch pipe, as sized from Table 9-6.

A $1/2$-inch pipe will adequately serve the 6 cwsfu demand of the laundry tray and two automatic clothes washers located in the basement laundry room.

STEP 10. The hot water supply fixture unit demand of the water heater is 26 hwsfu (as calculated in Step 2), which requires a 1-inch cold water supply.

STEP 11. At the point where the cold water supply to the water heater is taken from the cold water main pipe, the cold water supply fixture unit demand increases to 53 cwsfu. The table indicates that the cold water main pipe should increase to $1^1/4$-inch pipe at this point. However, this pipe does not need to increase in size above the originally selected 1-inch building supply pipe. This fact explains why the rest of the cold water main piping back to the water meter is all 1-inch size.

STEP 12. Referring to Figure 9-5 (which illustrates the sizing of the hot water piping in the four-unit apartment building of this example), the sizing begins in apartment 1, which is the most remote apartment from the water heater. A $1/2$-inch size branch pipe will adequately serve the three fixtures in this apartment that require a hot water supply (A ½-inch pipe serves the other three individual apartments' hot water needs as well.)

At the point where the hot water supply to apartments 1 and 2 joins the hot water main, the piping is increased to ¾-inch size. This same ¾-inch size pipe is also large enough to supply the combined 16 hwsfu demand of apartments 1 and 2 plus the basement laundry room.

The 6 hwsfu demand of the laundry room fixtures requires only a ½-inch size branch pipe.

The sizing of the hot water supply piping to apartments 3 and 4 is identical to that for apartments 1 and 2.

At the point where the hot water supply pipe for apartments 1, 2, and the laundry room joins the supply to apartments 3 and 4, the pipe is increased to 1-inch size. This 1-inch size pipe is continued back to the hot water outlet of the heater.

Sizing the Water Pipe in a Public Building

Figures 9-7 and 9-8 illustrate the cold and hot water supply piping of a one-story public building. In this building, water must be supplied to three toilet rooms, a service sink located in a janitor's closet, and a kitchen sink located in an employees' lunch area, as well as a water heater and a sill cock.

For the purpose of this example the following conditions will be assumed:

1. The water closets all use flushometer valves.

2. The two urinals in the men's toilet room are the stall-type with a demand of 5 wsfu each.

3. The developed length of the water supply piping is 200 feet from the street water main to the most remote fixture (the water closet at the far right in the women's toilet room).

4. The elvation difference is insignificant because this is a one-story building.

5. The available pressure at the street water main is 50 psi.

To size the water supply piping in this public building by the method of this chapter, the following steps would be taken:

STEP 1B. The cold water supply fixture unit (cwsfu) demand is calculated as follows from the "Public Use" column of Table 9-2:

Women's toilet room:

first water closet	40 cwsfu
second water closet	30 cwsfu
third water closet	20 cwsfu
fourth water closet	15 cwsfu
fifth water closet	10 cwsfu
sixth water closet	10 cwsfu
4 lavatories @2 cwsfu each . . .	8 cwsfu
Total cwsfu demand women's toilet room	133 cwsfu

Men's toilet room:

first water closet	40 cwsfu
second water closet	30 cwsfu
third water closet	20 cwsfu
fourth water closet	15 cwsfu
2 urinals @5 cwsfu each	10 cwsfu
2 lavatories @2 cwsfu each	4 cwsfu
Total cwsfu demand men's toilet room	119 cwsfu

Small toilet room at the front of the building: first water closet . .	40 cwsfu
lavatory	2 cwsfu
Total cwsfu demand small toilet room	42 cwsfu

Women's toilet room total	133 cwsfu
Men's toilet room total	119 cwsfu
Small toilet room total	42 cwsfu
Total cwsfu demand of all toilet rooms	294 cwsfu

plus: 1 service sink	4 cwsfu
1 kitchen sink	4 cwsfu
1 sill cock	5 cwsfu
Total cwsfu demand of building	307 cwsfu

STEP 1C. The hot water supply fixture unit (hwsfu) demand is calculated as follows from the "Public Use" column of Table 9-2:

7 lavatories @ 2 hwsfu each	14 hwsfu
1 kitchen sink	4 hwsfu
1 service sink	4 hwsfu
Total hwsfu demand	22 hwsfu

STEP 2. The developed length is given as 200 feet from the water main to the most remote outlet.

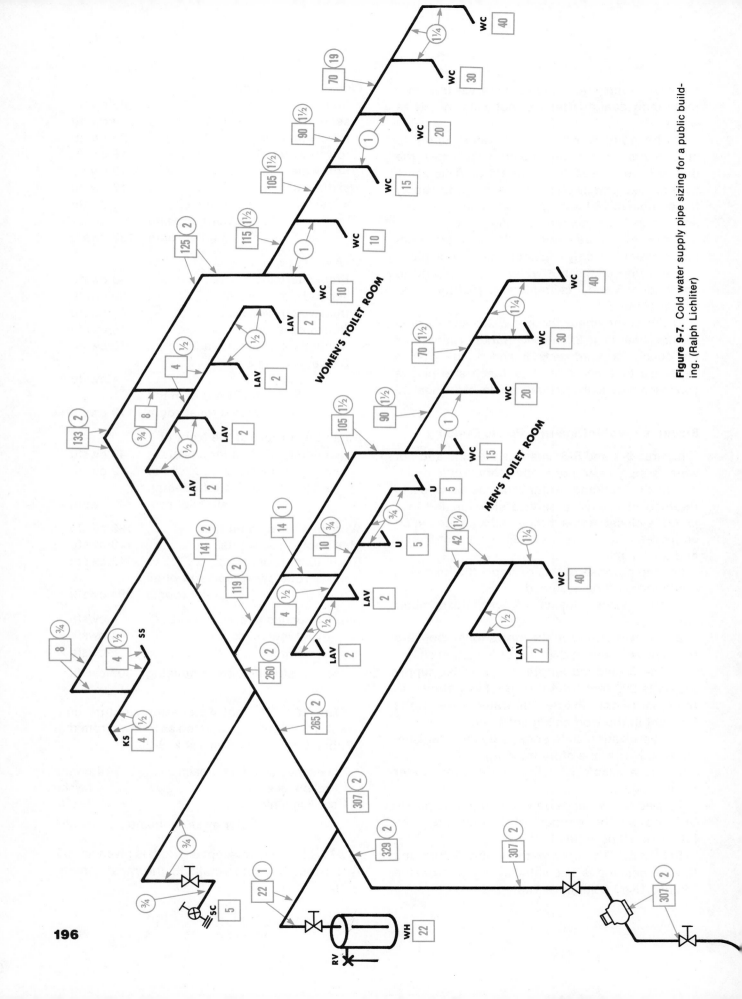

Figure 9-7. Cold water supply pipe sizing for a public building. (Ralph Lichliter)

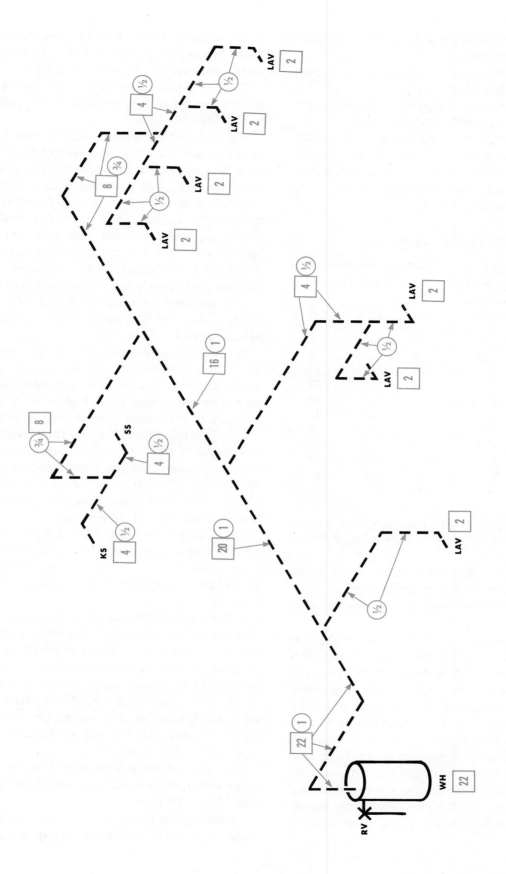

Figure 9-8. Hot water supply piping for a public building.

STEP 3. The elevation difference is insignificant for this example.

STEP 4. The pressure of the street water main for this example is 50 psi.

STEP 5. The pressure range table to be used is the 46 to 60 psi portion of Table 9-7.

STEP 6. The water piping for this entire building will be sized from the 200-foot length column of the 46 to 60 psi pressure range portion of Table 9-7. This portion of Table 9-7 is reproduced in Figure 9-9.

PRESSURE RANGE—46 TO 60 psi

Meter and Water Service (in inches)	Building Supply and Branches (in inches)	MAXIMUM ALLOWABLE LENGTH IN FEET		
		150	200	250
3/4	1/2	5	4	3
3/4	3/4	14	11	9
3/4	1	28	23	21
1	1	30	25	23
1	1 1/4	52	44	39
1 1/2	1 1/4	66	52	44
1 1/2	1 1/2	128	105	90
2	1 1/2	150	117	98
1 1/2	2	272	240	220
2	2	368	318	280
2	2 1/2	535	500	470

Figure 9-9. A reproduction of the portion of Table 9-7 that applies to the sizing of the water pipe illustrated in Figures 9-7 and 9-8.

STEP 7. The water service and water meter for this public building are 2-inch size.

STEP 8. The building supply pipe is also 2-inch size.

STEP 9. The cold water pipe sizing starts at the most remote cold water fixture outlet from the water meter, which is the water closet on the far right in the women's toilet room. This water closet, which is rated at a demand of 40 cwsfu, must be supplied with a 1 1/4-inch fixture branch pipe. The second water closet in this group, which is rated at a demand of 30 cwsfu, is also supplied with 1 1/4-inch fixture branch pipe. (Although the fixture branch pipe serving these two water closets is 1 1/4-inch size, the fixture supply pipe to the flushometer valves would be sized as 1-inch pipe.)

The section of pipe that supplies the water to these two water closets serves a demand of 70 cwsfu and must be 1 1/2-inch size pipe.

The third, fourth, and fifth water closets in this toilet room are all supplied with 1-inch individual fixture branch pipes.

The sixth water closet is also supplied with a 1-inch fixture branch pipe even though it is rated at a demand of 10 cwsfu and the sizing table indicates a 3/4-inch pipe would supply this demand. The reason this pipe must be sized 1-inch pipe is that in Step 1B it was stated that, "The piping supply for a flushometer valve may never be smaller in size than the size of the valve inlet." Table 9-6 indicates that a water closet flushometer valve has a 1-inch pipe size inlet.

The sections of water pipe that supply the third, fourth, and fifth water closets in this group are all sized 1 1/2-inch pipe.

At the point where the sixth water closet supply is taken off the cold water supply pipe, the cwsfu demand increases to 125 cwsfu and the pipe size increases to 2 inches in size. This 2-inch size continues back to the water meter.

The 8 cwsfu demand of the four lavatories in the women's toilet room is supplied with a 3/4-inch pipe taken from the 2-inch cold water pipe. At the point where this 3/4-inch cold water supply pipe divides to serve two lavatories on each side, it reduces to 1/2-inch size pipe.

The 8 cwsfu demand of the kitchen sink and service sink is supplied with a 3/4-inch pipe. This pipe reduces to 1/2-inch size to provide the individual fixture branches to each of these two fixtures.

In the men's toilet room, the sizing again starts at the most remote outlet. The first water closet, which is rated at a demand of 40 cwsfu, is supplied with a 1 1/4-inch fixture branch pipe. The second water closet, which is rated at a demand of 30 cwsfu, is also supplied with the 1 1/4-inch pipe. (Although, as was stated previously, the fixture supply pipe to the flushometer valves would be 1-inch pipe.)

The section of piping supplying these two water closets serves a demand of 70 cwsfu and is sized 1 1/2-inch pipe.

The third and fourth water closets in the men's toilet room, with their individual demands of 20 and 15 cwsfu respectively, may each be supplied with 1-inch size fixture branch pipes.

The cold water supply pipe to the four water closets in this toilet room is sized 1½-inch pipe, since the total demand of these closets is 105 cwsfu (which is the maximum demand that can be supplied with a 1½-inch pipe in this column of Table 9-7).

The two urinals in the men's toilet room are each supplied with ¾-inch individual fixture branch pipes. The section of pipe that serves both urinals is also sized ¾-inch pipe, since it serves a demand of 10 cwsfu.

The two lavatories in the men's toilet room with a combined demand of 4 cwsfu are supplied with a ½-inch pipe.

At the point where the ½-inch cold water supply pipe for the lavatories joins the ¾-inch urinal supply pipe, the demand increases to 14 cwsfu and the pipe must be increased to 1-inch size.

At the point where this 1-inch pipe serving the lavatories and urinals joins the pipe serving the four water closets in this toilet room, the demand increases from 105 cwsfu to 119 cwsfu, and the cold water supply pipe must be increased to 2-inch size.

The sill cock is served with a ¾-inch branch pipe.

In the small toilet room at the front of the building, the sizing again starts at the most remote water supply outlet, which is the water closet. This individual water closet is rated at a demand of 40 cwsfu and requires a 1¼-inch size supply pipe.

The lavatory in this toilet room is served with a ½-inch fixture branch pipe taken off the 1¼-inch pipe that supplies the cold water supply to the water closet. The 2 cwsfu demand of the lavatory plus the 40 cwsfu demand of the water closet increases the demand on the cold water supply pipe to this toilet room to 42 cwsfu, but the size of this pipe remains at 1¼-inch because a pipe of this size will supply 52 cwsfu.

STEP 10. The water heater must serve a demand of 22 hwsfu (as calculated in Step 1C), so it is supplied with a 1-inch cold water pipe.

STEP 11. At the point where the cold water supply to the water heater joins the cold water main pipe, the demand on the main increases to 329 cwsfu. However, it is not necessary to increase the size of the cold water main above the 2-inch size originally selected, even though the table indicates a larger size.

STEP 12. Figure 9-8 illustrates the sizing of the hot water supply piping for this public building. The sizing of this piping begins at the lavatories in the women's toilet room, since they are the most remote fixtures from the water heater. These four lavatories are all served by ½-inch individual fixture branch pipes. The sections of the hot water piping that supply two lavatories (or 4 hwsfu) in this toilet room are also piped with ½-inch pipe. At the point where the hot water supplies for all four lavatories join together, the hot water supply pipe is increased to ¾-inch size to serve the 8 hwsfu demand.

A ¾-inch size hot water pipe is necessary to serve the 8 hwsfu demand of the kitchen sink and service sink. However, the individual branch piping to each of these fixtures is sized as ½-inch pipe.

At the point where the hot water supply to the women's toilet room and the pipe serving the kitchen sink and service sink join, the demand increases to 16 hwsfu and the pipe increases to 1-inch size. This size continues back to the water heater.

The two lavatories in the men's toilet room are supplied with a ½-inch size pipe taken from the 1-inch hot water main.

The single lavatory in the small toilet room at the front of the building is also supplied with a ½-inch branch pipe.

⊙ REVIEW QUESTIONS

1. Define the term *flow rate* as applied to water supply in a building.

2. How is the *total demand* determined for a building water supply system?

3. Define the term *available pressure*.

4. Why is the water supply fixture unit (wsfu) system used in determining pipe sizes in a building? Where is the water supply fixture unit table found?

5. What does *pressure loss by friction* have to do with the sizing of pipe in a building water supply system?

6. In figuring pressure loss in the water system, how is the friction loss for valves and fittings calculated?

7. What is the source of Table 9-6, "Maximum Sizes of Fixture Water Branch Pipe"?

8. From what point in the building water supply system do you start when you determine pipe sizing? Why?

9. Make a drawing of the water supply piping in your home or apartment building and size it.

10. Make a drawing of the water supply piping in the men's and women's toilet rooms in your school and size it.

10

Plumbing Fixtures and Appliances

The majority of the piping installed by plumbers serves one of two uses: it either supplies water to a plumbing fixture or drains the waste water from a plumbing fixture after the fixture has been used.

A *plumbing fixture* is a receptacle for wastes, which are ultimately discharged into the sanitary drainage system.

A *plumbing appliance* is a special class of plumbing fixture intended to perform a special function.

Plumbing fixtures are manufactured from durable, corrosion-resistant, nonabsorbent materials that have smooth, easily cleaned surfaces. The materials currently used in the manufacture of plumbing fixtures are vitreous china, enameled cast iron, enameled pressed steel, stainless steel, fiberglass, plastic, terrazzo, and cement.

Modern plumbing fixtures and appliances are not only functional and well constructed, but are also attractively designed to add to the decor of a room. Plumbing fixtures and appliances can be obtained in a large variety of colors to harmonize with the color scheme chosen by the building designer.

In the following sections of this chapter, a general description of some of the more common plumbing fixtures and appliances and their associated trim will be presented. (Plumbing fixture *trim* consists of the water supply and drainage fittings that are installed on the fixture to control the flow of water into the fixture and the flow of waste water from the fixture to the sanitary drainage system.)

WATER CLOSETS

A *water closet* is a water-flushed plumbing fixture designed to receive human wastes directly from the user. (The term is sometimes also used to indicate the room or compartment in which the fixture is located.) Water closets are rated at 6 drainage fixture units of waste discharge and require a 3-inch waste pipe and a 2-inch vent.

Water closets are usually made of vitreous china. They are installed either directly on the floor (in which case they are called floor-set) or suspended from the wall on supporting chair carriers (called wall-hung water closets). Water closet bowls are manufactured in two different patterns: *elongated bowl* and *regular round front bowl* (which is also sometimes called the *plain bowl*). Elongated water closet bowls are generally 2 to 2³/₄ inches longer than round front bowls to allow the user of the fixture more room. In general, elongated water closet bowls are installed in all plumbing installations except those in residential or dwelling-type buildings.

Operating Principle of a Water Closet Bowl

The quality of a water closet bowl is judged by the efficiency with which it eliminates or flushes the wastes deposited in it. The flushing action, which is that of a syphon, is produced by a decrease in atmospheric pressure on the outlet side of the trap. When a water closet is in neutral position (that is, not being used), it contains a seal of water, and the atmospheric pressure is the same on both sides of the trap.

The construction of a water closet trap, as shown in Figure 10-1A, appears to be the same as that of an ordinary trap, with the inlet and outlet passageways of the same diameter. However, this is not the case. The diameter of the outlet passageway is smaller than that of the inlet passageway. It is built with short offsets and turns so that the flow of water through it is retarded, causing a head of water to build up in the fixture bowl. When the closet is flushed, the water passing through the outlet passageway eliminates the air in the outlet passageway, producing a partial vacuum. Atmospheric pressure and the head of water retained on the inlet side of the trap force the collected organic solids from the fixture.

Syphon Jet Water Closet. Figure 10-1 illustrates the flushing action of a syphon jet water closet, and this action is described in the following paragraphs:*

*Courtesy of the Kohler Company.

Syphon Jet Water Closet Flushing Operation:

1. Figure 10-1A illustrates a typical front-jet syphon jet water closet.

Water area: the size of the surface of the water in the bowl of a closet. Usually expressed as width × length.

Depth of seal: the height of water in a closet retained after proper refill. Measured from the surface of the water (or dam) to the top of the inlet of the passageway.

Passageway: the channel that connects the bowl to the outlet. In effect, it is where the syphonic action is created. Because of the configuration of the passageway (not necessarily round), the passageway is measured by the largest diameter ball it will pass.

Jet: the orifice that directs water into the upleg of the passageway to help create the syphonic flushing action.

2. Water starts to enter the closet (Figure 10-1B). Water enters the closet by way of the rim holes and jet. It rises in the bowl and flows over the dam, but no syphon or flushing action has started.

3. More water enters the passageway (Figure 10-1C). As soon as the water is accelerated, it will be directed over the dam to the other side of the passageway, and then redirected to the opposite side where it will create a curtain through the passageway. This curtain prevents air from entering the passageway.

4. The passageway water level begins to rise (Figure 10-1D). As the water in the down leg of the passageway rises, air from the passageway mixes with the water leaving the closet.

5. The passageway fills (Figure 10-1E). When the passageway is completely filled with water, a good flush or syphon action is created.

6. The syphon is broken (Figure 10-1F). As soon as the level of the water in the bowl drops to the level where air is again introduced into the passageway, the syphon is broken and the flush stops.

7. Possibility of sewer gas entering the room (Figure 10-1G). If the seal is not restored with refill water, sewer gas will enter the room.

In addition to the syphon jet water closet, plumbing fixture manufacturers offer water closets with reverse trap, washdown, and blowout flushing actions.

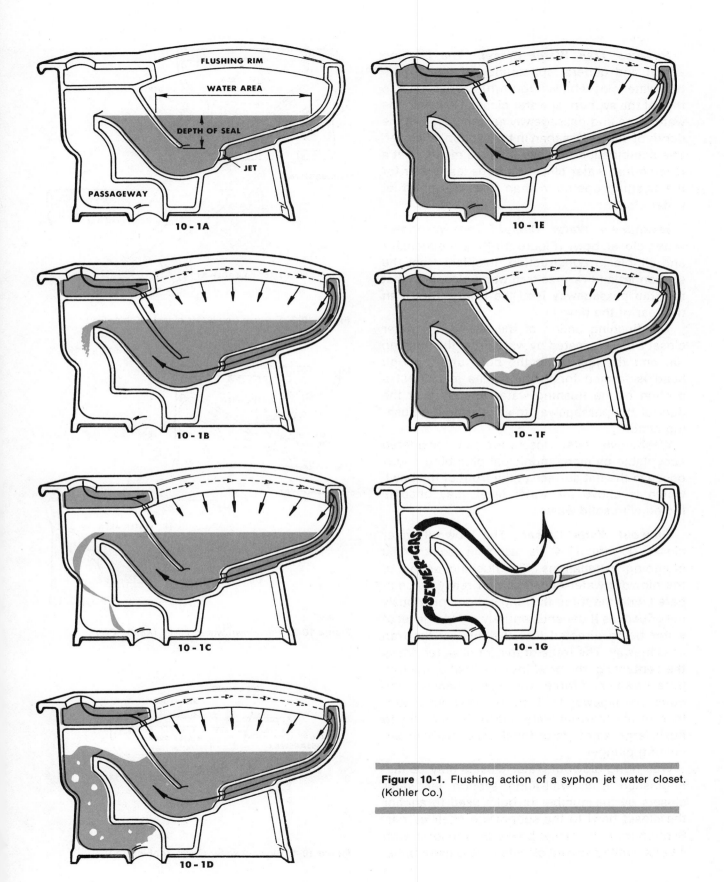

Figure 10-1. Flushing action of a syphon jet water closet. (Kohler Co.)

Reverse Trap Water Closet. The flushing action and general appearance of the reverse trap water closet bowl (Figure 10-2) are similar to that of the syphon jet water closet. However, the water area and passageway are smaller, and the depth of seal is less than in the syphon jet closet. The combination of these features means that a reverse trap water closet requires less water for the flushing operation than does a syphon jet water closet.

Washdown Water Closet. The washdown water closet bowl (Figure 10-3) is inexpensive and simple in construction. It differs from the syphon jet and reverse trap water closets in that the trap passageway is at the front rather than the rear of the bowl.

The flushing action of the washdown water closet bowl is created by water from the flushing rim and the jets filling the bowl until enough head is formed for flushing. The jets direct a portion of the flushing water directly over the dam of the passageway to accelerate the flushing action.

Washdown water closets are not considered acceptable by many municipal plumbing codes because the flat surface at the front of the bowl, which is above the water level, may become fouled with solid wastes.

Blowout Water Closet. The blowout water closet (Figure 10-4) is used in commercial plumbing installations. The flushing action of the blowout water closet cannot really be compared with the three flushing actions previously described, as it depends entirely on a large jet of water being directed into the inlet of the trap passageway. The force of this jet of water draws the contents of the bowl into the inlet of the trap passageway and forces (or blows) them into the outlet passageway to flush the bowl. An advantage of the blowout water closet is its ability to flush large amounts of toilet paper without becoming plugged.

Wall-hung blowout water closets may be distinguished from wall-hung syphon jet water closets by the number of bolts used to anchor the closet bowl to the supporting chair carrier. Syphon jet water closet bowls are anchored with 4 bolts, while blowout closets always use 3 bolts.

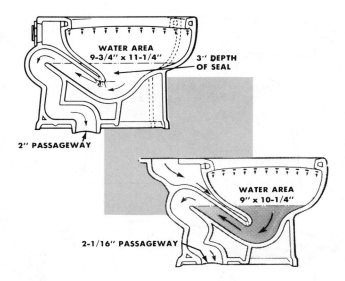

Figure 10-2. Reverse trap water closet. (Kohler Co.)

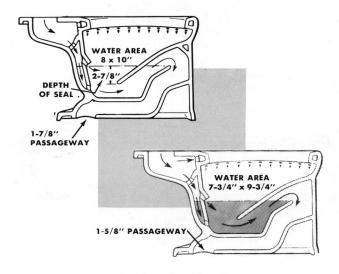

Figure 10-3. Washdown water closet. (Kohler Co.)

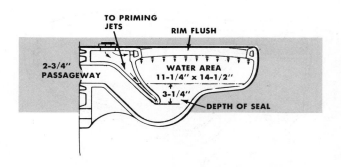

Figure 10-4. Blowout water closet. (Kohler Co.)

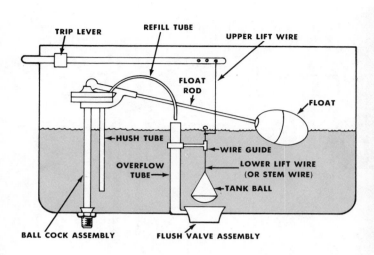

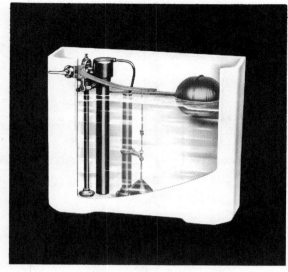

Figure 10-5. Component parts of a water closet flush tank. (Kohler Co.)

Flush Devices for Water Closets

The water to flush a water closet is supplied by either a flush tank (Figure 10-5) or a flushometer valve (Figures 10-6 and 10-7).

Flush tanks are commonly used on water clos-

ets in residential construction. In addition to being less expensive than flushometer valves, they have the advantages of quieter operation, and they require less water pressure and a smaller volume of water. (A water closet flushometer

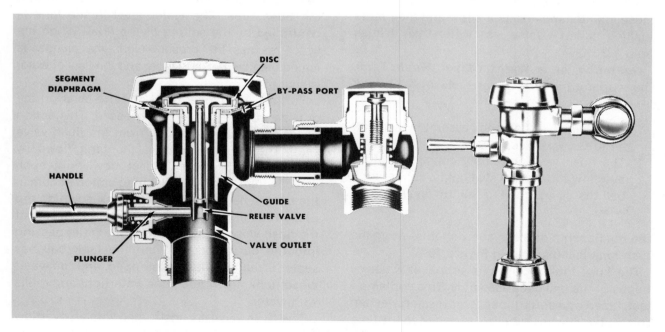

Figure 10-6. Diaphragm-type flushometer valve. (Sloan Valve Co.)

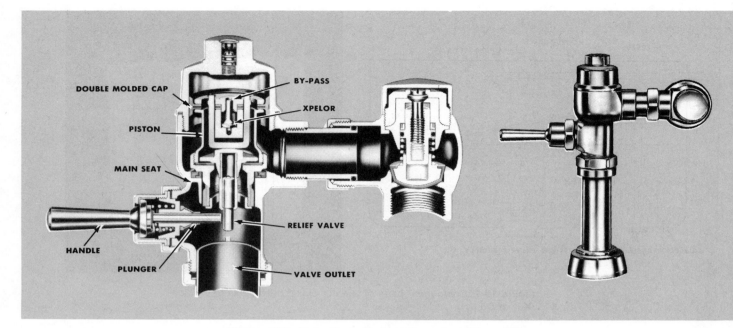

Figure 10-7. Piston-type flushometer valve. (Sloan Valve Co.)

valve requires a minimum of 1-inch size water supply, while a flush tank may be supplied with a ½-inch pipe, which lowers the cost of installation.)

On the other hand, *flushometer valves* are used on commercial plumbing installations because maintenance costs are low and their rapid flushing action wastes less water than a flush tank.

Operation of a Water Closet Flush Tank. The working parts of a water closet flush tank consist of:

1. A float valve or ball cock assembly connected to the water supply.
2. The flush valve assembly consisting of an overflow tube, valve seat, and a rubber tank ball that is attached to the trip lever with lift wires.

The relationship of these components within the flush tank is illustrated in Figure 10-5.

The flush tank receives its supply of water by means of the ball cock assembly. This contains a seat, faced by a soft rubber compression washer. The rubber washer, which is fastened to the float valve plunger, moves up and down over the

brass seat. The plunger is connected to the brass float rod, which has the float attached to the end.

Pressure on the trip lever starts the tank in operation. The float drops in the tank as the water recedes. This lifts the plunger from the seat, and water enters the closet tank through the hush tube. The water level in the tank is controlled by the setting of the float. When the water reaches the proper height, the plunger is forced onto the ball cock seat and the flow of water is shut off.

To flush the closet bowl, the trip lever on the outside of the flush tank is pressed. This action raises the rubber tank ball from the flush valve seat. The water stored in the tank flows by gravity into the water closet bowl with such velocity that a syphonic flushing action occurs in the bowl. The tank ball floats back down on top of the receding water. It is drawn onto the seat of the flush valve by the action of the water passing into the closet bowl. When the tank ball has seated itself, the flush is stopped, and the water closet tank again becomes watertight and refills with water.

The flush valve is constructed with an overflow tube, which is by-passed into the closet bowl below the seat of the flush valve. The overflow allows any leakage of the ball cock to enter the closet bowl rather than overflow the tank.

The refill tube is piped into the overflow tube. Since water closets are flushed by a syphonic action that draws practically all of the water seal from the trap, it is the function of the refill tube to replace this water in the closet bowl and so prevent the passage of sewer gas through the water closet trap. The flow of water through the refill tube stops when the float reaches the proper height to shut off the flow of water through the ball cock assembly.

Although nearly every plumbing fixture manufacturing company has its own peculiar design of internal component parts within a water closet flush tank, the operating principles of all flush tanks, as just described, are the same.

Types of Water Closet Flush Tanks. Water closet flush tanks are manufactured either as a close-coupled tank and bowl or as a one-piece water closet (see Figure 10-8). On a *close-coupled water closet,* the tank and bowl are separate, but the flush tank is bolted directly to the water closet bowl with a rubber gasket.

On a *one-piece water closet,* the flush tank and closet bowl are a single piece of vitreous china. Figure 10-8 illustrates a one-piece, low-silhouette water closet tank and bowl combination.

Flushometer Valves. A *flushometer valve* is a device that discharges a predetermined quantity of water to fixtures for flushing purposes and is actuated by direct water pressure. It is a very efficient device for flushing water closets, urinals, and some types of hospital fixtures. A flushometer valve is commonly installed in place of a flush tank for commercial plumbing installations because of its rapid flushing action and ease of service.

A flushometer valve can be operated at 6- to 10-second intervals. The efficiency of the valve is due to its small number of internal working parts. The principle involved in its operation is a simple one—equalization of pressure on both sides of a relief valve contained within the body of the valve proper.

Two types of flushometer valves are in use today: the *diaphragm* (Figure 10-6) and the *pis-*

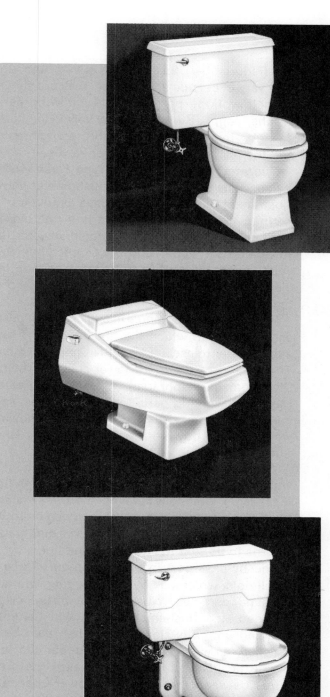

TOP — FLOOR SET CLOSET WITH CLOSE-COUPLED TANK AND BOWL.

MIDDLE — ONE-PIECE FLOOR SET CLOSET TANK AND BOWL COMBINATION.

BOTTOM — WALL-HUNG SIPHON JET WATER CLOSET WITH A CLOSE-COUPLED TANK AND BOWL.

Figure 10-8. Water closets used in residential construction. (Kohler Co.)

ton (Figure 10-7) flushometer valves. Although they vary in outside appearance, they are similar in their working principle. The following paragraphs, supplied by the Sloan Valve Company, describe the operation of both the diaphragm- and the piston-type flushometer valves.

Operation of a Diaphragm-Type Flushometer Valve. The diaphragm-type flushometer valve, illustrated in Figure 10-6, operates as follows:*

When the flushometer valve is in the closed position, the *segment diaphragm* divides the valve into an upper and lower chamber with equal water pressures on both sides of the diaphragm. The greater pressure area on top of the diaphragm holds it closed on its seat.

Movement of the *handle* in any direction pushes the *plunger,* which tilts the *relief valve* and allows water to escape from the upper chamber. The water pressure in the lower chamber (below the segment diaphragm), now being greater, raises the working parts (relief valve, disc, segment diaphragm, and guide) as a unit, allowing water to flow down through the *valve outlet* to flush the fixture.

While the valve is operating, a small amount of water flows through the *by-pass port* in the diaphragm, gradually refilling the upper chamber and equalizing the pressures once more. As the upper chamber gradually fills, the diaphragm returns to its seat to close the valve.

The by-pass port in the diaphragm is a very small hole, which can become plugged with dirt or sand particles. When this by-pass port is plugged, the flushometer valve will continue flushing until it is cleared, for when the port is blocked, no water can enter the upper chamber of the flushometer valve to equalize the pressures between the upper and lower chambers and stop the flushing action. For this reason, if dirty or sandy water is a problem in the building in which flushometer valves are to be installed, piston-type flushometer valves (which are described next) are preferred to diaphragm-type flushometer valves.

Operation of a Piston-Type Flushometer Valve. The piston-type flushometer valve, illustrated in Figure 10-7, operates as follows:*

*Courtesy of the Sloan Valve Company.

The *double-molded cap* divides the valve into an upper and lower chamber, with equal water pressures in both chambers when the valve is in the closed position. The greater pressure area on top of the *piston* holds the valve closed on the *main seat.*

A slight movement of the *handle* in any direction pushes the *plunger,* which tilts the *relief valve* and allows water to escape from the upper chamber. The water pressure in the lower chamber below the cup, which now is greater, raises the piston from the main seat, allowing water to flow down through the *valve outlet* to flush the fixture.

While the valve is operating, a small amount of water flows through the *by-pass* and chamber of the *xpelor,* gradually filling the upper chamber and equalizing pressures once more. As the upper chamber gradually fills, the piston returns to the seat to close the valve.

Water Closets Used in Residential Construction

There are many models of water closets presently manufactured for use in residential construction. Figure 10-8 pictures some of the more commonly used models, and the features of each type are described below.

The type most commonly used in residential construction is an inexpensive, *floor-set water closet* with a close-coupled flush tank and bowl combination. This closet has a syphon jet flushing action. Water closets of this type are also available in water-saving models. The flushing action of water-saving closets has been designed to ensure a positive flush with $3\frac{1}{2}$ gallons of water, compared with the $5\frac{1}{2}$ gallons required to flush other water closets.

A second type is the modern *one-piece water closet* flush tank and bowl combination. This floor-set closet features a low silhouette and an elongated pattern bowl. Since the water height in the flush tank is not adequate to give a proper rim flush, a connection is made from the ball cock assembly directly to the flushing rim. This closet requires a minimum of 35-40 pounds of water supply pressure for proper operation.

The third type shown in Figure 10-8 is a *wall-hung syphon jet water* closet with a close-coupled tank and bowl. The closet is anchored to the wall with a concealed supporting carrier. The off-the-floor design of this closet makes cleaning both the fixture and the floor much easier.

Water Closets Used in Commercial Construction

Figure 10-9 pictures some of the more common types of water closets used in commercial construction (that is, any construction that is nonresidential). The features of each type of water closet are described below.

One type of *wall-hung syphon jet water closet* has an elongated bowl. This closet uses a flushometer valve to supply the water for flushing.

Another *wall-hung syphon jet water closet* has a *concealed flushometer valve.* The flushometer is concealed behind the wall in the pipe space to prevent vandalism and/or tampering with the valve.

One type of *wall-hung blowout water closet* has an exposed flushometer. Blowout water closets are also available for use with a concealed flushometer valve.

The figure also shows a *floor-set syphon jet water closet* with an elongated bowl, which uses a flushometer valve. Closets of this type are manufactured in a variety of different bowl heights for different uses. (Bowl height is the height of the top of the flush rim of the closet bowl above the floor.) The standard bowl height is 15 inches, while 18 inches is the bowl height used for handicapped and elderly persons, and either a 10- or 13-inch bowl height is used for closets designed for small children (for example, in schools).

Another type of *floor-set syphon jet water closet* has a close-coupled flush tank. This closet also has an elongated bowl. Closets of this type are manufactured with a 15-inch bowl height for normal commercial use and with an 18-inch bowl height for use by handicapped persons.

Water Closet Seats

A water closet seat is the device installed around the rim of a water closet to support the person using the closet. For the comfort of the user and for cleanliness, water closet seats are manufactured from smooth, nonabsorbent materials. At the present time, most water closet seats are manufactured from solid plastic or pressed wood fibers with either a sprayed or a dipped-paint finish. The pressed wood seat is less expensive than solid plastic. Water closet seats are available in black, white, and a variety of colors to harmonize with or accent the bathroom plumbing fixtures.

Water closet seats (Figure 10-10) are manufactured with a closed front seat and cover for residential use, and with open fronts and no cover for public use.

10-9A. WALL HUNG SIPHON JET WATER CLOSET

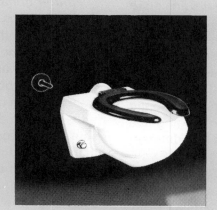

10-9B. WALL HUNG SIPHON JET WATER CLOSET WITH CONCEALED FLUSHOMETER VALVE

10-9C. FLOOR SET SIPHON JET WATER CLOSET

Figure 10-9. Water closets used in commercial construction. (Kohler Co.)

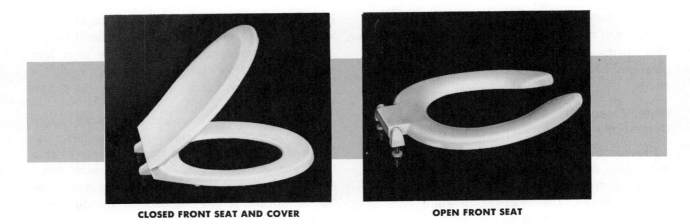

CLOSED FRONT SEAT AND COVER   OPEN FRONT SEAT

Figure 10-10. Water closet seats. (Kohler Co.)

⭕ URINALS

A urinal is a water-flushed plumbing fixture designed to receive urine directly. Urinals are manufactured from vitreous china or enameled cast iron. Urinals, like water closets, are available with a variety of different flushing actions: washout, syphon jet, and blowout.

Washout Urinals. In a urinal with a washout flushing action, the flushing water enters the top of the urinal, flows out into the fixture through openings in the top rim of the urinal, spreads across the back of the urinal, and flows by gravity out through the urinal trap into the sanitary drainage system. The trap of a washout urinal is not actually flushed, but rather the waste is washed out of it. Washout urinals are characterized by restricted openings over the inlet of the trap at the bottom of the urinal. This restricted opening may be either a beehive strainer (see Figure 10-11) or small openings cast in the china. The purpose of these small openings is to prevent any debris that is dropped or thrown into the urinal from plugging the fixture trap. Figure 10-11 illustrates four different styles of washout urinals.

A *stall urinal* is set into the floor of the toilet room. The beehive strainer waste outlet is connected with a caulked joint to a separate P-trap, which is installed below the floor level. Stall urinals are rated at 3 drainage fixture units of waste discharge and require a 2-inch P-trap and a 1 1/4-inch vent pipe.

A *wall-hung washout urinal* has an integral (built-in) 2-inch P-trap. The outlet openings to this trap consist of several small holes cast into the china at the bottom of the urinal. Wall-hung washout urinals with an integral trap are rated at 4 drainage fixture units of waste discharge and require a 2-inch waste pipe and a 1 1/4-inch vent pipe.

Another type is the *bottom outlet, wall-hung washout urinal.* The drain opening in this urinal is a beehive strainer, which is attached to a separate, exposed 1 1/2-inch trap. Bottom outlet, wall-hung washout urinals are rated at 2 drainage fixture units of waste discharge and require a 1 1/4-inch vent pipe.

Trough urinals are available in 3-, 4-, 5-, and 6-foot lengths. The flushing water enters a trough urinal through the perforated flush pipe attached to the back edge of the fixture (Figure 10-11). The waste flows into a beehive strainer and out the bottom to a separate 1 1/2-inch P-trap. A trough urinal is rated at 2 drainage fixture units of waste discharge and requires a 1 1/4-inch vent. Trough urinals are the only style of urinal manufactured in enameled cast iron.

Syphon Jet Urinals. The flushing action of a syphon jet urinal flushes the fixture trap clear of wastes in the same manner as the flushing action of a syphon jet water closet. These urinals have a large opening over the trap inlet in the bottom of the urinal. Several syphon jet urinals are pictured in Figure 10-12.

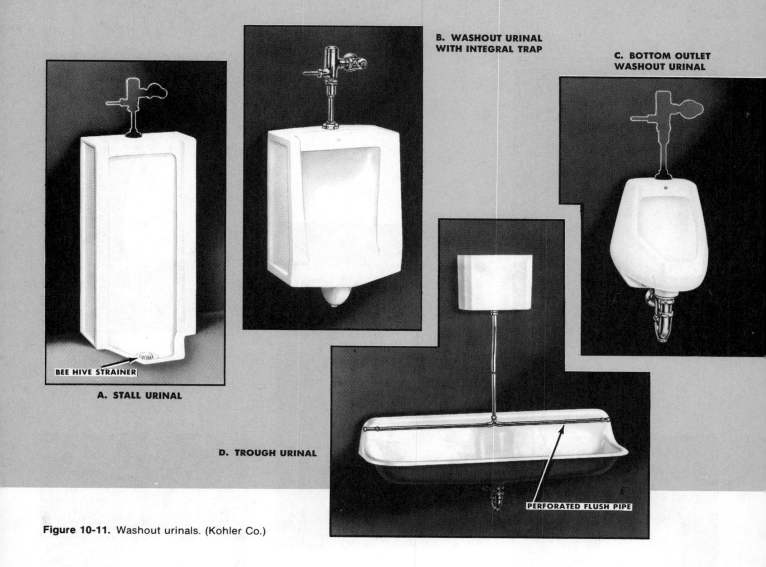

B. WASHOUT URINAL
WITH INTEGRAL TRAP

C. BOTTOM OUTLET
WASHOUT URINAL

BEE HIVE STRAINER

A. STALL URINAL

D. TROUGH URINAL

PERFORATED FLUSH PIPE

Figure 10-11. Washout urinals. (Kohler Co.)

Two *wall-hung syphon jet urinals* with a 2-inch waste outlet are shown. These urinals, which have an integral 2-inch trap, are rated at 4 drainage fixture units of waste discharge and require a 1¼-inch vent pipe.

A *pedestal syphon jet urinal* is set on the floor and has an integral trap. Pedestal urinals are rated at 6 drainage fixture units of waste discharge and require a 3-inch waste pipe and a 2-inch vent pipe.

Women's urinals are differently designed so that they can be straddled for use. One type of women's urinal is floor set, while another is wall-hung from a supporting chair carrier. Both of these urinals have integral traps, are rated at 6 drainage fixture units of waste discharge, and require a 3-inch waste pipe and a 2-inch vent pipe.

Blowout Urinals. A blowout urinal is flushed with a large jet of water that actually blows the water out of the fixture trap in the same manner as the flushing action of a blowout water closet. Blowout urinals have a large opening over the trap inlet at the bottom of the fixture. Since the flushing action of blowout urinals is very noisy, they are used only in installations where this noise will not create a problem.

Two different styles of blowout urinals are pictured in Figure 10-13. Both urinals have integral P-traps. One style is rated at 4 drainage fixture units of waste discharge and requires a 2-inch waste and 1¼-inch vent. The other is available with either a 2-inch or a 3-inch waste outlet. If the urinal has a 2-inch waste, it is rated at 4 drainage fixture units of waste discharge and requires a 2-inch waste pipe and a 1¼-inch

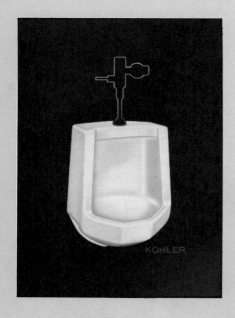

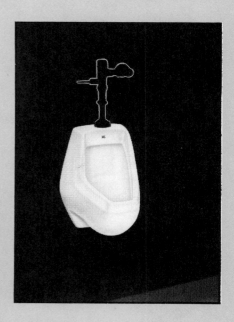

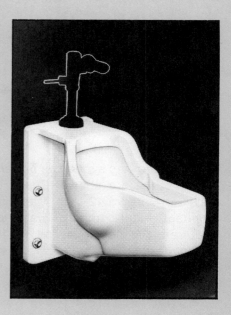

TOP: WALL-HUNG, SIPHON JET URINALS WITH A 2-INCH WASTE.

MIDDLE: FLOOR-SET, SIPHON JET URINAL WITH A 3-INCH WASTE.

BOTTOM: WOMENS URINALS; LEFT, FLOOR-SET; RIGHT, WALL-HUNG. BOTH REQUIRE A 3-INCH WASTE.

Figure 10-12. Syphon jet urinals. (Kohler Co.)

vent pipe. When supplied with a 3-inch waste outlet, this same urinal is rated at 6 drainage fixture units of waste discharge and requires a 3-inch waste pipe and a 2-inch vent pipe. Wall-hung blowout urinals with a 3-inch waste outlet usually are supported on concealed chair carriers.

Flush Devices for Urinals

A urinal must be equipped with an adequate flushing device to completely remove the urine from the fixture after it is used, to prevent the spread of disease, fouling of the urinal, and offensive urine odor. Urinals may be flushed with either a hand valve (Figure 10-14) or a flushometer valve (Figure 10-15).

The flushing of urinals in public buildings presents a problem. The user of a urinal in such buildings does not always take the time to flush the fixture after he uses it. For this reason, plumbers install flush devices that will automatically flush the urinal(s) at predetermined time intervals. Two types of automatic flushing devices installed by plumbers on urinals are the *automatic flushometer valve* (Figure 10-16) and the *urinal flush tank* (Figure 10-17). However, these automatic flush devices waste water because they flush the urinal at the predetermined time interval whether it has been used or not.

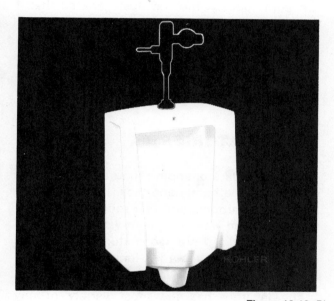

Figure 10-13. Blowout urinals. (Kohler Co.)

Figure 10-14. Urinal hand flush valve. (Chicago Faucet Co.)

Figure 10-15. Diaphragm-type flush-ometer valve for flushing a washout urinal. (Sloan Valve Co.)

Figure 10-16. Automatic flusho-meter valve. (Sloan Valve Co.)

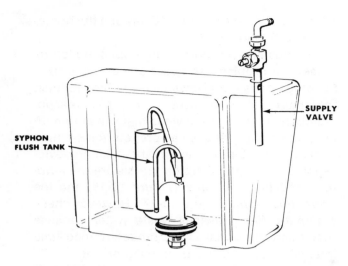

SUPPLY VALVE

SYPHON FLUSH TANK

Figure 10-17. Component parts of a urinal flush tank. (Kohler Co.)

A *hand valve* for flushing washout urinals is pictured in Figure 10-14. To flush a urinal with a valve of this type, the user of the fixture presses and holds the push button on the valve, allowing water to enter the fixture. When released, the valve slowly closes itself, shutting off the flow of water. Self-closing hand valves are also available with different styles of handles.

A *flushometer valve* for flushing urinals is pictured in Figure 10-15. Although the valve pictured is a diaphragm-type flushometer valve, piston-type flushometer valves (see Figure 10-7) also may be used to flush urinals.

Flushometer valves for urinals differ from those for water closets in two ways:

1. Since urinals normally require less water for flushing than water closets, the internal parts of urinal flushometer valves are adjusted to deliver a smaller volume of water.

2. The water outlet of a flushometer valve for a urinal (see Figure 10-15) is 3/4-inch size for washout urinals and 1 1/4- or 1 1/2-inch size for syphon jet and blowout urinals. (Water closet flushometer valves always use a 1 1/2-inch water outlet.)

An *automatic flushometer valve* is pictured in Figure 10-16. This flushometer valve has an electric motor operator that activates the flushing mechanism of the valve in place of the handle assembly (see Figure 10-15). This electric

motor operator is connected to an electric timer mechanism, which activates the operator to flush the urinal at predetermined time intervals. The timer can be adjusted so that the urinals are flushed only once each hour during periods of little or no use such as evenings and weekends.

A *urinal flush tank,* which is used to flush a washout urinal, is pictured in Figure 10-17. Since urinal flush tanks are mounted on the wall above the urinal, they are sometimes called *high tanks.* Flush tanks may be used to flush more than one washout urinal in the same toilet room. There are two components contained within a urinal flush tank—a supply valve and a syphon flush valve (see Figure 10-17).

The *supply valve* is adjusted to allow water to run continuously into the flush tank so that it flushes frequently enough to cleanse the urinal.

The *syphon flush valve* is designed so that as the water rises within the flush tank, a syphon is created that draws the water from the tank and discharges it into the urinal(s) to be flushed.

◎ LAVATORIES

A *lavatory* is a plumbing fixture designed for washing of the hands and face. It is commonly found in bathrooms and rest rooms. Lavatories are also called wash basins or vanity lavatories. They are rated at 1 drainage fixture unit of waste discharge and require a 1 1/4-inch waste pipe and a 1 1/4-inch vent pipe.

The lavatory has the most variety of any plumbing fixture with respect to color, size, and shape. Lavatories are made from vitreous china, enameled cast iron, enameled pressed steel, stainless steel, and plastic. They are available in wall-hung, vanity, and vanity-top models.

A *wall-hung lavatory* is a lavatory that is supported on a stamped steel or cast iron bracket fastened to the wall. This bracket is supplied with the lavatory by the plumbing fixture manufacturer.

A wall-hung lavatory is normally installed with its flood level rim 31 inches above the floor. This height permits the user of the fixture to stand slightly bent over the basin, so that water used

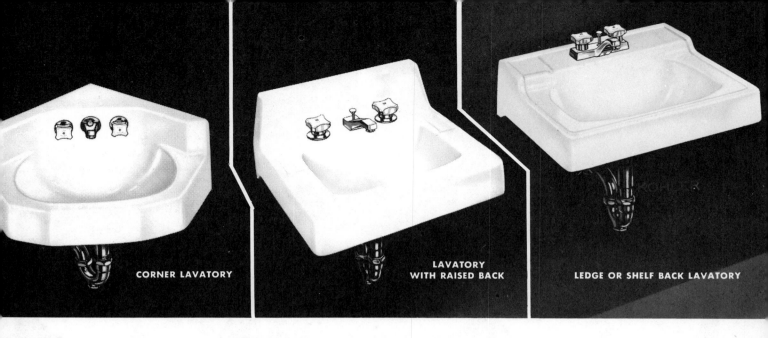

Figure 10-18. Wall-hung lavatories. (Kohler Co.)

for washing tends to drain from the arms back into the basin.

Several different styles of wall-hung lavatories are pictured in Figure 10-18. These are a corner lavatory, a raised back lavatory, and a ledge or shelf-back lavatory. These styles of wall-hung lavatories are manufactured from vitreous china and enameled cast iron.

Figure 10-19 shows a wall-hung *wheelchair lavatory*, made from vitreous china, that is installed in the rest rooms of public buildings for the convenience of handicapped persons. This lavatory is supported from a concealed-arm-type support.

Figure 10-19 also shows a sketch of a handicapped person using a wheelchair lavatory. To permit a wheelchair to slide under the lavatory, as pictured, the fixture must be installed with the flood level rim 34 inches above the floor.

Wheelchair lavatories use a special faucet that has long blade handles and a high spout, both of which extend toward the user of the fixture. In addition, the drain fittings below the lavatory are offset back toward the wall to move the trap away from the user's legs.

A variety of *vanity lavatories* are pictured in Figures 10-20 and 10-21. (A vanity is a bathroom fixture consisting of a lavatory set into the top of a cabinet.) Vanity lavatories are also called counter-top lavatories. They are manufactured from enameled cast iron, vitreous china, enameled pressed steel, and stainless steel.

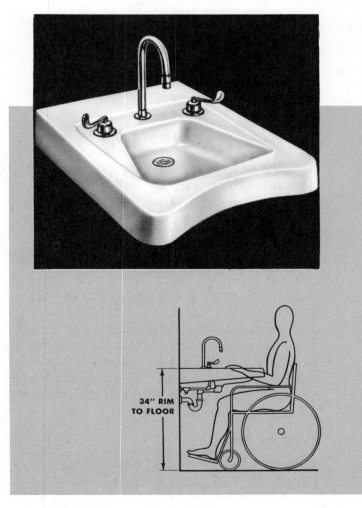

Figure 10-19. Wheelchair lavatory. (Kohler Co.)

Plumbing Fixtures and Appliances **215**

UNDER-COUNTER VANITY LAVATORY

SELF-RIMMING VANITY LAVATORY

RIM-TYPE VANITY LAVATORY

Figure 10-20. Types of vanity lavatories. (Kohler Co.)

Figure 10-20 illustrates the three installation styles of vanity lavatories:

1. *Rim type* lavatory set in a stainless steel rim, which is fastened to the counter top.
2. *Self-rimming* lavatory, which is set directly on the counter top.
3. *Undercounter* lavatory, which is mounted below an acrylic plastic or marble counter top.

In addition to the oval shaped vanity lavatories pictured in Figure 10-20, vanity lavatories are made in round and rectangular shapes (Figure 10-21).

A *vanity-top* lavatory is pictured in Figure 10-22. A vanity-top lavatory is a one-piece wash basin and counter top that is set on top of a vanity cabinet. Vanity-top lavatories are made from marbleized acrylic plastic in several different colors and sizes.

Lavatory Trim

Lavatory trim consists of a faucet (or lavatory fitting), which controls the supply of cold and hot water to the lavatory, and the lavatory drain fitting through which the water drains from the

Figure 10-21. Round and rectangular vanity lavatories. (Kohler Co.)

lavatory. Lavatory faucets may be classified as either compression faucets or port control faucets.

A *compression faucet* is a faucet in which the flow of water is shut off by means of a washer that is forced down (or compressed) onto its seat. Compression faucets operate in much the same manner as the globe valve, which was described in Chapter 3.

Figure 10-22. Acrylic plastic vanity-top lavatory. (Kohler Co.)

Most lavatory compression faucets are combination faucets, consisting of the cold and hot water compression valves joined together in one faucet body with a common mixer spout. This faucet permits the user of the fixture to mix the cold and hot water to the desired temperature.

A *port control faucet* is a single-handle, noncompression faucet that contains within the faucet body ports for both cold and hot water, and some method of opening and closing these ports. This control may be accomplished either by a cartridge or a disk. The cartridge (or disk) opens or closes one or both ports in the faucet body to mix the water to the desired temperature and volume. The working principle of one type of cartridge faucet is illustrated in Figure 10-23.

Lavatory drain fittings are installed in the bottom of the lavatory. The most common type of lavatory drain fitting is the pop-up waste fitting. This fitting consists of a brass waste outlet into which a sliding metal or plastic stopper is fitted. A lever that passes out the side of the drain fitting is connected to a lift rod on top of the lavatory. This rod is lifted to lower the stopper and allow the lavatory to fill. It is depressed to raise the stopper and drain the lavatory.

Lavatory drain fittings are also available with a perforated grate over the drain opening (Figure 10-24) for use in public buildings where it is not desirable to allow the user of the fixture to fill the lavatory with water.

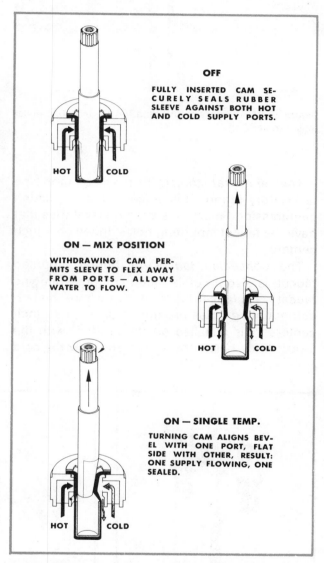

OFF

FULLY INSERTED CAM SECURELY SEALS RUBBER SLEEVE AGAINST BOTH HOT AND COLD SUPPLY PORTS.

HOT   COLD

ON — MIX POSITION

WITHDRAWING CAM PERMITS SLEEVE TO FLEX AWAY FROM PORTS — ALLOWS WATER TO FLOW.

HOT   COLD

ON — SINGLE TEMP.

TURNING CAM ALIGNS BEVEL WITH ONE PORT, FLAT SIDE WITH OTHER, RESULT: ONE SUPPLY FLOWING, ONE SEALED.

HOT   COLD

Figure 10-23. Working principle of a single-handle lavatory faucet. (Kohler Co.)

Although the previous paragraphs describe lavatory faucets and waste trim as separate items, they are usually supplied as a combination fitting, which includes a lavatory faucet and a pop-up drain assembly. Figure 10-25 illustrates some of the more common types of these fittings.

Figure 10-24. Lavatory drain fitting for use in public buildings. (Kohler Co.)

The *center-set fitting* is the most common type of lavatory fitting. This faucet is a combination compression faucet. It is used on lavatories that have the faucet mounting holes drilled on 4-inch centers.

The *concealed faucet*, or bottom-mounted faucet, is also a combination compression type faucet. This faucet fits lavatories where the faucet mounting holes are drilled on 8- or 12-inch centers. It is installed on the lavatory with the faucet body below the fixture and only the cold

and hot faucet handles and the spout on the top of the lavatory.

Figure 10-25 also shows the *single-handle faucet*. This faucet and drain assembly are also used on lavatories that have the faucet mounting holes drilled on 4-inch centers.

◯ BATHTUBS

A bathtub is a receptacle for water that is shaped to fit a human body and is used for bathing. Bathtubs are rated at 2 drainage fixture units of waste discharge, and require a $1\frac{1}{2}$ inch waste pipe and a $1\frac{1}{4}$-inch vent pipe.

Bathtubs are identified as right-hand or left-hand, according to the location of the tub waste opening. A *right-hand bathtub* has the drain on the right end of the tub as you face the length of the tub. A *left-hand bathtub* has the drain on the left.

Bathtubs are manufactured from enameled cast iron, enameled pressed steel, and fiberglass in a variety of sizes, shapes, and colors. The two most common types of bathtubs, the recessed bathtub and the bath-shower module, are pictured in Figure 10-26.

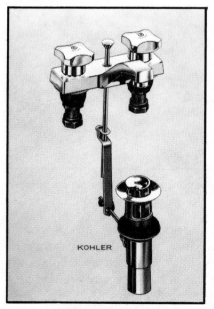

CENTERSET FITTING

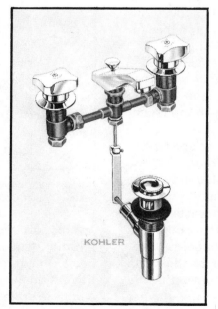

CONCEALED FAUCET

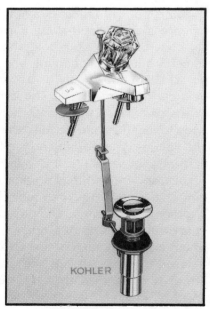

SINGLE HANDLE FAUCET

Figure 10-25. Lavatory faucets with pop-up drain assemblies. (Kohler Co.)

◄ BATH-SHOWER MODULE

RECESSED BATHTUB ◄

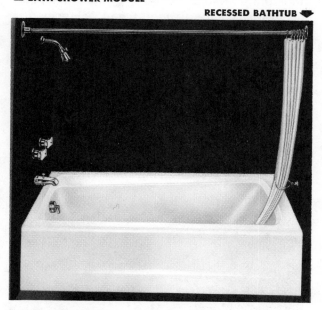

Figure 10-26. Bathtubs. (Kohler Co.)

A *recessed*, or *built-in*, *bathtub* is permanently attached or built in to the walls and floor of the bathroom. Recessed bathtubs are made from enameled cast iron and enameled pressed steel.

A *bath-shower module*, also shown in Figure 10–26, is an integral bathtub and shower wall combination that is made from fiberglass. Bath modules are usually made in one piece, although some models are made with the walls separate from the tub. This type of bathtub is especially popular in residential construction because the fiberglass shower walls eliminate the need for ceramic tile in the shower area.

Bathtub Faucets

Water is supplied to a bathtub through an *overrim bathtub fitting*. The overrim bathtub fitting consists of a faucet assembly and a mixing spout. These are mounted in the wall on the drain end of the bathtub with the spout above the flood level rim of the bathtub (hence the name *overrim*). The bathtub fitting, like the lavatory fitting, may be either a combination compression faucet or a single-handle faucet (Figure 10–27).

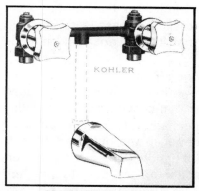

COMPRESSION FAUCET BATHTUB FITTING

SINGLE HANDLE BATHTUB FITTING

Figure 10-27. Overrim bathtub fitting. (Kohler Co.)

In many plumbing installations, the bathtub is used both as a bathtub and as a shower bath. In these installations, the plumber installs a combination bath and shower fitting (Figure 10–28). On

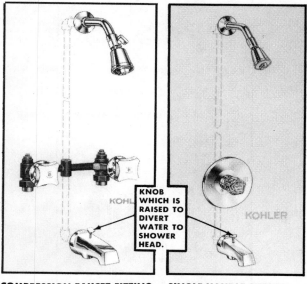

KNOB WHICH IS RAISED TO DIVERT WATER TO SHOWER HEAD.

COMPRESSION FAUCET FITTING **SINGLE HANDLE FITTING**

Figure 10-28. Combination bath and shower fittings. (Kohler Co.)

this fitting, the bathtub mixing spout contains a *diverter* attached to an external knob. This knob is raised to divert the flow of water from the bathtub spout to the shower head. The two combination bath and shower fittings illustrated in Figure 10–28 are the compression-faucet-type and the single-handle faucet.

Bathtub Drain Fitting

The drain fitting used on bathtubs is a *combination waste and overflow fitting*. This fitting provides both the outlet for the bathtub drain and an overflow to allow excess water to drain from the tub so that it does not spill onto the bathroom floor.

The *lift waste fitting* (Figure 10–29) is a type of waste and overflow fitting commonly installed on bathtubs. Within the overflow tube of this fitting is a lifting mechanism that is connected by a lever to the stopper in the outlet of the tub. Turning the control handle (located on the overflow) in one direction raises the stopper and allows the water to drain from the bathtub. Turning the handle in the opposite direction drops the stopper and allows the tub to fill. All the parts of this fitting are accessible through the tub overflow and the waste outlet.

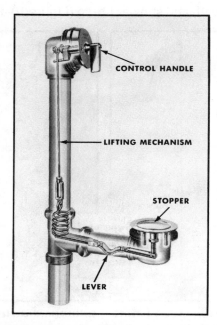

Figure 10-29. Lift-type bathtub waste and overflow fitting. (Kohler Co.)

SHOWER BATHS

A shower bath is a bath in which the bathing water is showered on the user's body from above. Shower baths are rated at 2 drainage fixture units of waste discharge and require a $1\frac{1}{2}$-inch waste pipe and a $1\frac{1}{4}$-inch vent pipe. However, most shower baths are supplied with a 2-inch waste outlet.

There are many types of shower baths. They may consist of one shower head installed in a small enclosed space in a residence or a series of heads installed in a large shower room in industrial buildings, schools, gymnasiums, etc.

There are two distinct types of residential shower baths. One consists of a shower enclosure, and the other is constructed of glazed ceramic tile.

Figure 10-30. One-piece fiberglass shower enclosure. (Kohler Co.)

A *shower enclosure* is pictured in Figure 10–30. This enclosure is constructed of fiberglass in one piece to completely waterproof the shower bath area. Other shower enclosures of this type are manufactured in fiberglass and painted steel, with the walls and base as separate units that the plumber assembles on the job site.

Ceramic tile shower baths may be built in almost any shape to fit the available space within the bathroom. In a ceramic tile shower, the plumber installs either a shower base (constructed of fiberglass or terrazzo) or a waterproof membrane. The tile setter then completes the installation of the shower enclosure. A typical fiberglass shower base is pictured in Figure 10–31.

Shower Bath Water Supply Valve

Shower baths, like lavatories and bathtubs, use either a *combination compression valve* or a *port control valve* (Figure 10–32) to control the volume and temperature of water flowing from the shower head.

A *thermostatic-mixing shower valve* (Figure 10–32) is also sometimes installed on shower baths. This shower valve contains a thermostatic device within the valve body that maintains the selected water temperature to prevent scalding of the person using the shower.

Figure 10-31. Fiberglass shower base. (Powers-Fiat, div. of Powers Regulator Co.)

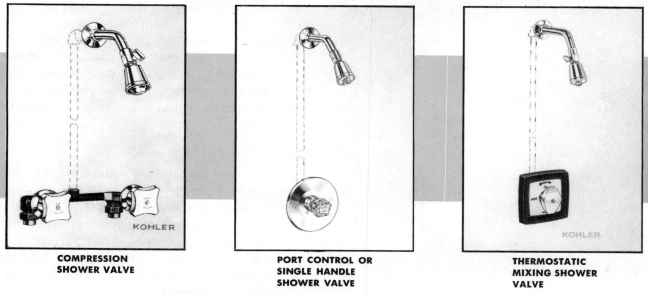

COMPRESSION
SHOWER VALVE

PORT CONTROL OR
SINGLE HANDLE
SHOWER VALVE

THERMOSTATIC
MIXING SHOWER
VALVE

Figure 10-32. Shower bath water supply valves. (Kohler Co.)

Shower Bath Drains

Shower bath enclosure (Figure 10–30) and shower bases (Figure 10–31) discharge their wastes into a 2-inch P-trap that is caulked into the strainer fitting in the bottom of the shower bath. The strainer fitting is supplied with either of these two types of showers by the manufacturer.

Shower baths with a floor constructed of ceramic tile use floor drains of the types pictured in Figure 10-45.

⊙ BIDETS

A bidet (Figure 10–33) is a low-set bowl equipped with cold and hot running water, which is used especially for bathing the external genitals and posterior parts of the body. The bidet is widely used in European bathrooms and is becoming increasingly popular in modern bathrooms in the United States.

Figure 10-33. A bidet. (Kohler Co.)

To use a bidet, the user sits astride the fixture, facing the cold and hot water faucets, to wash the genital and anal areas. The water supply to the bidet is controlled by the faucet mounted on the back of the bidet, which operates in the same way as a lavatory faucet to control both water temperature and volume. Fresh water enters the bidet through either an upward water spray in the center of the bowl or a flushing rim, which helps maintain bowl cleanliness. A stopper retains water in the bowl if desired.

⊙ KITCHEN SINKS

A kitchen sink is a shallow, flat-bottomed plumbing fixture that is used in cleaning dishes and in connection with food preparation. A kitchen sink is rated at 2 drainage fixture units of waste discharge and requires a 1½-inch waste pipe and a 1¼-inch vent pipe.

Although kitchen sinks are available in a large variety of sizes and shapes, the most popular is the *double-compartment kitchen sink* installed on a kitchen cabinet counter top. Double-compartment kitchen sinks are manufactured from enameled cast iron, enameled pressed steel, and stainless steel. Figure 10–34 pictures three double compartment kitchen sinks: a *self-rimming enameled cast iron sink*, a *rim-type enameled cast iron sink,* and a *stainless steel sink.* (Enameled pressed steel sinks are not pictured, but they are very similar in appearance to the enameled cast iron sinks shown in Figure 10–34.)

Kitchen Sink Faucets

Kitchen sink faucets are available either as combination compression faucets or as port control faucets with swing spouts. Another common feature of kitchen sink faucets is the flexible hose and spray attachment, which is used for rinsing dishes. Figure 10–35 illustrates both a *compression faucet* and a *port control faucet* with a hose and spray.

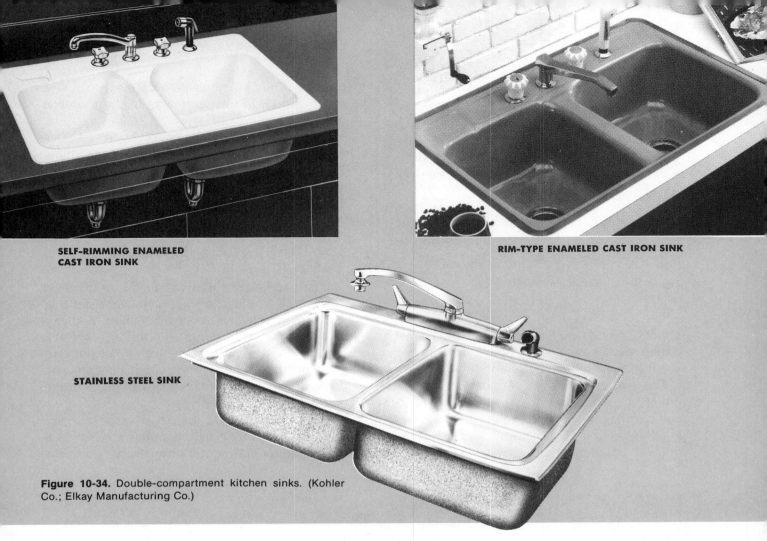

SELF-RIMMING ENAMELED CAST IRON SINK

RIM-TYPE ENAMELED CAST IRON SINK

STAINLESS STEEL SINK

Figure 10-34. Double-compartment kitchen sinks. (Kohler Co.; Elkay Manufacturing Co.)

Kitchen Sink Drain Fittings

Since kitchen sinks are used for cleaning dishes and in the preparation of food, the plumber installs a special drain fitting in each compartment of the sink to keep the solid food waste particles out of the drainage piping.

This fitting, pictured in Figure 10–36, is called a *basket strainer* (or *duo strainer*). It consists of a strainer body attached to the drain opening at the bottom of the sink compartment and a removable basket (also called a *crumb cup*). The strainer body has a strainer across its waste

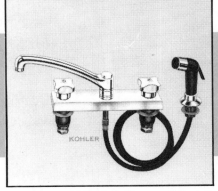

COMPRESSION VALVE SINK FAUCET

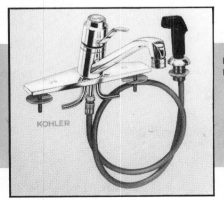

PORT CONTROL OR SINGLE HANDLE SINK FAUCET

Figure 10-35. Kitchen sink faucets with spray and hose attachment. (Kohler Co.)

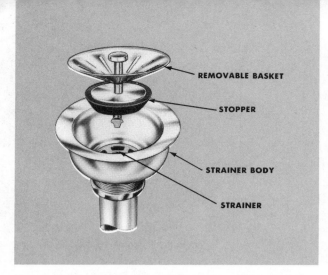

Figure 10-36. Kitchen sink basket strainer. (Elkay Manufacturing Co.)

outlet opening (see Figure 10–36). The basket or crumb cup is also the stopper used to retain water in the sink compartment.

The two methods of connecting the waste outlets from the basket strainers in a double-compartment kitchen sink to the drainage piping are illustrated in Figure 10–37. In the first method, each basket strainer waste is individually

trapped. The two traps then connect to a wye fitting, which in turn connects to the drainage piping.

In the other method, the basket strainer wastes are tied together with a continuous waste fitting. This continuous waste fitting discharges into a single P-trap, which drains into the drainage piping.

⊙ GARBAGE DISPOSALS

A garbage disposal, also called a food waste disposer, is an electric grinding device used with water to grind food wastes into pulp and discharge these wastes into the drainage system. Figure 10–38 pictures a typical domestic garbage disposal, which is mounted beneath one compartment of the kitchen sink in place of the basket strainer assembly.

Although domestic garbage disposals are rated at 2 drainage fixture units of waste discharge, this figure is not normally added to the

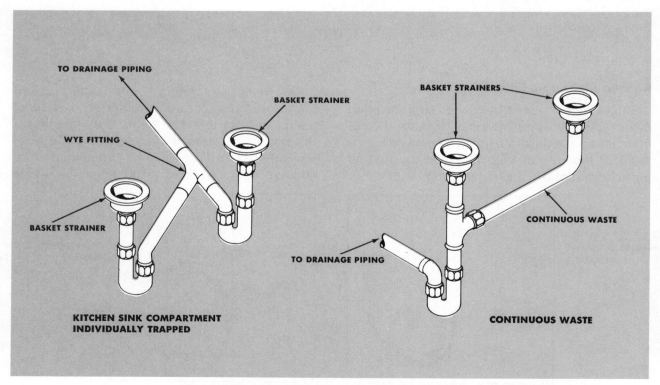

Figure 10-37. Methods of connecting kitchen sink drains to the drainage system.

Figure 10-38. Domestic garbage disposal. (In-Sink-Erator, div. of Emerson Electric Co.)

drainage fixture unit load of the kitchen sink drainage piping. A garbage disposal may discharge its waste into a separate P-trap in a piping arrangement similar to that illustrated in Figure 10–37. A garbage disposal may also discharge its waste into a continuous waste fitting (Figure 10–39). When a disposal drains into a continuous waste fitting, as illustrated in Figure 10–39, the tee connection joining the disposal waste and the waste for the other sink compartment must contain an internal baffle to prevent the disposal waste from backing up into the other sink compartment.

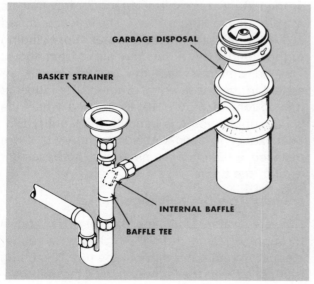

Figure 10-39. Continuous waste piping for a garbage disposal unit beneath a double-compartment kitchen sink.

The plumber should remind the homeowner that it is very important to run a large quantity of water into the garbage disposal when using it, so that the food waste pulp will be washed out of the drainage piping and to avoid a stoppage of the drain pipe.

DOMESTIC DISHWASHERS

A domestic dishwasher (Figure 10–40) is an electric appliance for washing dishes. A domestic dishwasher is rated at 2 drainage fixture units of waste discharge and requires a $1\frac{1}{2}$-inch waste pipe and a $1\frac{1}{4}$-inch vent pipe if it drains its wastes by gravity into the drainage piping. However, most of the domestic dishwashers installed by plumbers do not have a gravity drain, but their wastes are pumped out of the machine.

When a plumber installs a domestic dishwasher with a pumped-out waste, he may connect this waste with a rubber hose or copper tubing either to a dishwasher tailpiece beneath the kitchen sink basket strainer, or into the dishwasher drain connection on the garbage disposal, if the sink is equipped with this appliance (Figure 10–41).

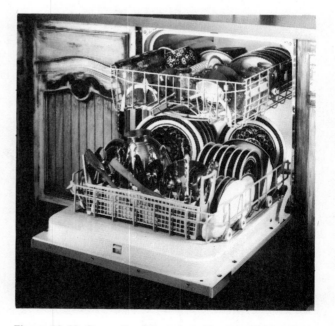

Figure 10-40. Domestic dishwasher. (General Electric Co.)

Plumbing Fixtures and Appliances **225**

In either of these installations, the plumber must make a high loop in the pump discharge piping as it travels from the dishwasher to the drain connnection. The purpose of this loop is to allow the dishwasher to drain properly and to prevent wastes from the kitchen sink and/or garbage disposal from backing up into the dishwasher.

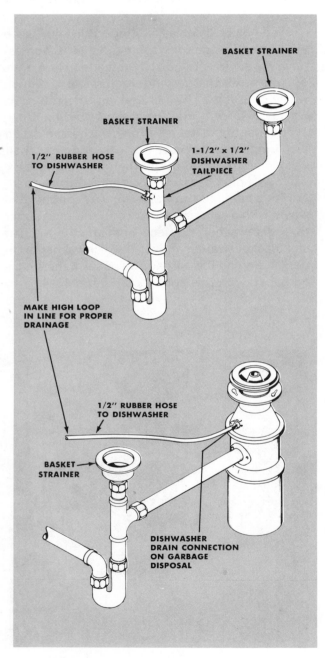

Figure 10-41. Dishwasher drain connections.

 LAUNDRY TRAYS

A laundry tray is a fixed tub, installed in the laundry room of a home, that is supplied with cold and hot water and a drain connection. The laundry tray is used for washing clothes and other household items. It is also used to receive the waste from the automatic clothes washer and to store the water from this machine, if it is equipped with a water reuse cycle.

Laundry trays (sometimes called tubs) are rated at 2 drainage fixture units of waste discharge and require a 1½-inch waste pipe and a 1¼-inch vent pipe.

Laundry trays are manufactured at present from fiberglass or plastics. They are available as either single-compartment or double-compartment tubs in wall-hung and floor-set models. A wall-hung laundry tray is mounted on a bracket supplied with the fixture, while a floor-set model stands on steel legs, also supplied with the fixture. Figure 10-42 pictures both a *floor-set* and a *wall-hung double-compartment laundry tray.*

Laundry Tray Faucets

The faucet used on laundry trays is a two-handle compression faucet with a swing spout that has a hose connection on the spout. The laundry tray faucet is mounted on the back ledge of the fixture. Two styles of laundry tray faucets are pictured in Figure 10–43. The faucet pictured at the top is used when the water supply piping to the laundry tray enters the faucet from above. This is common when the laundry tray is installed in the basement of a home. The faucet at the bottom of Figure 10–43 is used when the water supply piping is beneath the laundry tray. This faucet is used on laundry trays that are installed in laundry rooms above the basement level of the home.

Laundry Tray Drain Fittings

Waste water from the laundry tray flows through a strainer fitting, which uses a rubber stopper to retain the water in the tub. This fitting is supplied with the laundry tray by the manufacturer. A single strainer fitting is used on a single-compartment laundry tray. A double-

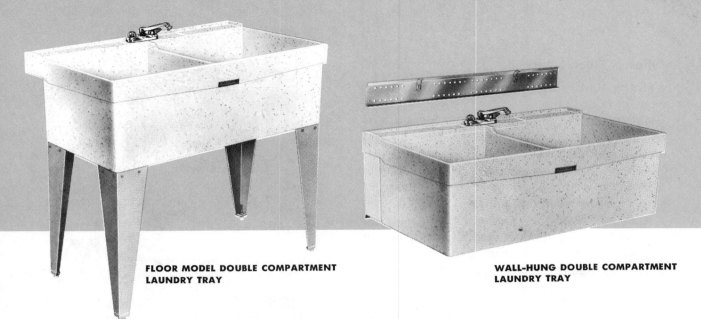

FLOOR MODEL DOUBLE COMPARTMENT LAUNDRY TRAY

WALL-HUNG DOUBLE COMPARTMENT LAUNDRY TRAY

Figure 10-42. Laundry trays. (E. L. Mustee and Sons, Inc.)

compartment laundry tray uses a wye fitting (Figure 10–44) that ties the two compartments of the laundry tray together into a common waste opening that discharges into a single P-trap.

Figure 10-43. Laundry tray compression faucets. Top: Type used when the water supply piping is above the laundry tray; Bottom: Type used when the water supply piping is below the laundry tray. (Union Brass)

Figure 10-44. Double-compartment laundry tray drain fitting. (E. L. Mustee and Sons, Inc.)

 FLOOR DRAINS

A floor drain is a receptacle used to receive water that is to be drained from the floor into the drainage system. Floor drains are considered plumbing fixtures.

The drainage fixture unit rating of a floor drain depends on the outlet size of the drain:

1. A 2-inch floor drain is rated at 2 drainage fixture units.
2. A 3-inch floor drain is rated at 3 drainage fixture units.
3. A 4-inch floor drain is rated at 4 drainage fixture units.

As a rule, the drainage fixture unit load of a floor drain is added to the sanitary drainage piping into which the floor drain discharges its waste; it is not added to the vent piping. In fact, many plumbing codes do not require individual floor drains to have a vent pipe if they are installed within 25 feet of a vented drainage pipe line.

Types of Floor Drains

Figure 10–45 pictures the three most common types of floor drains:

A floor drain with an integral trap. Floor drains of this type are usually installed below ground level in buildings.

A floor drain body for use with a separate P-trap. This type of drain is usually installed above ground level in buildings.

A dry pan floor drain. Dry pan floor drains are installed in areas of buildings that are subject to freezing, such as entryways and garages. The water collected by the dry pan drain is piped to a trap that is located in a heated area of the building.

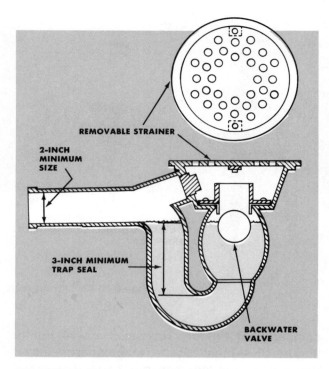

Figure 10-46. Features of a floor drain.

1. The minimum outlet size of a floor drain is 2 inches.

2. The floor drain strainer or grate must be removable.

3. The combined free area of the holes in the strainer (grate) must equal the drain outlet size.

4. Floor drain traps must be a deep seal trap with a 3-inch minimum trap seal.

5. Floor drains installed below ground level must be equipped with a backwater valve.

Uses of Floor Drains

Plumbers install floor drains to receive the water used to wash floors and in any location where there is a possibility of overflow, leakage, and/or spillage of liquid waste onto the floor. Some typical locations for floor drains are:

1. Home laundry and utility rooms.
2. Basements of all buildings.
3. Public rest rooms.
4. Janitors' closets.
5. Entrances and exits to large shower rooms.
6. Building entryways.
7. Garages.
8. Restaurant kitchens.
9. Food markets.

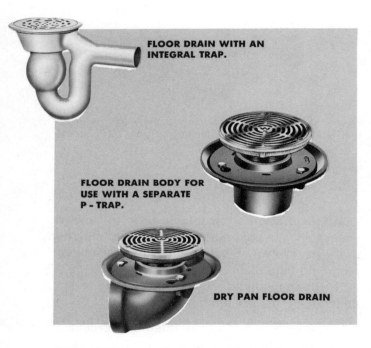

Figure 10-45. Types of floor drains. (Wade Division, Tyler Pipe)

Features of Floor Drains

The three types of floor drains pictured in Figure 10–45 have several features in common. Figure 10–46 pictures a cross sectional view of a floor drain with an integral P-trap; the following section lists the common features of a floor drain. (It should be noted that these features are also plumbing code minimum requirements for floor drains.)

In addition to being used for washing the floor and providing a receptacle for water spills, floor drains are also installed in restaurant kitchens and food markets to receive the liquid waste from the equipment used to prepare and store food. Floor drains installed for this purpose are called *indirect waste receptors.* (An *indirect waste* is a waste pipe that does not connect directly with the drainage system, but conveys liquid wastes by discharging them into a plumbing fixture or receptacle that is connected directly to the drainage system.) Plumbing codes usually require that all plumbing fixtures and equipment used in the preparation and/or storage of food have an indirect waste.

Plumbing codes also require that floor drains used as shower bath drains, janitors' receptors, slop sumps, or any such receptor with a built-up curb or threshold and a water supply, must be equipped with a regular seal P-trap, and that this trap must be vented according to the standard trap-to-vent distance table (Table 6–7).

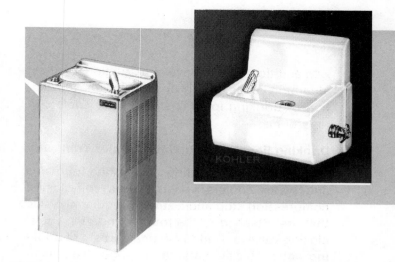

Figure 10-47. Wall-hung drinking fountain and a wall-hung water cooler. (Kohler Co.; Elkay Manufacturing Co.)

DRINKING FOUNTAINS AND WATER COOLERS

A drinking fountain is a fixture that delivers a stream or jet of drinking water through a nozzle. This stream of water is delivered at an upward angle so that the user of the fixture may drink from the fountain without using a cup. Drinking fountains are manufactured from vitreous china, enameled cast iron, and stainless steel.

A water cooler is an electric appliance that combines a drinking fountain with a water-cooling unit. The water-cooling unit cools the drinking water entering the appliance to the desired temperature before delivering it to the nozzle. (Most authorities believe 50°F (10°C) to be the ideal temperature for drinking water.)

Drinking fountains and water coolers are both rated at 1 drainage fixture unit of waste discharge.

Although drinking fountains and water coolers are available in a variety of styles, the style most commonly installed by plumbers is the wall-hung fixture. Figure 10–47 pictures a *wall-hung viterous china drinking fountain* and a *wall-hung water cooler.*

Sanitary Features of Drinking Fountains and Water Coolers

A typical plumbing code requires that all drinking fountains and water coolers be constructed with the following sanitary features (see Figure 10–48):

1. The bowl of the fountain must be constructed of a nonabsorbent material.

2. A mouth protector must be provided over the nozzle.

3. The drinking water must be delivered at an angle so that no water can fall back onto the nozzle.

4. The nozzle must be located above the spill line (flood level rim) of the fountain.

5. The water supply to the fountain must be entirely separate from the waste.

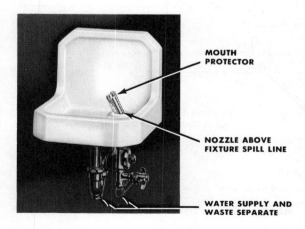

MOUTH PROTECTOR

NOZZLE ABOVE FIXTURE SPILL LINE

WATER SUPPLY AND WASTE SEPARATE

Figure 10-48. Features of drinking fountains and water coolers. (Kohler Co.)

In addition to the above construction features, plumbing codes usually *do not* allow the installation of drinking fountains and water coolers in rest rooms.

Drinking Fountain and Water Cooler Trim

The water supply fitting for drinking fountains and water coolers consists of a self-closing compression stop and a stream-regulating valve that are attached to the fountain nozzle. A self-closing valve is used to eliminate waste of drinking water. This stop and regulating-valve assembly is normally installed on the water cooler or drinking fountain by the manufacturer.

The waste outlet for a water cooler or a drinking fountain is a strainer fitting installed in the drain opening at the bottom of the fountain bowl, which connects to a 1¼-inch trap. This

strainer fitting also is usually installed on the fixture by the manufacturer.

Drinking Fountain and Water Cooler Waste Connections

Plumbing codes usually permit drinking fountains and water coolers to connect to the drainage system either directly or indirectly, as illustrated in Figure 10–49.

A drinking fountain or water cooler that drains directly into the drainage system must have a 1¼-inch P-trap, and both a 1¼-inch waste pipe and a 1¼-inch vent pipe.

When a drinking fountain or water cooler discharges its wastes indirectly into the drainage system (that is, its waste discharges into a floor drain or some other receptacle, as shown in Figure 10–49), only a 1¼-inch P-trap and a 1¼-inch waste pipe are necessary. A vent is not usually required.

○ SERVICE SINKS AND MOP BASINS

Service sinks and mop basins are fixtures that are installed in janitors' closets and building maintenance areas for use by the building maintenance personnel.

A service sink, also called a *slop sink,* is a sink with a deep basin to accommodate a scrub pail. It is used for the filling and emptying of scrub pails, the rinsing of mops, and the disposal of cleaning water. Service sinks (Figure 10–50) are normally made of enameled cast iron and are supplied with a P-trap standard.

A mop basin (or mop receptor) is a floor-set service sink. Mop basins are somewhat more convenient to use because mop pails do not have to be lifted in and out of the fixture. Mop basins may be made of enameled cast iron (Figure 10–51), terrazzo, fiberglass, or even of ceramic tile.

Both service sinks and mop basins are rated at 3 drainage fixture units of waste discharge and require a 2-inch waste pipe and a 1½-inch vent pipe. Although most service sinks and mop basins are manufactured with 3-inch waste outlets, they are still rated at 3 drainage fixture units of waste discharge.

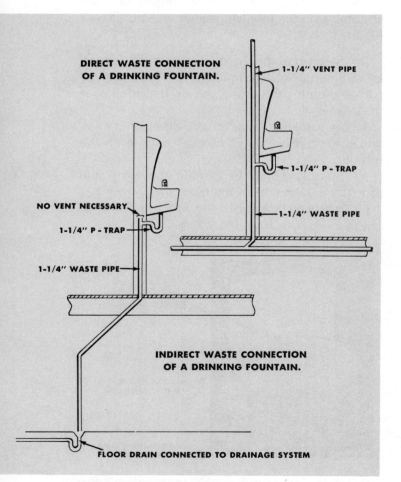

DIRECT WASTE CONNECTION OF A DRINKING FOUNTAIN.

1-1/4" VENT PIPE

1-1/4" P - TRAP

1-1/4" WASTE PIPE

NO VENT NECESSARY.

1-1/4" P - TRAP

1-1/4" WASTE PIPE

INDIRECT WASTE CONNECTION OF A DRINKING FOUNTAIN.

FLOOR DRAIN CONNECTED TO DRAINAGE SYSTEM

Figure 10-49. Direct and indirect waste connections for drinking fountains.

Figure 10-50. Enameled cast iron service sink with a P-trap standard. (Kohler Co.)

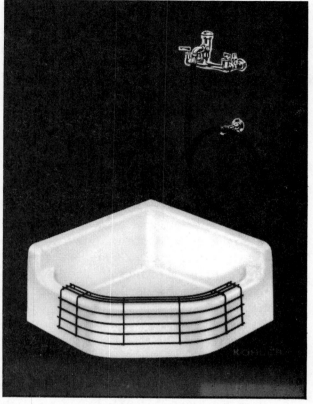

Figure 10-51. Enameled cast iron mop basin. (Kohler Co.)

Water Supply Fitting

Service sinks and mop basins use a combination compression faucet with a hose thread and a pail hook on the spout, like the faucets pictured in Figures 10–50 and 10–51. Since it is common practice for the users of these fixtures to hang a scrub pail on the faucet spout when filling it with water, service sink faucets are often supplied with a wall brace (Figure 10–52) to support this weight.

Drain Fittings

Service sinks of the type pictured in Figure 10–50 are supplied by the manufacturer with a strainer fitting for the waste outlet and a P-trap standard.

Mop basins are supplied with a strainer fitting that is normally caulked into a P-trap installed below the floor. Mop basins that are constructed of ceramic tile use a floor drain of the types illustrated in Figure 10–45.

The apprentice is reminded that when a floor drain is used for a mop basin drain, it must have a regular seal P-trap and a vent.

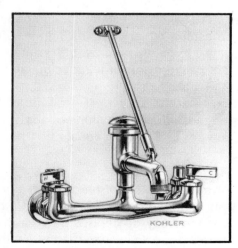

Figure 10-52. Service sink faucet with a wall brace. (Kohler Co.)

⦿ WATER SOFTENERS

One of the most frequent problems encountered in water supplies is the condition known as "hard water." Hard water is water that contains excessive amounts of the minerals calcium and magnesium. These minerals are dissolved by water as it passes through underground limestone strata.

These are some of the objections to hard water:

1. The calcium and magnesium minerals curdle soaps and some detergents, making them hard to lather (hence the term *hard water*).

2. Clothes and fabrics clog with soap curd. This curd dulls colored fabrics, grays white fabrics, and shortens the fabric's life.

3. Soap curd tends to stick on a person's skin after bathing in hard water, and it leaves a dull film on hair.

4. Dishes, glasses, and silverware are water-spotted and eventually may become etched by the minerals.

5. A scum or ring is left in bathtubs, lavatories, and sinks.

6. When hard water is heated, calcium tends to precipitate (or drop out) of the water. This calcium builds up a scale in pots and pans, water heaters, hot water piping, and in hot-water heating boilers and radiators, reducing their efficiency.

Water hardness is measured in either grains per gallon or parts per million (ppm). One grain is equal to 0.002 ounces of calcium or magnesium per U. S. gallon of water. (For comparison, a common aspirin tablet contains 5 grains. Therefore, water 5 gains hard would contain that amount of calcium and/or magnesium dissolved in one U.S. gallon of water.)

One part per million equals 1 pound of calcium or magnesium per 1 million pounds of water. One grain per U. S. gallon equals 17.1 parts per million.

Water Softening Process

To solve the problem of hard water, plumbers install an appliance called a *water softener*. A water softener may be defined as an appliance that removes dissolved water-hardness minerals (calcium and magnesium) from water by the process of ion exchange.

To understand this process, it is necessary for the apprentice to picture how the hardness minerals behave when they are dissolved in water.*

When any mineral is dissolved in water, it breaks down into two or more components known as *ions.* These are individual atoms or groups of atoms that carry an electrical charge. In any solution, the number of positively charged ions must equal the number of negatively charged ions. Dissolved calcium and magnesium, the objectionable ingredients in hard water, are both positive ions. These are the particles that react with soap to form curd and scum.

The basic procedure for getting rid of them in a domestic water supply system is known as *ion exchange*. This is simply a process of swapping the "hard" calcium and magnesium ions for "soft" sodium ions.

The steps of the water softening process are shown in Figure 10–53. The process takes place on the surface of an ion exchange medium, called a zeolite. The most common zeolite is a synthetic resin bead.

Zeolites have the property of being able to take hardness-causing ions out of solution and to replace them with sodium ions. This interaction takes place on the granular particles of the zeolite in the softener.

In actual practice, the zeolite is precharged with sodium ions. As the hard water flows through the softener, the zeolite pulls the calcium and magnesium ions out of solution and replaces them with sodium ions. This process occurs literally billions of times every second when a softener is in use.

After a while, the surface of the zeolite becomes relatively saturated with the *hard* ions and relatively free of *soft* sodium ions. As this action occurs, the softening process gradually stops, and the water emerging from the unit remains hard.

To restore the softener's effectiveness, it must be stripped of its hard calcium and magnesium ions and then given a new charge of exchange-

---

*This description of the water softening process was taken from an article published in *The Journal of Plumbing, Heating and Air Conditioning* entitled "The Purification and Treatment of Domestic Water Supplies" by Richard Weikart and Allan Stahl.

READY TO START	OPERATION	EXHAUSTION	REGENERATION

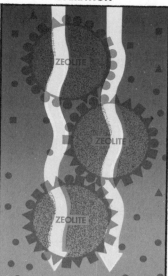

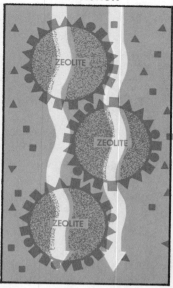

LEGEND: CALCIUM IONS ▲   MAGNESIUM IONS ■   SODIUM IONS ●

PRECHARGED ZEOLITE WITH SOFT (NEGATIVE) SODIUM IONS.

HARD WATER FLOWS THROUGH, AND HARD (POSITIVE) CALCIUM AND MAGNESIUM IONS ARE EXCHANGED FOR SOFT SODIUM IONS. HARD IONS STICK TO THE ZEOLITE. WATER BECOMES SOFT.

ZEOLITE IS EXHAUSTED (CAN ACCEPT NO MORE HARD IONS). NO FURTHER SOFTENING CAN TAKE PLACE.

ZEOLITE IS REGENERATED BY RUNNING SODIUM SOLUTION THROUGH UNIT. SOFT SODIUM IONS ARE AGAIN ATTACHED TO ZEOLITE. HARD IONS ARE FLUSHED AWAY.

Figure 10-53. What goes on inside a water softener.

able sodium ions. This process is known as *regeneration.*

Regeneration (Figure 10–53) is nothing more than operating the softener in reverse. A fairly concentrated solution of sodium chloride, which is ordinary table salt, is run through the zeolite. This concentrated solution of sodium chloride is called a *brine solution.* The brine solution, rich in sodium ions, forces the calcium and magnesium off the zeolite surface and forces the sodium ions onto the same surface. The unwanted calcium and magnesium ions, together with the excess salt solution, are then flushed down the drain.

The frequency with which a softener must be regenerated depends on a number of factors: the hardness of the incoming water, the type of equipment involved, and the level of water consumption.

Automatic Water Softeners

Automatic water softeners are the most common type of water softener installed in homes. An automatic water softener is one with a time clock control to automate the service (softening) and regeneration cycles of the appliance. Homeowners prefer them because the only mainte-

nance they require is the addition of salt to the brine tank.

There are two styles of automatic water softeners installed by plumbers: the *two-tank softener* (Figure 10–54) which has separate zeolite and

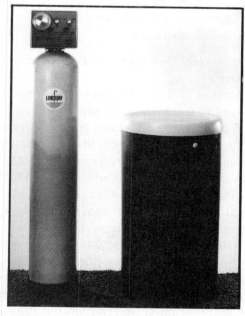

Figure 10-54. A two-tank automatic water softener. (Ecodyne Corporation, the Lindsay Division)

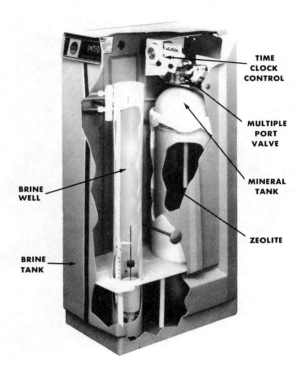

Figure 10-55. A self-contained automatic water softener. (Ecodyne Corporation, the Lindsay Division)

brine tanks, and the *self-contained softener* (Figure 10–55) in which the zeolite tank is contained within the brine tank. There is no difference in the operation of these two styles.

The common features of these two styles of automatic water softener, as illustrated in Figure 10–55, are listed below:

1. A *mineral tank* that contains the zeolite.

2. The *zeolite.*

3. A *brine tank.* Salt stored in the brine tank is dissolved in water to make the brine solution.

4. A *brine well* containing the brine valve assembly. Brine is drawn from the brine well by the brine valve assembly to regenerate the zeolite.

5. A *multiple port valve,* which is used to divert the water and brine solutions in and out of the mineral tank in proper sequence for service (water softening) and regeneration cycles.

6. A *time clock control,* which automates the operations of the multiple port valve. This control is a seven-day time clock that is programmed with the number of times per week the water softener must be regenerated in order to supply an adequate amount of soft water. The

regeneration cycle is normally timed to take place between midnight and 5:00 a.m., a time when water is not normally being used in a home.

Installation of an Automatic Water Softener

An automatic water softener is normally piped into the water supply system of a home in such a way as to provide soft water to all the plumbing fixtures and appliances except the outside sill cocks and the kitchen sink cold water. Figure 10–56 illustrates the piping of a typical automatic water softener in a home. (See Chapter 12 for a detailed description of the installation of the hard and soft water piping in a home.)

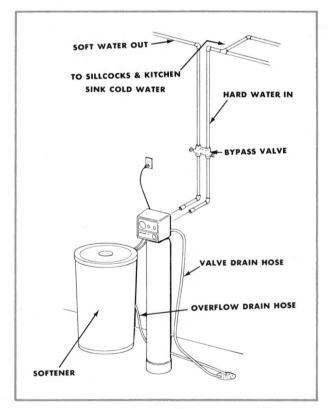

Figure 10-56. Piping of a water softener. (Ecodyne Corporation, the Lindsay Division)

As illustrated in Figure 10–56, a by-pass is provided on the water softener piping so that any repairs to the water softener do not require the interruption of the water supply to the entire household. The by-pass may be either a by-pass valve (as pictured in Figure 10–56) or it may be

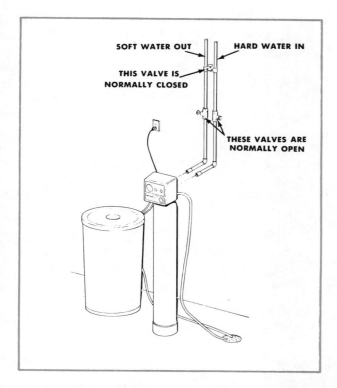

SOFT WATER OUT   HARD WATER IN

THIS VALVE IS
NORMALLY CLOSED

THESE VALVES ARE
NORMALLY OPEN

Figure 10-57. A water softener by-pass constructed with three gate valves and 2 tees.

constructed with three valves and two tees as pictured in Figure 10–57.

In addition to piping the hard water in and the soft water out of the water softener, the plumber must provide a valve drain hose. This hose, which is extended to a floor drain, carries the discharge from the softener during the regeneration cycle. An overflow drain hose must also be provided from the brine tank to a floor drain.

 ## WATER HEATERS

A water heater is an appliance for heating and supplying the hot water used within a building for purposes other than space heating. Although there are a number of different types of water heaters, this text will present only the two most commonly installed types—automatic gas and electric storage tank water heaters.

An automatic storage tank water heater—whether gas or electric—heats and stores hot water at a thermostatically controlled temperature for delivery on demand.

Automatic Gas Storage Tank Water Heater

An automatic gas storage tank water heater utilizes the heat produced by the burning of gas (either natural gas or liquid petroleum gas) and transfers this heat to the cold water contained within the storage tank through the tank bottom and flue.

Figure 10–58 illustrates the components of a typical automatic gas storage tank water heater.

Jacket and top cover. The jacket and top cover form the outer shell of the water heater. Both pieces are constructed of steel finished in baked enamel.

Insulation. To prevent the loss of heat from the water contained within the storage tank, a layer of insulation is provided between the tank and the outer shell.

Storage Tank. The storage tank pictured in Figure 10–58 is commonly referred to as a center flue-type tank. The design of this type of water heater storage tank allows the heated flue gases to cover the entire bottom of the tank before they enter the vertical flue inside the tank. This flue is fitted with a flue baffle to retard the flow of the hot combustion gases and so allow maximum heat transfer.

To protect the water heater storage tank from the corrosive effects of the hot water it contains, the inside surfaces of the tank are covered with a glass coating. This coating is fused to the steel tank. Water heaters with this glass coating on their storage tanks are commonly called *glass-lined.*

The top of the storage tank has three openings:

1. *Cold water inlet.* This is the opening that is connected to the cold water supply pipe. It contains the inlet or dip tube.

2. *Hot water outlet.* The hot water supply pipe is connected to this opening. In this particular model of water heater, a magnesium anode rod is incorporated into the hot water outlet fitting. (An anode rod is a sacrificial rod which is installed in water heaters to prevent electrolytic deterioration of the water heater.)

3. *Relief valve opening.* This opening (which is not visible in Figure 10–58) is provided in water heaters for the installation of a temperature and

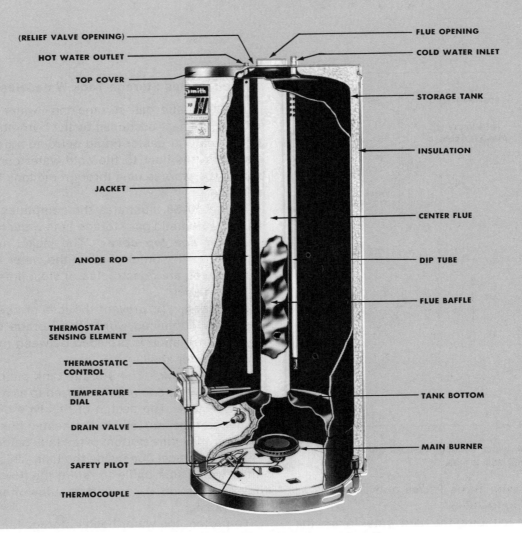

Figure 10-58. Sectional view of an automatic gas storage tank water heater. (A. O. Smith)

pressure relief valve. (The purpose of a relief valve will be discussed later.)

The side of the storage tank has two openings, one of which accommodates the water heater control and the other the tank drain valve. The drain valve opening is located as close to the tank bottom as possible.

Dip tube. All water heaters with the cold water inlet at the top of the tank have a dip tube extending toward the bottom of the tank. The purpose of this dip tube is to carry the incoming cold water through the stored water to the bottom of the tank. In this way, the cold water does not mix with the hot water and is delivered to the bottom of the heater where it can be rapidly heated.

Each dip tube has a small opening, called an anti-syphon hole, located near the top end of the

tube. This hole prevents the hot water in the tank from being syphoned out during an interruption in the cold water supply.

Thermostatic Control. The operation of an automatic gas storage tank water heater is regulated by a control that contains a thermostatic sensing element immersed in the water contained in the storage tank. This control has a temperature dial which permits the selection of the desired hot water temperature.

When hot water is drawn from the storage tank, it is replaced with cold water. This cools the thermostatic sensing element, which then turns on the gas to the main burner to heat the water. This same thermostat also shuts off the gas to the main burner when the temperature of the water in the storage tank reaches the selected temperature.

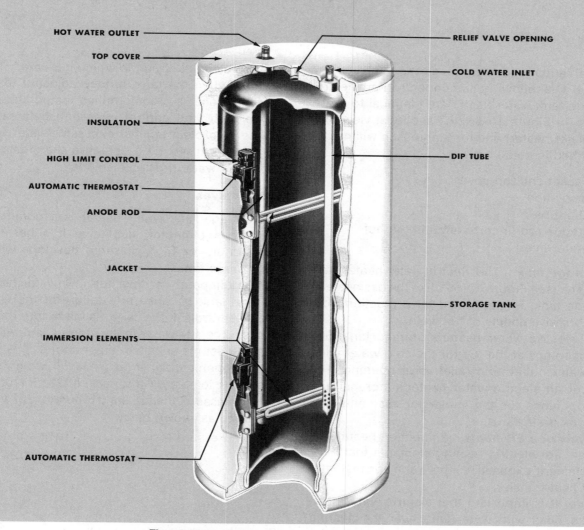

HOT WATER OUTLET

TOP COVER

INSULATION

HIGH LIMIT CONTROL

AUTOMATIC THERMOSTAT

ANODE ROD

JACKET

IMMERSION ELEMENTS

AUTOMATIC THERMOSTAT

RELIEF VALVE OPENING

COLD WATER INLET

DIP TUBE

STORAGE TANK

Figure 10-59. Sectional view of an automatic electric storage tank water heater. (A. O. Smith)

The thermostat will also turn on the gas to the main burner if the temperature of the water contained within the storage tank cools below the selected temperature.

Main Burner. The main burner is a device for the final conveyance of the fuel gas or a mixture of the fuel gas and air to the combustion chamber, where it is burned to produce heat.

Safety Pilot. The safety pilot ignites the gas at the main burner when the thermostat turns it on. It is called a safety pilot because if it goes out, the gas supply to the water heater is automatically shut off.

Thermocouple. The safety pilot also generates a small amount of electricity because its flame heats an ingenious device called a thermocouple. A thermocouple is a tiny electric generator made of two different metals joined firmly.

When these metals are heated, a small electric current is generated. Although the electric energy is small, it is enough to hold the safety shut-off gas valve open. If the pilot flame becomes too small, or if it goes out, the thermocouple then does not produce enough electricity, and a spring closes the gas valve. Because of this device, the gas automatic water heater not only works by itself, but is entirely safe to use.

Automatic Electric Storage Tank Water Heater

An automatic electric storage tank water heater utilizes the heat produced by the flow of electricity through a resistance wire contained in the heating elements, which are immersed in the storage tank, to heat the cold water contained in the tank.

Figure 10–59 illustrates the components of a

typical automatic electric storage tank water heater. The following components of an electric storage tank water heater are identical to those shown in Figure 10–58 (the sectional view of a gas water heater) and for that reason will not be described in detail:

Jacket and top cover
Insulation
Dip tube
Anode rod (incorporated in the hot water inlet)

Storage tank. The electric water heater storage tank (see Figure 10–59), like the gas heater's storage tank, is glass-lined for protection from the corrosive effects of hot water.

An electric water heater's storage tank also has openings at the top for the cold water inlet, hot water outlet, and relief valve opening. The side of an electric water heater's storage tank has openings for the immersion elements and the tank drain valve.

Immersion Elements. Immersion heating elements are electric heating elements that have the element exposed to the water for fast, efficient heat transfer.

Note: It is important that electric water heaters be filled with water whenever the heating elements are operated. This is because "dry firing" will cause the elements to burn out, as there is no water to carry the heat away from the element.

Each immersion element has an automatic thermostat.

Automatic Thermostat. The water heater thermostat is the primary device that starts and stops the flow of electricity to the heating elements. It may be adjusted to give the desired hot water temperature. The thermostat in this heater senses the outside surface temperature of the storage tank. Some other types of electric water heaters use a thermostat that has a remote sensing bulb located in an immersion well in the element flange so that it is exposed to the water. Either type of thermostat will effectively control the water temperature.

High Limit Control. All electric water heaters have a high limit control. The purpose of this control is to stop the flow of electricity to the

heating element circuit in which it is wired. This is done when the tank surface adjacent to the device reaches a predetermined temperature.

The high limit control is essentially a safety device to protect against excessive water temperatures caused by a defective thermostat or grounded water heating element.

Relief Valves

A storage tank water heater is a potentially dangerous plumbing appliance if either high temperature or high pressure develops within the storage tank.

High temperature develops in a water heater storage tank because of a failure of the water heater controls. If the controls fail to turn off the heat source, the water in the storage tank will be heated past the selected temperature.

In an open container (at sea level), water boils and turns to steam (vaporizes) at 212°F (100°C). Under pressure, the boiling temperature of water increases as shown below:

Gauge pressure (psi)	Temperature °F
0	212
30	274
60	307
90	331

Since the water contained within the water heater storage tank is normally under pressure, it can be heated to a very high temperature, or *superheated.* (Superheated water is water that is heated above its boiling point without vaporization.) Superheated water is still a liquid, but if the pressure is reduced, some of the water will flash to steam to establish the temperature-pressure relationship shown in the table above.

The pressure within the water heater storage tank can be reduced in this way if someone uses a hot water plumbing fixture. Superheated water would come out of the plumbing fixture supply valve and, on being released in the atmosphere, this water would immediately flash to steam. It would expand to approximately 1,700 times its original volume. This escaping steam might damage the plumbing fixture and/or injure the user. If the bottom of the water heater storage tank were to rupture while it contained superheated water, the tank could explode. In fact, there are cases on record in which water heater explosions have sent the heater through the roof of the home!

High pressures can develop within a water heater storage tank from high water service pressure, water hammer, or thermal expansion.*

High water service pressure can be eliminated as a problem by the installation of a pressure-reducing valve on the water service (see Figure 3-24).

Water hammer is simply the banging noise in pipes caused by a quick-closing valve. A momentary pressure rise is created, and the effect is transmitted to all parts of the water supply system. Water hammer can be controlled by the installation of a water hammer arrestor of the type illustrated in Figure 10–60 on the supply pipe near the fixture with the quick-closing valve.

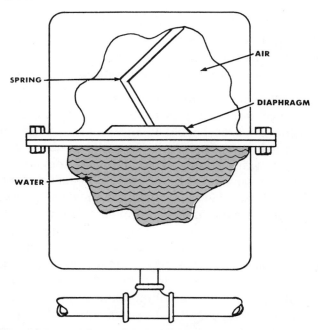

Figure 10-60. Water hammer arrestor. (A. O. Smith)

Thermal expansion, or heat expansion, is caused by the fact that when water is heated, it expands about 2½ percent in volume for every 100°F of temperature rise. If, for some reason, the hot water cannot expand within the water supply system, such an increase in pressure within the system could cause damage.

*A.O. Smith, "Residential Gas Water Heater Service Manual."

Of the two conditions just described—high temperature and high pressure—high temperature is by far the more dangerous in a water heater. If a water heater storage tank ruptures from high pressure, a stream of water will pour from the tank until the pressure is relieved. However, if the tank ruptures from high temperature, it may go through the roof of the building.

To prevent the development of high temperature and/or high pressure within a water heater storage tank, the plumber must install a temperature and pressure (T and P) relief valve in the opening provided in the storage tank (see Figure 10–59). A T and P relief valve is a safety valve designed to protect against the development of dangerous conditions by relieving high temperature and/or high pressure from the water heater. Figure 10–61 illustrates a typical temperature and pressure relief valve.

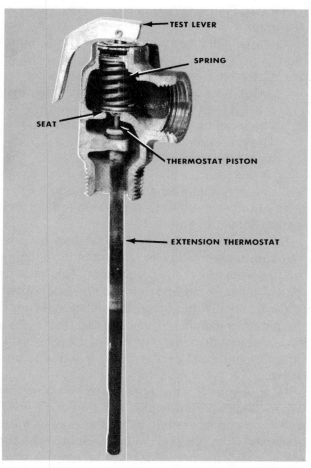

Figure 10-61. A temperature and pressure relief valve. (Watts Regulator Co.)

Pressure and temperature relief valves are selected on the basis of temperature, pressure, and relief capacity:

Temperature. The temperature relief portion of the valve must open before the water temperature reaches the atmospheric boiling point. Normally this means that the relief valve is set to open at 210°F.

Pressure. The pressure relief portion of the valve must open before the pressure within the system exceeds the working pressure of the lowest pressure-rated component in the system. Most relief valves are set to open at a minimum of 125 psi.

Relief Capacity. The valve must be able to relieve high temperature water at a rate equal to or greater than the ability of the water heater to generate it.

In addition, a temperature and pressure relief valve (see Figure 10–61) must have the following features:

1. A name plate or label showing this information:
   The valve is approved by the AGA (American Gas Association) or the ASME (American Society of Mechanical Engineers).
   The temperature setting of the valve.
   The pressure setting of the valve.
   Its relief capacity.

2. A test lever.

3. ³/₄-inch minimum size inlet and outlet.

4. Automatic self-closing (function of the spring shown in Figure 10–61).

5. An extension thermostat installed with the temperature-sensing element in the top 6 inches of hot water in the storage tank. This may be either in a tank tapping or in the hot water outlet.

6. A full-size drain from the relief valve must be piped to within 18 inches of the floor or to a safe place of disposal. The end of the relief valve drain piping must not have either a thread or a valve.

It is helpful for the apprentice to know that when the temperature relief function of a T and P relief valve is operating, a large volume of water escapes from the valve drain piping. On the other hand, the pressure relief function is characterized by water dripping from the valve drain pipe.

 INSTALLATION OF PLUMBING FIXTURES AND APPLIANCES

The plumber must take extra care to install the plumbing fixtures and appliances properly, because the customer will judge the quality of workmanship of the entire plumbing installation by their appearance, as this is the only part of the plumbing installation he can see.

The proper installation of plumbing fixtures and appliances takes place in two steps or phases: *rough-in* and *finishing*.

Rough-in (also called roughing-in or rough plumbing) is the installation of all parts of the plumbing system that can be completed prior to the installation of the fixtures. This includes the installation of:

1. The drainage piping.
2. The vent piping.
3. The water supply piping.
4. Any necessary fixture supports.

Plumbing fixture manufacturers provide a rough-in drawing for each plumbing fixture and appliance as an aid to the plumber. With this drawing, he may properly install the waste, vent, and water supply piping so that they fit the fixture or appliance when it is installed. These rough-in drawings show the physical dimensions of each fixture or appliance as well as the dimensions for the waste and water supply pipes and any necessary fixture supports (or backing).

Finishing is the term applied to the actual installation or setting of the plumbing fixtures and appliances. Plumbing fixtures and appliances are installed after the rooms in which they are to be installed have been decorated and are ready to be occupied. (The bathtub, shower base, and fiberglass tub enclosures and fiberglass shower baths, because they are *built-in fixtures,* are exceptions to this rule.)

Each type of plumbing fixture and appliance has its own peculiar installation procedure. However, the following steps generally apply to the installation of all plumbing fixtures and appliances (although not necessarily in the order given below):

1. Uncrate and inspect the fixture and its trim for damage and/or defects.

2. Check the measurements for the waste and water supply piping with the rough-in drawing for the fixture.

3. Attach the fixture wall or floor supports.

4. Attach the fixture trim (if necessary).

5. Hang or set the fixture.

6. Align, level, and/or plumb the fixture.

7. Secure the fixture fasteners.

8. Connect the water supply and waste piping to the fixture.

9. Purge the water supply piping to relieve any air and remove any dirt in the piping so that it will not affect the operation of the fixture.

10. Test the water supply and waste connections.

11. Caulk or grout the fixture where it meets the wall and/or floor to make a watertight connection that will prevent the accumulation of dirt and make the fixture easier to clean.

12. Clean and inspect the fixture.

It should be emphasized that plumbing fixtures must be handled carefully by the plumber. They do not stand abuse, and any roughness on the part of the plumber usually results in breakage. (This caution should also be heeded by the users of the fixtures.) The materials used to manufacture plumbing fixtures, although they are durable, are on a par with those in fine chinaware and should be treated with care. It is unwise to use scratchy, gritty soaps for cleaning purposes or harsh chemical compounds for the removal of accumulated dirt and scale.

REVIEW QUESTIONS

1. Define: a plumbing fixture; a plumbing appliance.

2. Name at least five materials of which plumbing fixtures are made.

3. Explain the flushing action of a syphon jet water closet and the action of a flush tank mechanism.

4. How does a flushometer valve operate?

5. What is the standard size drain connection for a water closet? What is the size of the vent?

6. How does a single-handle or port control faucet function?

7. Describe two ways a double kitchen sink may be trapped. What size pipe is required for each kind of trap connection?

8. Name the two ways to connect the waste of a dishwasher to the drainage piping described in this chapter.

9. Which plumbing fixture does not require a vent?

10. What is an indirect waste? With what fixtures may it be used? With what fixtures is it required?

11. Give five sanitary requirements for a drinking fountain.

12. Explain the water softening process called ion exchange.

13. Diagram the parts of an automatic storage tank hot water heater, either gas or electric.

14. List at least 10 important steps in setting, connecting, and testing plumbing fixtures.

CHAPTER 11

Testing and Inspecting the Plumbing System

The plumbing system must be subjected to adequate tests and inspections in a manner that will disclose all leaks and defects in the work or the material. It should also be noted that the testing and inspection of the plumbing system is a requirement of all plumbing codes.

Testing the Plumbing System

The practice of testing the plumbing system assures the building owner of a safe, leak-free, plumbing system, and, to a large degree, certifies the plumber's installation. To permit any concealed pipe (whether it be drainage, vent, or water supply pipe) to be enclosed in the building partitions or buried in a trench without the formality of testing it is too great a risk for the plumber to assume. Without complete testing, faulty workmanship (which is sometimes unavoidable) and defects in materials will become apparent only after the plumbing system is put into operation and the building is occupied. Your employer—the plumbing contractor—is the person who will be held responsible, regardless

of whether the defect is in materials or workmanship, and he may be held liable for any damage.

Inspection of the Plumbing Systems

Not only must the plumbing systems be tested, but an inspection by some person of authority who is familiar with plumbing practices is essential. The individual best suited for this responsibility is one who has had previous plumbing installation experience and who possesses complete knowledge of the principles involved in plumbing systems. Often the responsibility of inspection is placed in the hands of building inspectors—men whose knowledge of building crafts is general rather than specific. These men, when inspecting a plumbing installation, look for leaks and other obvious faults, but do not always recognize improperly installed piping. Neither the plumbing contractor nor the building owner is properly served by an inspection of this type.

Most states have plumbing departments, directed by competent plumbers who are responsible for sanitation standards. Field men, called plumbing inspectors, are responsible for designated territories, and their duty is to make inspections and offer installation advice in their territory. Larger cities often employ a staff of full-time plumbing inspectors, and it appears that this practice is a most efficient one.

Hopefully, a universal plumbing code specifying minimum standard practices will be established soon, with each state adopting the code and adding to it whatever regulations would be necessary to cover existing local conditions. Based on this plumbing code, each state could then develop a policy of inspection of plumbing installations to be carried out under local supervision.

Order of Tests and Inspections

To protect both the building owner and the plumbing contractor from the hazards of faulty workmanship and defective materials and to certify that the installation meets the requirements of the plumbing code, all parts of the plumbing system must be tested and inspected. Normally, this means that the following plumbing systems would be tested and inspected as they are installed:

1. The building sewer.
2. The water service.
3. The sanitary drainage and vent pipe system.
4. The storm drainage system.
5. The potable water supply and distribution system.
6. The plumbing fixture installation.

Several of the above systems may be tested and inspected at the same time (such as the building sewer and the building water service) when they lie in the same trench. On the other hand, on larger plumbing installations some of the systems may be tested in sections and never as a whole unit. This is particularly true of the sanitary drainage and vent piping in multistory buildings or in buildings that cover a large area.

PLUMBING SYSTEM TESTS

Air Test of Sanitary Drainage and Vent Piping

The air test of sanitary drainage and vent piping is made by attaching an air compressor to any suitable opening and closing all other inlets and outlets to the system with the appropriate pipe cap or plug, or with testing plugs. Air is then forced into the system until it reaches a uniform pressure of 5 pounds per square inch on the system (or portion thereof) being tested. The pressure must remain constant for 15 minutes without the addition of more compressed air.

To apply an air test to a sanitary drainage and vent piping system, the plumber will need the following items:

1. An air compressor.
2. Materials to close all inlets and outlets of the system.
3. A test gauge assembly.
4. A soap solution to check for leaks.

Figure 11-1. Hand pump for applying air test to sanitary drainage and vent piping.

Air Compressor. The source of compressed air for testing sanitary drainage and vent piping on an installation such as a single-family dwelling may be either a hand pump (Figure 11-1) or a small air compressor (Figure 11-2). To test larger installations, a large gasoline motor-driven compressor of the type shown in Figure 11-3 will be required. In addition to the air compressor, the plumber will need enough air hose to pipe the compressed air from the compressor to the test opening.

Pipe Caps and Plugs. To contain the compressed air within the sanitary drainage and vent pipe system, all the openings must be closed. Where possible, this is done with the appropriate pipe cap or plug. For example, to close iron pipe (threaded) openings, iron pipe caps and plugs are used; to close copper tubing openings, soldered caps are used; to close plastic pipe openings, plastic caps and plugs are used; etc.

It is the usual procedure to close the tees left for the drains for wall-hung or counter-set sinks and lavatories with a nipple and cap that extends *past* the finished wall line. This prevents the

Figure 11-2. Small air compressor. (ITT Pneumotive)

drain opening from being covered over when the rough wall is enclosed.

Test Plugs. Some of the openings in the sanitary drainage and vent system cannot be sealed with a pipe cap or plug. These openings are sealed with either a mechanical or inflatable rubber test plug, as illustrated in Figures 11-4, 11-5, and 11-6.

Figure 11-3. Gasoline motor driven air compressor.

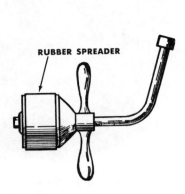

Figure 11-4. Mechanical test plug.

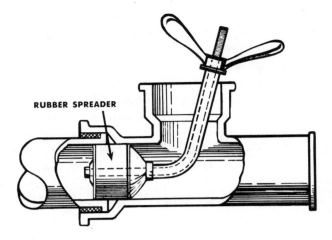

Figure 11-5. Bent or angle mechanical test plug inserted in a cleanout tee.

The *mechanical test plug* (Figure 11-4) consists of a capped stem of 1/2-inch pipe, equipped with a running thread. The pipe is provided with two flanges, the lower one of which is stationary. The top flange is moved up or down by a large wing nut and cast iron body. A heavy rubber spreader placed between the flanges completes the test plug. The rubber portion of the plug may be inserted into the inside diameter of the pipe and expanded by turning the wing nut to the right. Where more than a few pounds of pressure exist, it often becomes necessary to wire or block the plug to prevent it from being blown out of the opening. The test tee must be sealed in similar fashion, except that a *bent* or *angle test plug* is used, as shown in Figure 11-5.

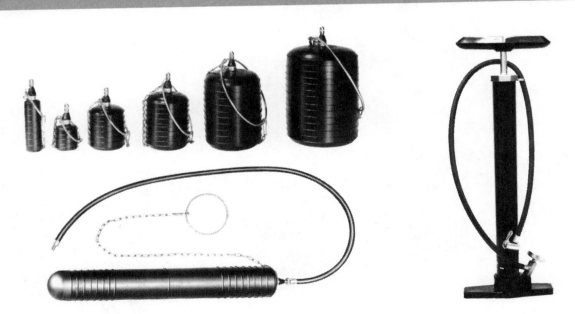

Figure 11-6. Inflatable rubber test plugs and a tire pump for inflating them. (Cherne Industrial, Inc.)

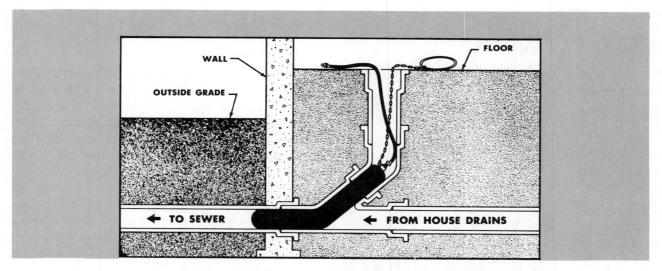

Figure 11-7. Long inflatable rubber test plug being used to seal the building drain at the front main cleanout opening. (Cherne Industrial, Inc.)

Inflatable *rubber test plugs* are shown in Figure 11-6. These test plugs are inserted into the opening and inflated with a tire pump, which is also illustrated in Figure 11-6. Figure 11-7 shows a long rubber test plug being used to seal the building drain at the front main cleanout opening. Figure 11-8 pictures a test tee blocked with a short rubber test plug, while Figure 11-9 illustrates the use of a short rubber test plug to seal a closet bend opening.

Plaster of Paris. Another method commonly used to seal closet bend openings and floor drains is plaster of Paris. A piece of shaped cardboard or a wad of oakum tied with a wire is pushed into the opening. Approximately 1 inch of plaster of Paris, which is mixed with water to a pouring consistency, is then poured over the cardboard or oakum and allowed to harden.

Plaster of Paris can also be used to seal roof vent terminals, but this practice is prohibited by some plumbing codes because plumbers sometimes neglect to remove it. If the vent terminal is left sealed, it naturally destroys the effectiveness of the entire vent system.

However, plaster of Paris is a very effective sealant for the previously mentioned closet bends and floor drains because it can be left in these openings until the building is ready to be occupied and the fixtures set. While these openings are thus sealed, they are protected from

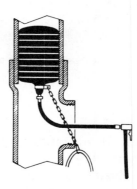

Figure 11-8. Test tee blocked with a short rubber test plug. (Cherne Industrial, Inc.)

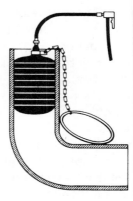

Figure 11-9. Closet bend opening sealed with a short rubber test plug. (Cherne Industrial, Inc.)

Figure 11-10. Test gauge assembly inserted in a cleanout tee.

having dirt, cement, and other debris dropped or pushed into them.

Test Gauge Assembly. A test gauge assembly is necessary to indicate to the plumber (and the plumbing inspector) that the system does indeed contain 5 pounds per square inch of air pressure and that it is holding this pressure. A common test gauge assembly inserted in a cleanout tee is pictured in Figure 11-10. Test gauge assemblies are usually adapted to a cleanout tee or a drain opening for a sink or lavatory. The test gauge assembly is also the opening through which the compressed air is admitted into the system.

Soap Solution. To finds leaks in the sanitary drainage and vent piping when an air test is being applied usually requires the use of a soapy water solution. A soapy water solution is made by mixing common household liquid dishwashing detergent with water in a small container. The soapy water solution is then brushed on the joints with a small paint brush. This mixture will bubble quite vigorously if the pipe joint leaks.

Water Test of Sanitary Drainage and Vent Piping

The water test may be applied to the sanitary drainage and vent piping system either in its entirety or in sections. If the test is applied to the entire system, all openings in the piping must be tightly closed, except the highest vent opening above the roof. The system is then filled with water to the point of overflow above the roof.

If the water test is applied to the system in sections, each opening must be tightly closed, except the highest opening of the section being tested. However, no section should be tested with less than a 10-foot head of water (which is equivalent to 4.34 psi of air pressure).

In testing successive sections, at least the upper 10 feet of the preceding section must be retested so that no joint or pipe in the building will have been tested with less than a 10-foot head of water (except the uppermost 10 feet of the building piping).

The system, or portion of it under test, must be able to hold the 10-foot head of water pressure for 15 minutes without the addition of water.

Water Test Procedure. The procedure followed for a water test of the sanitary drainage and vent piping system is as follows:

1. Plug all openings, except the highest vent openings, as discussed previously for an air test.

2. Fill the system with water, either through the highest opening or by admitting water through a valve and hose connection placed in one of the fixture drain openings.

3. Check for water leaks, and if found, repair.

In tall buildings, the sanitary drainage and vent system is usually tested in sections because of the tremendous pressure that would be produced by the extreme height of the water column if the entire height of the pipe were to be filled with water. Remember, every 1 foot of head of water equals 0.434 psi of pressure; a 100-foot head would equal 43.4 psi of pressure. For this reason, it is usually not advisable to test more than five floors of piping at one time.

On completion of the water test, the water must be removed from the system. In homes or small buildings, this does not present much of a problem. The air is simply released from the inflatable rubber test plug placed in the front

Figure 11-11. Water removal plate. (Cherne Industrial, Inc.)

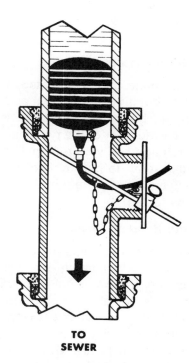

Figure 11-12. Small inflatable rubber test plug inserted in a test tee in a stack for a water test with a water removal plate installed. (Cherne Industrial, Inc.)

main cleanout, allowing the water to drain into the building sewer.

However, in tall buildings that are tested in sections by inserting an inflatable rubber test plug into a test tee, a large quantity of water will splash out of the test tee when the air is released from the plug. There is also a danger of losing the test plug down the pipe because it will tend to wash with the water rushing down the pipe.

To avoid this difficulty, a water removal plate of the type illustrated in Figure 11-11 is used to seal the test tee opening, as shown in Figure 11-12. This water removal plate is designed so that when it is attached to the test tee, the air may be released from the test plug through the front of the plate. When the air is released, the plug will drop onto the rod extended from the water removal plate and be caught, as illustrated in Figure 11-13.

The water test just described is preferred by many plumbers because it is very easy to find leaky joints and defects in the piping—there is water at the leak to indicate it. However, the water test cannot be used in colder climates during the months when the temperature could drop below freezing for fear of the water freezing in the pipes and bursting them. In cold weather conditions, an air test will have to be used.

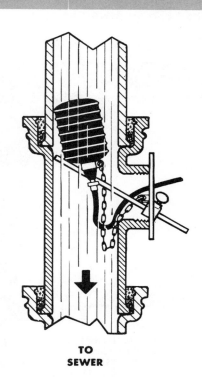

Figure 11-13. Use of a water removal plate to prevent spillage of water when removing water from a stack at the end of a water test. (Cherne Industrial, Inc.)

Tests Of Storm Drainage Piping

Storm drainage piping located within the building (the building storm drain and rainwater leader) should be tested with either a 5-pound air test or a water test of the type mentioned for sanitary drainage and vent piping. Exterior rainwater leaders that do not connect to the storm sewer system are not required by plumbing codes to be tested.

Tests of the Building Sewer

Plumbing code requirements for a test of the building sewer (either sanitary sewer or storm sewer) can vary from a visual inspection of the pipe after it is installed in the trench, to a 5-pound air test, to a water test with 10 feet of head pressure, or to a flow test of the sewer.

Air and water tests are applied to building sewer piping in much the same manner as these tests are conducted on other drainage piping. To block the end of the sewer pipe at its connection to the sanitary sewer, a partially filled test ball attached to a long rubber air hose is floated down the sewer. When the test ball reaches the end of the building sewer (as measured by the length of the air hose), it is inflated so that the building sewer can be tested with either air or water. To obtain the necessary 10 feet of head pressure on a building sewer, a 10-foot length of pipe could be set into the front main cleanout and filled with water.

A flow test of a building sewer is conducted by running a large volume of water through the pipe to see that it flows freely into the sewer main without leakage.

Air Test of Water Supply and Distribution Piping

An air test of water supply and distribution piping is administered in much the same manner as an air test on sanitary drainage and vent piping. The three differences between an air test on water pipe and an air test on drainage piping are:

1. The pressure at which the water supply system is tested is quite a bit larger. Water supply and distribution piping is normally tested at 1½ times the working pressure or 150 pounds per square inch, whichever is greater.

2. The test period is usually 12 to 24 hours.

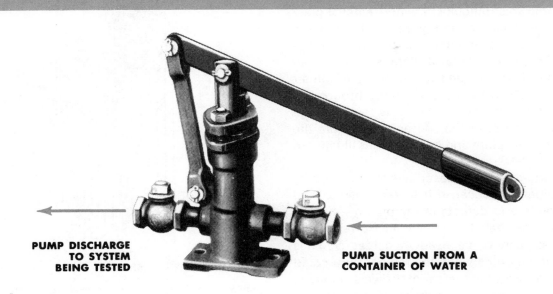

PUMP DISCHARGE TO SYSTEM BEING TESTED

PUMP SUCTION FROM A CONTAINER OF WATER

Figure 11-14. Hydrostatic test pump. (Crane Co.—Deming Pumps)

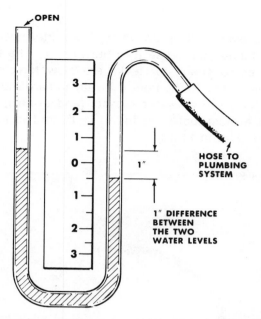

OPEN

3
2
1
0
1
2
3

HOSE TO
PLUMBING
SYSTEM

1"

1" DIFFERENCE
BETWEEN
THE TWO
WATER LEVELS

Figure 11-15. Manometer connected to finished plumbing system. It indicates a pressure equal to a 1-inch water column. (Ralph Lichliter)

3. Since the pressure and time of test are greater than for air tests of drainage piping, the openings in the water piping are sealed with the system control valves and the appropriate pipe caps and plugs. (Valves, caps, and plugs normally hold pressure better than testing plugs.)

Hydrostatic Test of Water Supply and Distribution Piping

A hydrostatic test is a test in which the pipe being tested is filled with water and then submitted to additional water pressure. A hydrostatic test is the common test for water main and water service piping.

The procedure for applying a hydrostatic test is to fill the pipe(s) being tested *completely* with potable water (after sealing all openings) and then to force additional water into the pipe(s) with a hydrostatic test pump of the type pictured in Figure 11-14. This pump has suction and discharge ends with check valves on each end to prevent the water pressure from forcing water back through the pump.

To test with a hydrostatic test pump, with hoses: (1) extend the suction end of the pump to a container of potable water, and (2) connect the discharge end to a test gauge assembly that is connected to the system being tested.

The pump handle is then stroked to force enough extra water into the system to raise the pressure within the system to the pressure required for the test. This usually does not involve the introduction of very much additional water, since water is practically incompressible. For example, the addition of only one-half cup of water forced into a piping system containing 30 gallons of water will raise the pressure from 0 to 50 pounds per square inch.

The key to a successful hydrostatic test is to eliminate all pockets of air within the system by complete flushing of the pipes with water. If the system contains air pockets, the water forced into the system will compress the air pockets rather than the water, and the test gauge pressure will not hold steady. If there *are* air pockets, just keep flushing.

Finished Plumbing Test

The finished plumbing test, which is sometimes called a *final air test,* is a test of the plumbing fixtures and their connections to the sanitary drainage system.

One plumbing code* describes the finished plumbing test as follows:

After all the plumbing fixtures have been set and their traps filled with water, their connections shall be tested and proven (sewer) gas and water tight by plugging the stack openings on the roof and the building drain where it leaves the building, and air introduced into the system equal to the pressure of a 1-inch water column. This pressure shall remain constant for the period of the inspection (usually 15 minutes) without the introduction of additional air.

The testing apparatus used to check the pressure for a finished plumbing test is a *manometer.* A manometer is a U-shaped, clear plastic tube with a section of ruler in the middle, as illustrated in Figure 11-15.

The procedure for applying a finished plumbing test is as follows:

1. Fill all fixture and floor drain traps with water.

2. Plug all stack openings on the roof.

*Minnesota Plumbing Code.

3. Plug the building drain at the front main cleanout.

4. Insert the manometer hose through the trap seal of a water closet bowl and blow any water out of this hose.

5. Attach the hose to the manometer, which has been previously filled with water to the "0" marks on the ruler. Set the manometer up on the open closet seal to rest.

6. With another hose inserted through the trap of the water closet, blow air into the system until a 1-inch differential of pressure can be read and maintained on the manometer, as illustrated in Figure 11-15.

If the finished plumbing test just described does not hold the required pressure of 1 inch of water column, the leaks must be found and fixed. Leaks in the finished plumbing system can be located with either freon gas, smoke, or oil of peppermint.

Freon Gas Leak Detection. One method of locating leaks in the finished plumbing is with freon gas (Figure 11-16) and a halide torch (Figure 11-17) equipped with a snifter hose. Freon is introduced into the finished plumbing system through the water closet trap. Any freon gas that seeps out from a leak is detected because the flame on the halide torch will turn green. The following procedure is used to apply a freon leak detection test:

1. Repeat steps 1 to 5 of the previously described "Finished Plumbing Test".

2. Attach a freon charging hose to a can of freon gas.

3. Insert the charging hose from the freon can through the trap seal of the water closet bowl, making sure that the hose passes completely through the trap.

4. Open the freon valve on the freon can to introduce freon gas into the plumbing system until a pressure differential of 1 inch of water column is indicated on the manometer.

5. Light the halide torch and allow the halide detector head to warm up until the head is a dull red color.

6. Pass the snifter hose around the fixtures and their connections to the drainage system. While doing this, watch the flame on the detector head carefully because any freon leaking out of

the system will be drawn up the snifter hose to the flame and cause it to burn green. The first places to check for leaks are at water closet connections to the floor (or wall) and around the flushing rim of water closets (and urinals) because most leaks are found in these two locations (Figure 11-18).

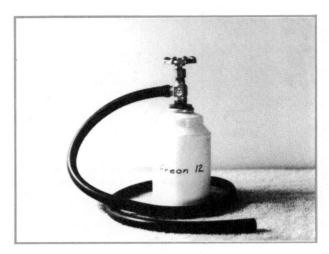

Figure 11-16. Freon gas container with charging hose.

Figure 11-17. Halide torch.

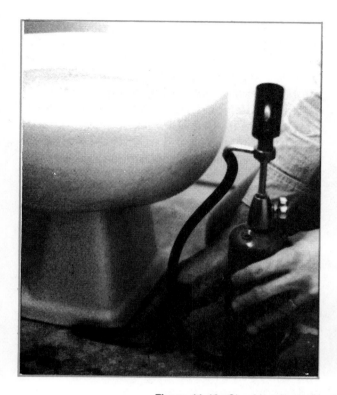
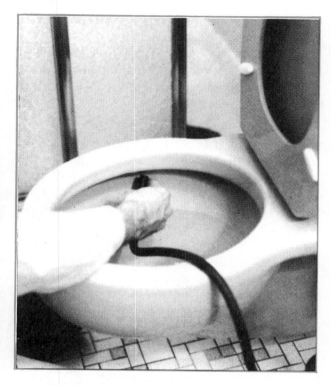

Figure 11-18. Checking for leaks with a halide torch. Left: At the water closet connection to floor; Right: At the water closet flush rim.

7. Fix all leaks found with the above method and reapply the test. (Water closets and urinals with leaks in the flushing rim will have to be replaced and sent back to the manufacturer, as this defect cannot be repaired.)

Smoke Leak Detection. A smoke test is conducted by filling the fixture traps with water, plugging the building drain at the front main cleanout, and introducing an odorous, thick smoke into the system. Smoke is produced by one or more smoke machines. When smoke appears at the stack opening(s) on the roof, it is closed and a pressure of 1 inch of water column is built up and maintained within the system for 15 minutes.

A smoke machine is just a hollow drum provided with a cover and an air pump attached to it. Oily rags or some other substance that will produce a dense smoke when burned are placed in the drum and ignited. The cover is then put on the drum and air is pumped into the drum. As the air passes through the drum it carries the smoke

with it into the plumbing system. Figure 11-19 illustrates the principle of a smoke machine.

Since the plumber using the smoke test must depend on his sense of smell to help locate any

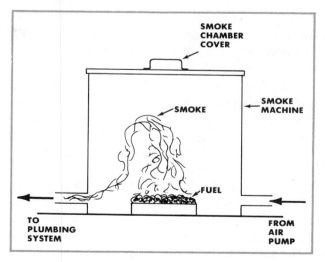

Figure 11-19. Principle of a smoke machine.

leaks, he must not have any of the smoke odor on his clothes. For this reason, the smoke machine should be located either outside of the building or in a room with no plumbing fixtures. It should be operated by a plumber not involved in finding the leaks.

Peppermint Leak Detection. When the peppermint test is used, the fixture traps are filled with water and the building drain plugged at the front main cleanout. Two ounces of oil of peppermint are then poured into each stack and the building drain. After closing these openings, air is applied to the system to a pressure of 1 inch of water column. To locate leaks, the plumber must depend on his ability to smell the peppermint.

When using this test, the plumber must be careful to avoid spilling any of the oil of peppermint on to his body or clothing, as he will carry this smell with him. It is best to have some person who will not be involved in locating leaks add the oil of peppermint to the system and then have this person leave the building.

Of the three leak detection methods just described (freon, smoke, and oil of peppermint), the freon test is probably the best for locating leaks in finished plumbing.

Some notes of caution are necessary when applying a finished plumbing test:

1. In all of the leak detection methods described, as well as the original test, the systems were pressurized to 1 inch of water column. The reason for this pressure limitation is that some plumbing fixture traps contain only 2 inches of water seal and any pressure exceeding the depth of water seal of the shallowest trap in the system will blow this trap.

2. During the period of the finished plumbing test, no large volumes of water can be added to the system quickly without blowing some fixture traps. In particular, no water closets or urinals may be flushed. For this reason, this plumbing test should be done at a time when no other tradesmen but plumbers are working in the building. If it is not possible to work in the building alone, the plumbers should turn off all flushometer valves and empty all water closet flush tanks during the period of the test to prevent the flushing of these fixtures.

3. If the period of the test is to cover more than one day, the test plug in the front main cleanout opening will have to be removed. When removing this plug, it is advisable to first remove the plug(s) from the roof vent openings. Since there will usually be some water in the building drain, this water when released will syphon trap seals if there is no other vent, and any trap seal syphoned will naturally have to be found and refilled.

 ## HOW TO APPLY A PLUMBING TEST

Since the testing of plumbing systems is a very important part of the plumber's work, the apprentice should become acquainted with the procedure that is followed for all plumbing tests.

However, before presenting the plumbing test procedure, the apprentice should be made aware of two basic requirements of all plumbing codes concerning the testing and inspection of plumbing systems. The first requirement is that the plumbing contractor must furnish the equipment, material, power, and labor necessary for each test and inspection. The second requirement is that no work may be covered (either with dirt or enclosed in a wall or ceiling) until it has been tested and inspected. In fact, the plumbing inspector will usually require any untested piping that is concealed to be uncovered.

The procedure for all plumbing tests follows the eight points listed below:

1. Assemble the testing apparatus.
2. Seal all openings.
3. Apply the test.
4. Check for leaks.
5. Fix defects.
6. Call for an inspection and test.
7. Assist the inspector at time of test.
8. Remove testing apparatus on completion of test and inspection.

Assemble Testing Apparatus. Since all plumbing tests involve subjecting the plumbing system being tested to pressure, first check to see that you have the proper test plugs, pipe caps, and/or plugs to seal all openings so that the pressure will be contained within the system being tested.

Next, see that you have the proper equipment to apply pressure. This may be in the form of an air compressor for an air test, or it may be a supply of potable water to fill the pipe for a hydrostatic test. In addition to a source of pressure, the plumber will need the proper hose to get the compressed air or water from the source to the system.

A method of checking the pressure being applied to the system is also needed. This may be either in the form of a test gauge assembly or a manometer.

Finally, you will need to have some method of checking for leaks in the system, if an air test is being applied. A container of soapy water and a small paint brush will be required for air tests. A manometer, freon gas and a halide torch, or a smoke generator, or oil of peppermint will be necessary for final air tests. (In a water test, the leaks will naturally be quite easily noticed by water at the point of leakage.)

Seal All Openings. Since plumbing tests are pressure tests, the pressure must be contained within the system. This involves closing all openings with either the proper pipe cap or plug, a test plug, or plaster of Paris. In a final air test, filling all the traps with the proper amount of water will seal the trap.

Apply Test. At this point in the testing procedure, the plumber simply follows the appropriate testing method for the type of test being applied to the system, as detailed earlier in the chapter. The plumber should find and identify leaking pipe joints and defective materials in the plumbing system being tested.

Fix Defects. Having located and identified the defective portions of the plumbing system being tested, the plumber must proceed to correct these defects. Leaks at joints must be repaired. Cracked and defective pipe and fittings must be removed and replaced.

Plumbing fixtures that will not pass the final air test must also be removed and stored carefully so that they can be returned to the manufacturer for credit.

If the plumber encounters a considerable amount of defective pipe and fittings, it is also wise to save this material because the plumbing contractor may be able to file a claim against the manufacturer of these items.

Call for an Inspection and Test. After the plumber is satisfied that the plumbing system he is testing meets the test requirements, he should notify the plumbing inspector that the work is ready for an inspection.

Plumbing inspectors usually require a minimum of one working day's notice before an inspection so that they may schedule their inspection load. Knowing this, plumbers are sometimes tempted to call the inspector before they are sure that a plumbing system will pass a test. This practice is not recommended because it wastes the time of the inspector, and it can be embarrassing for the plumber to have the inspector arrive on the job too early and find that a piping system will not pass a test.

Assist Inspector at the Time of Test. When the plumbing inspector arrives to test and inspect the work, the plumber must apply the test for him. While the test is in progress, the plumbing inspector may require that the plumber remove a cleanout or some other plug to see that the pressure has reached all parts of the system under test.

The plumber should also escort the inspector around the job so that he may visually inspect the work to see that it meets the requirements of the plumbing code.

Remove Testing Apparatus. On completion of the test and inspection, all testing apparatus should be removed. This usually involves removing the testing equipment and tools and sending these items back to the plumbing shop.

At this time, all the test caps and plugs should also be removed and any water or air that was used to test the piping system should be drained from the system. Test nipples and caps are normally left in water supply and sanitary drainage openings after testing to prevent the covering over of these openings when the finished walls are erected.

On completion of the final air test, the plumber must be particularly careful to remove all test caps and plugs placed on roof vent terminals, for if these openings are left sealed, the vent system of the building will be seriously impaired.

◉ REVIEW QUESTIONS

1. When should the inspection of the different parts of the plumbing system be made?

2. Why should the drainage piping of a multistory building be tested in sections?

3. Is it feasible or not feasible to test the drainage system separate from the vent system? Why, or why not?

4. List the pipes to be plugged, the trap seals to be filled, and the equipment to be used to make a water test of an entire drainage system.

5. At what point or points in the installation of the sanitary drainage system is the water test made?

6. What system or parts of the system for storm water drainage must be tested? What parts need not be tested?

7. Describe the test usually made on the water supply and distribution system of a building before final inspection.

8. To what pressure is the water supply and distribution system usually subjected? For how long is this pressure maintained?

9. Describe the "finished plumbing test" given the plumbing system of a building prior to inspection by a local or state plumbing inspector.

10. What is recommended for the disposition of plumbing pipe, fittings, and fixtures that fail in any plumbing test?

CHAPTER 12

Plumbing a House

To correlate the information given in previous chapters, this last chapter illustrates and explains the design of a typical plumbing system of a single-family home.

The home selected as an example is of medium price and of the type being constructed on suburban housing developments in various parts of the country. The house is an L-shaped rambler containing approximately 1,200 square feet of living space, with an attached double garage.

The plumbing in the home consists of a modern kitchen with a garbage disposal unit and a built-in automatic dishwasher. There is a full bath (a bathroom containing a water closet, a lavatory, and a bathtub) on the main floor, as well as a three-quarter bath (a bathroom containing a water closet, a lavatory, and a shower bath) off the master bedroom.

The plumbing in the basement consists of a laundry tray with a standpipe for the automatic clothes washer drain, a water heater, a water softener, and two floor drains, one beneath the laundry tray and the other by the water softener to receive the backwash from the water softener.

This drain also receives the condensate from the air-conditioning coil in the furnace.

A rough-in for a future three-quarter bath is also indicated in the basement. It is common practice to rough-in for future basement bathrooms at the time of the original construction of the home. This saves the homeowner the expense and inconvenience that would be involved in breaking up the basement floor to provide the plumbing for these fixtures at a later date.

In the following sections of this chapter, the blueprints for the home, the specifications, the rough-in drawings, the illustrations used, and the design and installation of the plumbing systems of the home will be explained in regular order.

Blueprints

The blueprints on which the illustrative example is based are designated and numbered as Sheets 1 to 8.

All drawings indicating floor plans and elevations were originally done according to the $1/4'' = 1'0''$ scale. The detail drawings, which appear on some of the plan and elevation plates, were drawn to the $\frac{1}{2}'' = 1'0''$ scale.

The blueprints were originally drawn full size, but they have been reduced to fit the page size of this book. Therefore, although the blueprints are all in correct proportion, they cannot be scaled.

A study of the various floor plans will show that the architect has indicated all the plumbing fixtures and their locations. The plans indicated by the architect should be followed exactly unless obstacles arise, in which case the architect should be consulted before any changes are made.

The blueprints should be studied so that the plumber will have a perfect visualization of the entire house and its structural details, all of which must be considered as the design of the plumbing system is being carried on.

In the working drawings or blueprints that architects prepare as a guide to the construction of houses, there are seldom, if ever, any details of the required plumbing system given, other than those that show the fixture symbols in their proper places. Designing the piping systems, therefore, must be done by the plumbing contractors before they can make estimates, bid on such jobs, or actually install the systems.

Specifications

The function of written specifications is to supplement the blueprints. In other words, the specifications give instructions and explanations that are not or cannot be shown on the blueprints. They should be studied as carefully as the blueprints, and every item noted with extreme care. For example, the blueprints show the locations of water closets, lavatories, etc., but only the specifications describe these items as to their quality, manufacturer, color, etc. The blueprints do not indicate any piping for the home, but the specifications will tell the plumber what *type* of pipe to use for each plumbing system.

The specifications for most homes are simply fill-in forms of the type illustrated in Figure 12-1. This form is supplied by the U.S. Government Printing Office for home builders who will be using FHA (Federal Housing Authority) or VA (Veterans Administration) appraisals for home loan financing. Other types of specification forms also are commonly used.

The items of particular interest to the plumber on these specifications are found on Figure 12-1, page 3, Section 22: "Plumbing." This lists the location, type, and color of all the plumbing fixtures in a chart. Also indicated in this section are the type of water supply and sewage disposal system, the type of pipe to be used for the house (building) drain and water piping, and the type of water heater.

Section 26: "Special Equipment," found on Figure 12-1, page 4, specifies special equipment of interest to the plumber, such as the dishwasher, garbage disposal unit, and water softener.

Some specifications are also shown on the blueprints in the form of notes. These notes are as important as any of the symbols, plans, or items in the written specifications.

FHA Form 2005
VA Form 26-1852
Rev. 4/73

U. S. DEPARTMENT OF HOUSING AND URBAN DEVELOPMENT
FEDERAL HOUSING ADMINISTRATION

For accurate register of carbon copies, form
may be separated along above fold. Staple
completed sheets together in original order.

Form Approved
OMB No. 63—R0055

DESCRIPTION OF MATERIALS

No. _____
(To be inserted by FHA or VA)

☐ Proposed Construction

☐ Under Construction

Property address __11432 Galtier Drive__ City __Burnsville__ State __Minn.__

Mortgagor or Sponsor _____
(Name) _____ *(Address)*

Contractor or Builder _____
(Name) _____ *(Address)*

INSTRUCTIONS

1. For additional information on how this form is to be submitted, number of copies, etc., see the instructions applicable to the FHA Application for Mortgage Insurance or VA Request for Determination of Reasonable Value, as the case may be.

2. Describe all materials and equipment to be used, whether or not shown on the drawings, by marking an X in each appropriate check-box and entering the information called for in each space. If space is inadequate, enter "See misc." and describe under item 27 or on an attached sheet. THE USE OF PAINT CONTAINING MORE THAN ONE PERCENT LEAD BY WEIGHT IS PROHIBITED.

3. Work not specifically described or shown will not be considered unless

required, then the minimum acceptable will be assumed. Work exceeding minimum requirements cannot be considered unless specifically described.

4. Include no alternates, "or equal" phrases, or contradictory items. (Consideration of a request for acceptance of substitute materials or equipment is not thereby precluded.)

5. Include signatures required at the end of this form.

6. The construction shall be completed in compliance with the related drawings and specifications, as amended during processing. The specifications include this Description of Materials and the applicable Minimum Property Standards.

1. EXCAVATION:
Bearing soil, type __Sand__

2. FOUNDATIONS:
Footings: concrete mix __1:3:5 2,000 lb. test__ strength psi _____ Reinforcing _____
Foundation wall: material __12" concrete block__ Reinforcing _____
Interior foundation wall: material _____ Party foundation wall _____
Columns: material and sizes __3" steel pipe__ Piers: material and reinforcing _____
Girders: material and sizes __8 WF 10 & 8 WF 12__ Sills: material __Sill seal__
Basement entrance areaway _____ Window areaways __Where required-corrugate metal__
Waterproofing __1 coat trowel mastic__ Footing drains _____
Termite protection _____
Basementless space: ground cover _____ ; insulation _____ ; foundation vents _____
Special foundations _____
Additional information: _____

3. CHIMNEYS:
Material __Metalbestos 5"__ Prefabricated *(make and size)* _____
Flue lining: material _____ Heater flue size _____ Fireplace flue size _____
Vents *(material and size)*: gas or oil heater _____ ; water heater __3" Gal. or Alum. to chimne__
Additional information: _____

4. FIREPLACES:
Type: ☐ solid fuel; ☐ gas-burning; ☐ circulator *(make and size)* _____ Ash dump and clean-out _____
Fireplace: facing _____ ; lining _____ ; hearth _____ ; mantel _____
Additional information: _____

5. EXTERIOR WALLS:
Wood frame: wood grade, and species __#2 fir__ ☐ Corner bracing. Building paper or felt __15 lb. on corner & rim joists__
Sheathing __strong board__ ; thickness __½"__ ; width __4"__ ☒ solid; ☐ spaced _____ o. c.; ☐ diagonal;
Siding __aluminum__ ; grade _____ ; type __lap__ ; size __9x12__ ; exposure __8"__ ; fastening __alum. nails__
Shingles _____ ; grade _____ ; type _____ ; size _____ ; exposure _____ ; fastening _____
Stucco _____ ; thickness _____ "; Lath _____ ; weight _____ lb.
Masonry veneer _____ Sills _____ Lintels _____ Base flashing _____
Masonry: ☐ solid ☐ faced ☐ stuccoed; total wall thickness _____ "; facing thickness _____ "; facing material _____
Backup material _____ ; thickness _____ "; bonding _____
Door sills _____ Window sills _____ Lintels _____ Base flashing _____
Interior surfaces: dampproofing, _____ coats of _____ ; furring _____
Additional information: _____
Exterior painting: material _____ ; number of coats _____
Gable wall construction: ☐ same as main walls; ☒ other construction __Common truss with web backing, sheathed as above__

6. FLOOR FRAMING:
Joists: wood, grade, and species __2x10 #2 fir__ ; other _____ ; bridging __Metal__ ; anchors __3 5/8"__
Concrete slab: ☒ basement floor; ☐ first floor; ☒ ground supported; ☐ self-supporting; mix __3000 # mix__ ; thickness __3 5/8"__ ;
reinforcing _____ ; insulation _____ ; membrane _____
Fill under slab: material __Sand__ ; thickness __4"__. Additional information: _____

7. SUBFLOORING: *(Describe underflooring for special floors under item 21.)*
Material: grade and species __½" CD Plyscore__ ; size __4 x 8__ ; type __Solid__
Laid: ☒ first floor; ☐ second floor; ☐ attic _____ sq. ft.; ☐ diagonal; ☒ right angles. Additional information: _____

8. FINISH FLOORING: *(Wood only. Describe other finish flooring under item 21.)*

LOCATION	ROOMS	GRADE	SPECIES	THICKNESS	WIDTH	BLDG. PAPER	FINISH
First floor	Living & Dining	#1	Com.Oak	25/32"	1½"	Red Resin	Sand, seal & wax
XXXXXXX	Bedrooms, Hall	#1	Com.Oak	25/32"	1½"	Red Resin	Sand, seal & wax
Attic floor				sq. ft.			
Additional information:							

Figure 12-1. Typical specifications (page 1).

9. PARTITION FRAMING:
Studs: wood, grade, and species _#2 Fir_ size and spacing _2x4 -16" OC_ Other _____
Additional information: _____

10. CEILING FRAMING:
Joists: wood, grade, and species _____ Other _____ Bridging _____
Additional information: _____

11. ROOF FRAMING:
Rafters: wood, grade, and species _____ Roof trusses (see detail): grade and species _____
Additional information: _Stress grade material 2x4 truss 24" o.c._

12. ROOFING:
Sheathing: wood, grade, and species _½" CD Plywood_ ; ☒ solid; ☐ spaced _____" o.c.
Roofing _Asphalt sealdown_ ; grade _1_ ; size _4 x 8_ type _____
Underlay _15 lb. felt_ ; weight or thickness _235#_ size _12x36_ ; fastening _Nails_
Built-up roofing _____ ; number of plies _____ ; surfacing material _____
Flashing: material _Galvanized_ ; gage or weight _26 ga._ ; ☐ gravel stops; ☐ snow guards
Additional information: _____

13. GUTTERS AND DOWNSPOUTS:
Gutters: material _aluminum_ ; gage or weight _.027_ ; size _5"_ ; shape _Box_
Downspouts: material _aluminum_ ; gage or weight _.027_ ; size _5"_ ; shape _Box_ ; number _____
Downspouts connected to: ☐ Storm sewer; ☐ sanitary sewer; ☐ dry-well. ☒ Splash blocks: material and size _Concrete 10"x24"_
Additional information: _____

14. LATH AND PLASTER
Lath ☐ walls, ☐ ceilings: material _____ ; weight or thickness _____ Plaster: coats _____ ; finish _____
Dry-wall ☒ walls ☒ ceilings: material _Gypsum_ thickness _½"_ ; finish _Sanded_
Joint treatment _3 coats perfatape & cement_

15. DECORATING: *(Paint, wallpaper, etc.)*

ROOMS	WALL FINISH MATERIAL AND APPLICATION	CEILING FINISH MATERIAL AND APPLICATION
Kitchen		
Bath		
Other _All Rooms_	Prime coat - flat latex 2nd coat flat - flat latex	flat latex texture sprayed

Additional information: _____

16. INTERIOR DOORS AND TRIM:
Doors: type _Flush_ ; material _Oak_ ; thickness _1-3/8"_
Door trim: type _Ranch_ material _Oak_ Base: type _2 member_ ; material _Oak_ ; size _2-3/4_
Finish: doors _Stain and seal_ ; trim _Stain and seal_
Other trim *(item, type and location)* _Basement door - pine trim, seal coated._
Additional information: _Closet trim in 2 member pine stained and sealed._

17. WINDOWS:
Windows: type _Casements_ ; make _____ ; material _Pine_ ; sash thickness _1-3/4"_
Glass: grade _SSB_ ; ☐ sash weights; ☐ balances, type _____ ; head flashing _Aluminum_
Trim: type _Ranch_ ; material _Oak_ Paint _Stain and seal_ ; number coats _3_
Weatherstripping: type _Compression_ ; material _Aluminum & vinyl_ Insulated Glass
Screens: ☒ full; ☐ half; type _Aluminum_ ; number _All_ ; screen cloth material _Gun metal_
Basement windows: type _Awning_ ; material _Pine_ ; screens, number _All_ ; Storm sash, number _All_
Special windows _____
Additional information: _____

18. ENTRANCES AND EXTERIOR DETAIL:
Main entrance door: material _Oak_ ; width _3'0"_ ; thickness _1-3/4"_ Frame: material _Pine_ ; thickness _5/4"_
Other entrance doors: material _Oak_ ; width _2'8"_ ; thickness _1-3/4"_ Frame: material _Pine_ ; thickness _5/4"_
Head flashing _26 GA GI_ Weatherstripping: type _Spring Bronze_ ; saddles _GI_
Screen doors: thickness _____" ; number _____ ; screen cloth material _____ Storm doors: thickness _____"; number _____
Combination storm and screen doors: thickness _1_"; number _1_ ; screen cloth material _Aluminum_
Shutters: ☐ hinged; ☒ fixed. Railings _____ Attic louvers _Soffit and roof_
Exterior millwork: grade and species _Clear Redwood_ Paint _Lead & oil_ ; number coats _2_
Additional information: _____

19. CABINETS AND INTERIOR DETAIL:
Kitchen cabinets, wall units: material _Oak_ ; lineal feet of shelves _See Plans_ shelf width _11"_
Base units: material _Oak_ ; counter top _Textolite_ ; edging _Textolite_
Back and end splash _Textolite_ Finish of cabinets _factory finished lacquer_ ; number coats _3_
Medicine cabinets: make _Vanity and linen cabinet, per plan_ model _mirror over vanity_
Other cabinets and built-in furniture _____
Additional information: _____

20. STAIRS:

STAIR	TREADS		RISERS		STRINGS		HANDRAIL		BALUSTERS	
	Material	Thickness	Material	Thickness	Material	Size	Material	Size	Material	Size
Basement										
Main	Fir	5/4"	Fir	1"	Fir	2x10	Oak	2"		
Attic							W.I.			

Disappearing: make and model number _____
Additional information: _____

2

Figure 12-1 (cont'd.). Typical specifications (page 2).

21. SPECIAL FLOORS AND WAINSCOT:

	Location	Material, Color, Border, Sizes, Gage, etc.	Threshold Material	Wall Base Material	Underfloor Material
Floors	Kitchen	Inlaid linoleum	None	Plywood	Plyscore
	Bath	Ceramic	Marble	Cove	Plyscore
	Entry	Inlaid linoleum	None	Plywood	Plyscore

	Location	Material, Color, Border, Cap. Sizes, Gage, etc.	Height	Height Over Tub	Height in Showers (From Floor)
Wainscot	Bath – Full:	Ceramic wall tile		4'0	
	– 3/4:	Ceramic wall tile	4.0		

Bathroom accessories: ☒ Recessed; material ___Ceramic___ ; number __3__ ; ☒ Attached; material __ceramic__ ; number __2__
Additional information: _____

22. PLUMBING:

Fixture	Number	Location	Make	Mfr's Fixture Identification No.	Size	Color
Sink	1	Kitchen	Elkay	LR 3372	33x22	Stainless S.
Lavatory	1	Full Bath	Kohler	Caxton K 2210C	17 x 14	Green
Water closet	1	Full Bath	Kohler	Wellworth K3512 PBA		Green
Bathtub	1	Full Bath	Kohler	Seaforth K746SA	5'0"	Green
Shower over tub △	None					
Stall shower △		3/4 Bath	Fiat	Cascade 48 CST	48 x 34	White
Laundry trays	1	Basement	Fiat	Fiberglass LTD	2 compt.	White
Lavatory	1	3/4 Bath	Kohler	Chesapeake K1746C	19 x 17	Blue
Water Closet	1	3/4 Bath	Kohler	Wellworth K3512 PBA		Blue

△ ☐ Curtain rod   ☒ ☒ Door   ☐ Shower pan: material _____
Water supply: ☒ public; ☐ community system; ☐ individual (private) system. ★
Sewage disposal: ☒ public; ☐ community system; ☐ individual (private) system. ★
★ Show and describe individual system in complete detail in separate drawings and specifications according to requirements.
House drain (inside): ☒ cast iron; ☐ tile; ☒ other _Inside drain above ground_   House sewer (outside): ☒ cast iron; ☐ tile; ☐ other _Hubless cast iron or equal._
Water piping: ☐ galvanized steel; ☒ copper tubing; ☐ other _____   Sill cocks, number __3__
Domestic water heater: type _Gas Automatic_ ; make and model _A.O. Smith PGDL 40_ ; heating capacity
___42.0___ gph. 100° rise. Storage tank: material _glass lined_ ; capacity __40__ gallons.
Gas service: ☒ utility company; ☐ liq. pet. gas; ☐ other _____   Gas piping: ☐ cooking; ☒ house heating.
Footing drains connected to: ☐ storm sewer; ☐ sanitary sewer; ☐ dry well. Sump pump; make and model _____
_____ ; capacity _____ ; discharges into _____

23. HEATING:

☐ Hot water. ☐ Steam. ☐ Vapor. ☐ One-pipe system. ☐ Two-pipe system.
☐ Radiators. ☐ Convectors. ☐ Baseboard radiation. Make and model _____
Radiant panel: ☐ floor; ☐ wall; ☐ ceiling. Panel coil: material _____
☐ Circulator. ☐ Return pump. Make and model _____ ; capacity _____ gpm.
Boiler: make and model _____ Output _____ Btuh.; net rating _____ Btuh.
Additional information: _____
Warm air: ☐ Gravity. ☒ Forced. Type of system _Gas – Perimeter_
Duct material: supply _Sht.metal Galv._ return _Galv. Sheet Metal_ Insulation _____ thickness _____. ☒ Outside air intake.
Furnace: make and model _GE LU 105 B2B_ Input _105,000_ Btuh.; output _84,000_ Btuh.
Additional information: _____
☐ Space heater; ☐ floor furnace; ☐ wall heater. Input _____ Btuh.; output _____ Btuh.; number units _____
Make, model _____ Additional information: _____
Controls: make and types _Honeywell, or equal_ _____
Additional information: _____
Fuel: ☐ Coal; ☐ oil; ☒ gas; ☐ liq. pet. gas; ☐ electric; ☐ other _____ ; storage capacity _____
Additional information: _____
Firing equipment furnished separately: ☐ Gas burner, conversion type. ☐ Stoker: hopper feed ☐; bin feed ☐
Oil burner: ☐ pressure atomizing; ☐ vaporizing
Make and model _____ Control _____
Additional information: _____
Electric heating system: type _____ Input _____ watts; @ _____ volts; output _____ Btuh.
Additional information: _____
Ventilating equipment: attic fan, make and model _____ ; capacity _____ cfm.
kitchen exhaust fan, make and model _GE JV 33_
Other heating, ventilating. or cooling equipment _____

24. ELECTRIC WIRING:

Min. of
Service: ☒ overhead; ☐ underground. Panel: ☐ fuse box; ☒ circuit-breaker; make _100 amp ser_ No. MP's _____ No. circuits __8__
Wiring: ☐ conduit; ☐ armored cable; ☐ nonmetallic cable; ☐ knob and tube; ☒ other _Greenfield_
Special outlets: ☐ range; ☐ water heater; ☐ other _____
☐ Doorbell. ☒ Chimes. Push-button locations _Front door_ Additional information: _____

25. LIGHTING FIXTURES:

Total number of fixtures _12*_ Total allowance for fixtures, typical installation. $ _____
Nontypical installation _* includes luminous light in ktchen and bath_
Additional information: _____

DESCRIPTION OF MATERIALS

Figure 12-1 (cont'd.). Typical specifications (page 3).

26. INSULATION:

Location	Thickness	Material, Type, and Method of Installation	Vapor Barrier
Roof			
Ceiling	5"	Mineral wool blown	Yes
Wall	2"	Blanket – fiberglass	Yes
Floor			

HARDWARE: (make, material, and finish.) Weiser

SPECIAL EQUIPMENT: (State material or make, model and quantity.   Include only equipment and appliances which are acceptable by local law, custom and applicable FHA standards.   Do not include items which, by established custom, are supplied by occupant and removed when he vacates premises or chattles prohibited by law from becoming realty.)

Dishwasher: G.E. Pot Scrubber
Garbage disposal unit: Insinkerator ISE 77
Other: Water Softener: Lindsay Imperial Probe

27. MISCELLANEOUS: (Describe any main dwelling materials, equipment, or construction items not shown elsewhere; or use to provide additional information where the space provided was inadequate.   Always reference by item number to correspond to numbering used on this form.)

1   Clothes chute
4   12 x 7 roof louvers
2   Special roof louvers
18 Garage service door: material – pine; Width – 2'8" thickness – 1 3/8"
    Luminous kitchen ceiling
    Aluminum facia

PORCHES:

TERRACES:

GARAGES:
    20 x 22 attached per plan

WALKS AND DRIVEWAYS:
Driveway: width 16'0"; base material Sand; thickness "; surfacing material Concrete; thickness 3 5/8"
Front walk: width 4'; material Concrete, thickness 3 5/8" Service walk: width _____; material _____; thickness _____"
Steps: material _____; treads _____"; risers _____". Cheek walls _____

OTHER ONSITE IMPROVEMENTS:
(Specify all exterior onsite improvements not described elsewhere, including items such as unusual grading, drainage structures, retaining walls, fence, railings, and accessory structures.)

LANDSCAPING, PLANTING, AND FINISH GRADING:
Topsoil 4" " thick: XX front yard; XX side yards; XX rear yard to lot line feet behind main building.
Lawns (seeded, sodded, or sprigged): ☐ front yard _____; ☐ side yards _____; ☐ rear yard _____
Planting: ☐ as specified and shown on drawings; ☐ as follows:
_____ Shade trees, deciduous. _____" caliper. | _____ Evergreen trees. _____' to _____', B & B.
_____ Low flowering trees, deciduous, _____' to _____' | _____ Evergreen shrubs. _____' to _____', B & B.
_____ High-growing shrubs, deciduous, _____' to _____' | _____ Vines, 2-year _____
_____ Medium-growing shrubs, deciduous, _____' to _____'
_____ Low-growing shrubs, deciduous, _____' to _____'

IDENTIFICATION.—This exhibit shall be identified by the signature of the builder, or sponsor, and/or the proposed mortgagor if the latter is known at the time of application.

Date_____ Signature_____

Signature_____

Figure 12-1 (cont'd.). Typical specifications (page 4).

Rough-In Drawings

The drawings shown in Figures 12-2 to 12-18 are what plumbers call *rough-in drawings* or *rough-in sheets.*

By studying these drawings, you can see that they show all the necessary dimensions for waste and water supply pipes going to and from the various fixtures. The purpose of the drawings is to show the plumber where to install the waste and water supply piping for the fixtures so that these will fit the fixtures when they are installed during the finishing operations (which are done after the wall coverings and floors have been placed).

Rough-in drawings are secured by plumbing contractors when they place their fixture orders with their suppliers.

It should be kept in mind that rough-in drawings usually apply only to the specific job for which the manufacturer issued them. Such drawings should be obtained with each order placed with the manufacturer, because some of the important dimensions change frequently as improvements and other design changes are made in the fixtures.

Plumbing Permits

Before starting to install any piping on a plumbing job, the plumber must be certain that the proper permits for that job have been obtained. Plumbing permits are required by most municipalities so that there is a record of all plumbing work being performed in the municipality.

Plumbing permits are usually obtained by the plumbing contractor. When the contractor applies for a plumbing permit, he may be required to submit blueprints and specifications of the plumbing work he proposes to install so that the plumbing inspector may see that they meet the local plumbing code requirements. On approval of the plumbing plans, the municipality will issue plumbing permits for the work to be performed.

Plumbing permits are usually required for water service connections, building sewer connections, and the plumbing to be installed within the building. The plumbing contractor is required to pay a fee for each permit. The purpose of this fee is to help defray the costs of the plumbing inspections necessary for the work.

As a note of caution: if the plumber starts to install any phase of the plumbing on a job before the proper plumbing permits have been obtained, the plumbing inspector may refuse to inspect the work until the permits have been obtained. In addition, in some municipalities, plumbing inspectors have the authority to double the usual permit fee for jobs on which the plumbing work is started before the proper permits are obtained.

Lustertone®
TWO COMPARTMENT SINK
NO. LR-3322

SPECIFICATION
#18 Gauge, type 302 (18-8), stainless steel ledge back sink with self rimming feature. Compartments are recessed 7/16" below faucet deck and other outside flanges. Compartments have 1-3/4" radius vertical and horizontal coved corners. Exposed surfaces are hand blended to a uniform LK-6K-H satin finish. Entire underside of sink is sound deadened.

SPECIFIED FITTINGS:

Counter Top Opening
32-3/8" X 21-3/8" — Corners, 1-1/2" Radius

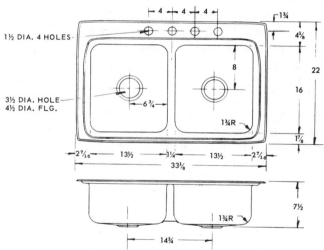

1½ DIA. 4 HOLES

3½ DIA. HOLE
4½ DIA. FLG.

INDICATE FAUCET DRILLINGS REQUIRED

☐ 3 Holes

☐ 4 Holes

JOB NAME _____ DATE: _____

CUSTOMER _____

ARCHITECT _____

ELKAY new concepts in stainless steel sinks

ELKAY MANUFACTURING CO., 2700 S. 17th AVE., BROADVIEW, ILL. 60153

This specification describes an Elkay product with design, quality and functional benefits to the user. When making a comparison of other producers' offerings, be certain these features are not overlooked.

Figure 12-2. Double-compartment kitchen sink rough-in sheet. (Elkay Mfg. Co.)

"CENTURA" Sink Mixing Faucets

K-6894   Lever handle, two water.

K-6895   Lever handle, two water with spray.

K-6896   Two water.

If wanted with ground joint inlets, add suffix P.

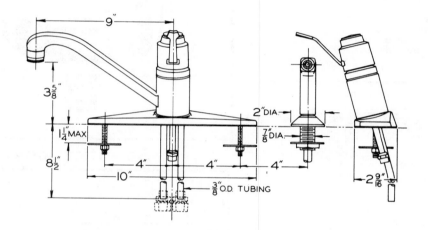

KOHLER | PLUMBING FIXTURES & FITTINGS
ENGINEERED FOR EACH OTHER

KOHLER CO. KOHLER, WISCONSIN 53044

K-6894 (1-76)

Figure 12-3. Kitchen sink faucet rough-in sheet. (Kohler Co.)

MODEL 77
1/2 HP Stainless Steel Garbage Disposer

Submittal Sheet

- 5-year warranty on parts, 1-year warranty on labor
- Whisper Quiet polystyrene sound barrier
- INSTANT ENERGY CAPACITOR MOTOR
- Stainless steel 2-piece easygrip stopper
- Stainless steel grind chamber
- Exclusive self-service "Wrenchette"
- "Quick lock" mounting
- Exclusive polypropylene corrosion shield
- Automatic reversing action
- Permanently lubricated upper and lower bearings
- Exclusive cushioned anti-splash baffle
- Sound-absorbing upper shell
- Dishwasher drain connection
- Overload protection manually reset

Dimensions

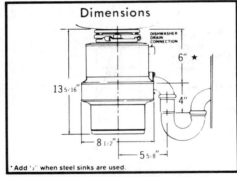

* Add ½" when steel sinks are used.

Sample Specifications

Garbage Disposer(s) shall be In-Sink-Erator stainless steel Model 77. Continuous feed, automatic reversing switch, with ½ H.P. capacitor start motor, corrosion protection shield, self-service wrenchette, unit lifetime corrosion warranty, 5-year warranty on parts, 1-year warranty on labor.*

* The complete ISE warranty and all parts covered, is included in the Care & Use Booklet packed with each unit.

Specifications

Type of Feed	Continuous	Unit Finish	Acrylic White Enamel Stainless Steel
On/Off Control	Wall Switch		
Motor	Instant Capacitor Start	Overall Height	13-5/16"
		Rough-In for Drain*	6"
HP	1/2		
Volts	115	Grind Chamber Cap.	5-1/8 cups **
HZ	60		
RPM	1725	Motor Protection	Manual reset Overload
Amp. (Avg. Load)	6.9		
Time Ratings	Intermittent	Average Water Usage	1-1/2 Gallons Per Person Per Day
Lubrication	Permanently Lubricated Upper & Lower Bearings	Average Electrical Usage	1/2 KWH Per Month
		Drain Connection	
Shipping Weight (Approx.)	23 lbs.	Dishwasher Drain Connection	

* Add ½" when steel sinks are used. Meets MIL-G-15840-B Type II.

** Unlimited capacity — continuous feed.

Job Specifications

Figure 12-4. Garbage disposal rough-in sheet. (In-Sink-Erator of Emerson Electic Co.)

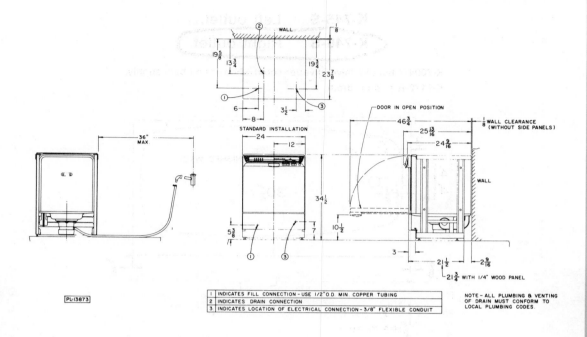

DETAILS AND CONNECTIONS

	INDICATES FILL CONNECTION - USE 1/2" O.D. MIN. COPPER TUBING
	INDICATES DRAIN CONNECTION
	INDICATES LOCATION OF ELECTRICAL CONNECTION - 3/8" FLEXIBLE CONDUIT

NOTE - ALL PLUMBING & VENTING OF DRAIN MUST CONFORM TO LOCAL PLUMBING CODES.

CAUTION

For installation when the dishwasher is to be left unused in freezing temperatures: Turn off electrical power and shut off water supply at the hand valve. Remove lower panel. Disconnect both inlet and outlet lines at the DRAIN valve. This permits water to drain from the dishwasher tank, pump and drain line. Disconnect both the inlet and outlet lines at the FILL valve. Make provisions to control water drained from unit. Reinstall lower panel. Turn on electrical power. Then, push door handle down to lock position and push "Normal Cycle" button. Let dishwasher run through first pre-rinse only (approximately four minutes) to drain all water trapped in the dishwasher. Turn off electricity. To complete installation later:

Connect drain and fill valves-both inlet and outlet.

Reinstall lower panel.

Turn on water and electrical supply. Then push button for cycle desired.

KITCHENAID DIVISION **HOBART** CORPORATION TROY, OHIO 45374

FORM 12106A (Rev. 4-76) (Supsds. F. 12106, 2-76)   PRINTED IN U.S.A.

Figure 12-5. Pot-scrubber dishwasher rough-in sheet. (Kitchen Aid Division, Hobart Corp.)

"SEAFORTH" Recess Bath

K-745-S Left outlet.

K-746-S Right outlet.

K-7004-T built-in valve, diverter spout, shower and bath supply.
K-7172-R pop-up drain.

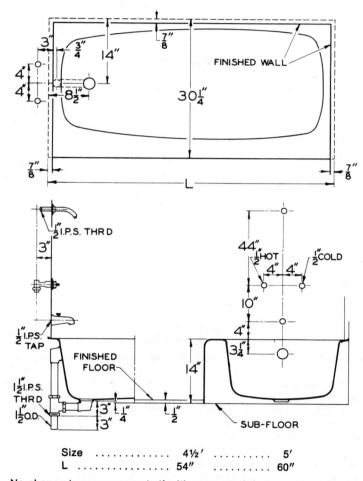

Size .... 4½' .... 5'
L .... 54" .... 60"

No change in measurements if with connected drain and overflow.

KOHLER | PLUMBING FIXTURES & FITTINGS ENGINEERED FOR EACH OTHER

KOHLER CO. KOHLER, WISCONSIN 53044

K-745-S (1-76)   Measurements may vary ½".

Figure 12-6. Bathtub rough-in sheet. (Kohler Co.)

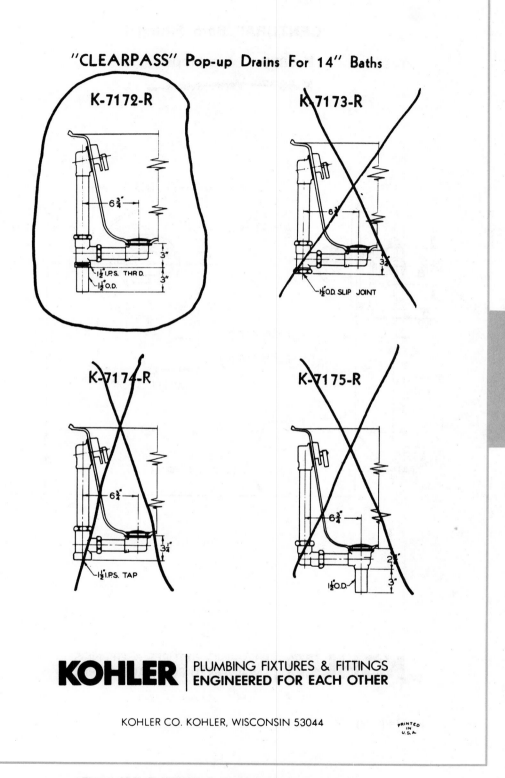

Figure 12-7. Bathtub drain rough-in sheet. (Kohler Co.)

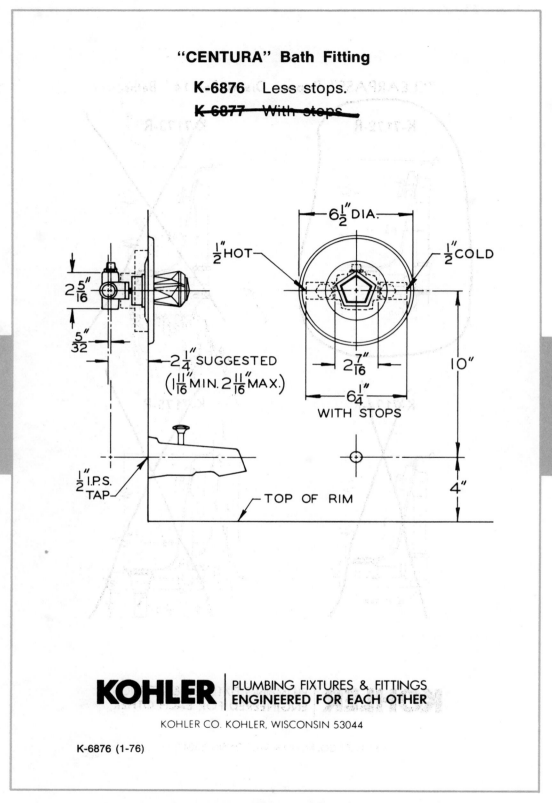

"CENTURA" Bath Fitting

K-6876 Less stops.

~~**K-6877** With stops.~~

KOHLER | PLUMBING FIXTURES & FITTINGS
ENGINEERED FOR EACH OTHER

KOHLER CO. KOHLER, WISCONSIN 53044

K-6876 (1-76)

Figure 12-8. Bathtub fitting rough-in sheet. (Kohler Co.)

WELLWORTH WATER - GUARD
Closet Combination

	X	L
K-3500-EB	12"	29"
K-3500-PB	12"	27"
K-3502-PB	10"	27"
K-3504-PB	14"	27"

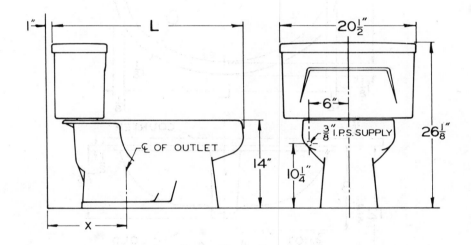

Measurements may vary ½".

KOHLER | PLUMBING FIXTURES & FITTINGS
ENGINEERED FOR EACH OTHER

KOHLER CO. KOHLER, WISCONSIN 53044

PRINTED IN U.S.A.

Figure 12-9. Water closet rough-in sheet. (Kohler Co.)

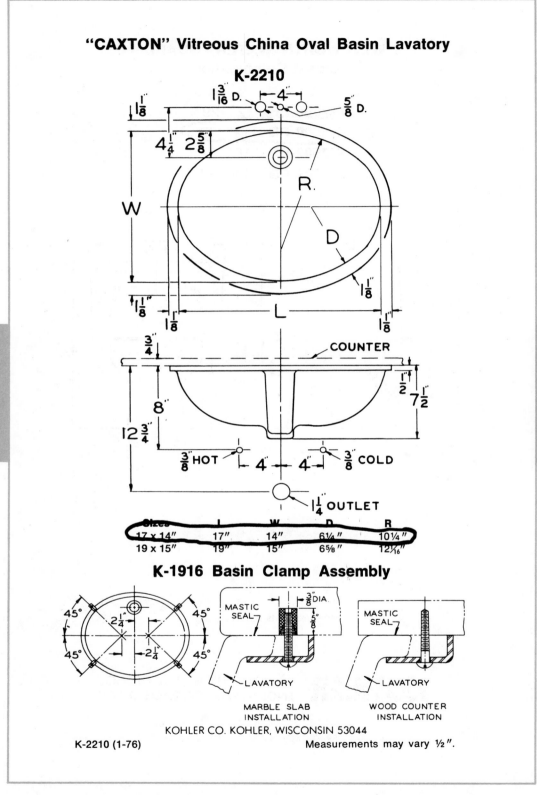

"CAXTON" Vitreous China Oval Basin Lavatory

K-2210

Sizes	L	W	D	R
17 x 14"	17"	14"	6¼"	10¼"
19 x 15"	19"	15"	6⅝"	12¹/₁₆

K-1916 Basin Clamp Assembly

MASTIC SEAL

⅜"DIA.

⅝"

LAVATORY

MARBLE SLAB INSTALLATION

MASTIC SEAL

LAVATORY

WOOD COUNTER INSTALLATION

KOHLER CO. KOHLER, WISCONSIN 53044

K-2210 (1-76)   Measurements may vary ½".

Figure 12-10. Vanity lavatory rough-in sheet. (Kohler Co.)

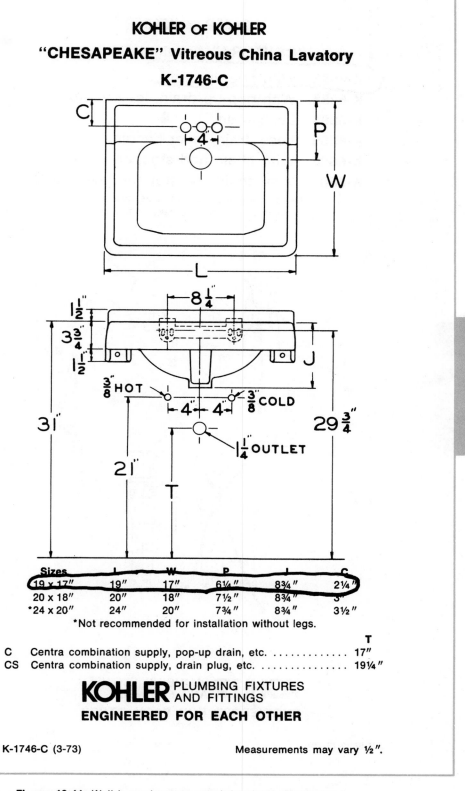

KOHLER of **KOHLER**

"CHESAPEAKE" Vitreous China Lavatory
K-1746-C

Sizes	L	W	P	J	C
19 x 17"	19"	17"	6¼"	8¾"	2¼"
20 x 18"	20"	18"	7½"	8¾"	3"
*24 x 20"	24"	20"	7¾"	8¾"	3½"

*Not recommended for installation without legs.

		T
C	Centra combination supply, pop-up drain, etc.	17"
CS	Centra combination supply, drain plug, etc.	19¼"

KOHLER PLUMBING FIXTURES AND FITTINGS

ENGINEERED FOR EACH OTHER

K-1746-C (3-73)   Measurements may vary ½".

Figure 12-11. Wall-hung lavatory rough-in sheet. (Kohler Co.)

KOHLER of KOHLER

"CENTURA" Lavatory Fittings

K-6882 With pop-up drain.

K-6883 With pop-up drain, lever handle.

K-6884 Less drain.

K-6885 Less drain, lever handle.

K-6886 With chain and stopper.

K-6887 With chain and stopper, lever handle.

If wanted with ground joint inlets, add suffix P.
If wanted with stream breaker, add suffix S.

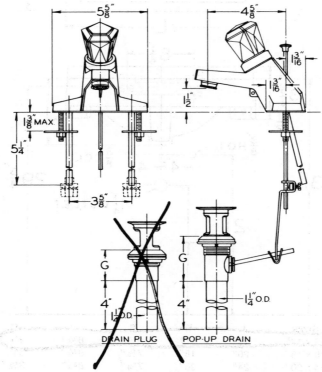

G Pop-up drain 2⅝". Drain plug 1¾".

KOHLER PLUMBING FIXTURES AND FITTINGS

ENGINEERED FOR EACH OTHER

K-6882 (3-73)

Figure 12-12. Lavatory faucet rough-in sheet. (Kohler Co.)

MODEL	NOMINAL SIZE		ACTUAL SIZE		ROUGHING IN DIMENSIONS			
No.	A	B	A	B	a	b	c	d
32CST	32	32	31-7/8	31-7/8	32-1/4	32-1/16	16-1/8	16-1/8
34CST	34	34	33-7/8	33-13/16	34-1/4	34	17-1/8	17-1/16
36CST	36	36	35-7/8	35-13/16	36-1/4	36	18-1/8	18-1/16
42CST	42	34	41-13/16	33-13/16	42-1/8	34	21-1/16	17-1/16
48CST	48	34	47-13/16	33-13/16	48-1/8	34	24-1/16	17-1/16
54CST	54	34	53-13/16	33-3/4	54-1/8	33-15/16	27-1/16	17-1/16
60CST	60	34	59-3/4	33-13/16	60-1/8	34	30-1/16	17-1/16
34CDT	34	34	33-7/8	33-7/8	34-1/16	34-1/16	17-1/8	17-1/8
36CDT	36	36	35-13/16	35-13/16	36	36	18-1/8	18-1/8
38CSTC	38	38	37-13/16	37-13/16	38	38	12-3/16	12-3/16
48ST*	48	32	47-9/16	31-5/8	47-7/8	31-3/4	23-15/16	15-7/8

*Not Illustrated.

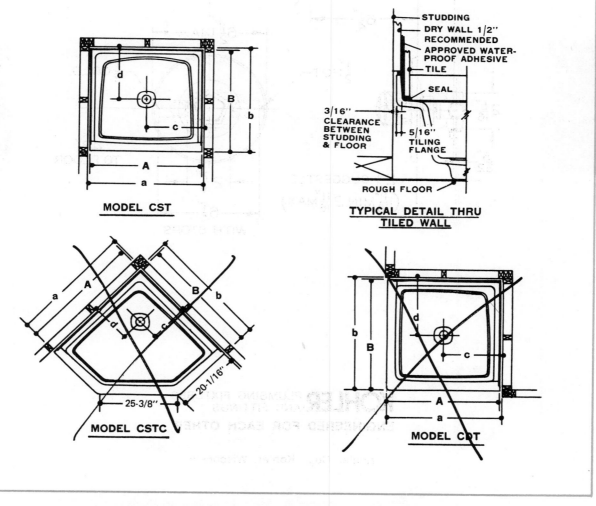

Figure 12-13. Shower bath base rough-in sheet. (Powers–Fiat Co.)

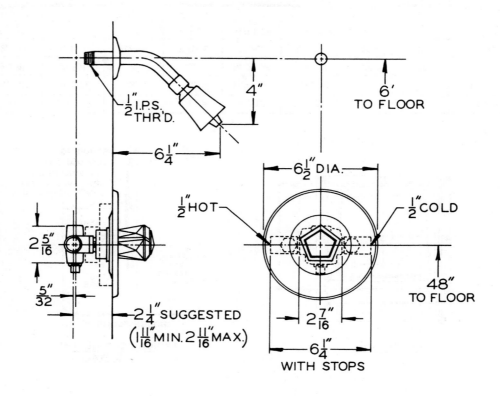

KOHLER OF KOHLER
"CENTURA" Shower Fitting
K-6874 Less stops.
K-6875 With stops.

½" I.P.S. THR'D.

4"

6' TO FLOOR

6¼"

6½" DIA.

½" HOT

½" COLD

2 5/16"

5/32"

2¼" SUGGESTED
(1 11/16" MIN. 2 11/16" MAX.)

2 7/16"

6¼"

48" TO FLOOR

WITH STOPS

KOHLER PLUMBING FIXTURES AND FITTINGS
ENGINEERED FOR EACH OTHER

Kohler Co.,   Kohler, Wisconsin

Figure 12-14. Shower faucet rough-in sheet. (Kohler Co.)

SERV-A-SINK®

MODEL L-1*

MODEL LTD
(U.S. Patent No. 3,605,456)

STANDARD SIZES: Dimensions shown conform to those specified in Federal Specifications WW-P-541/5A-3.2.7 (Laundry Tubs) and WW-P-541/5A-3.2.7.7 (Cabinet Type).

	Single Tub	Double Tub
Overall Outside Dimensions	23 x 21-1/2 x 13-7/16	45-1/8 x 21-3/8 x 15-3/16
Inside Between Walls and Bottom	20-1/2 x 17-1/4 x 13	21-1/2 x 16-1/2 x 12-3/4 ea. side
Deck Space With Integrally Molded Soap Dish	4″	4″
Capacity	20 Gallons	40 Gallons
Static Load (Wall Hung and Leg Mounted)	600 Pounds	600 Pounds
Overall Height From Floor	34-3/4	35-1/4

* U S Patent Pending

GENERAL: MOLDED-STONE® Laundry Tub shall be manufactured by Powers-Fiat. Molding shall be done in matched metal dies under heat and pressure, resulting in a homogeneous molded section. Strainer(s) shall be furnished with stopper. Capacity shall not be less than 20 gallons for L-1 and FL-1 or 40 gallons for LTD and FLTD models.

COLOR: No. 218 Confetti White is standard. Optional without cost: No. 233 Green Drift.

MODELS: L-1*, single tub wall mounted; FL-1, single tub leg mounted, L-2, single tub cabinet mounted; LTD, double tub wall mounted; FLTD, double tub leg mounted.

OPTIONAL ACCESSORIES: See reverse side.

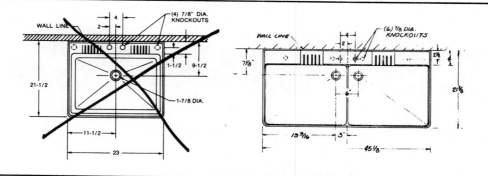

DIVISION OF POWERS REGULATOR COMPANY

Figure 12-15. Laundry tray rough-in sheet.

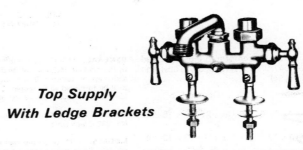

**Top Supply
With Ledge Brackets**

NUMBER	SPOUT	UNION CONNECTIONS	FINISH
50214	4¾"	DUAL ½" I.P.—½" C	SATIN

Figure 12-16. Laundry tray faucet rough-in sheet.
(Wolverine Brass)

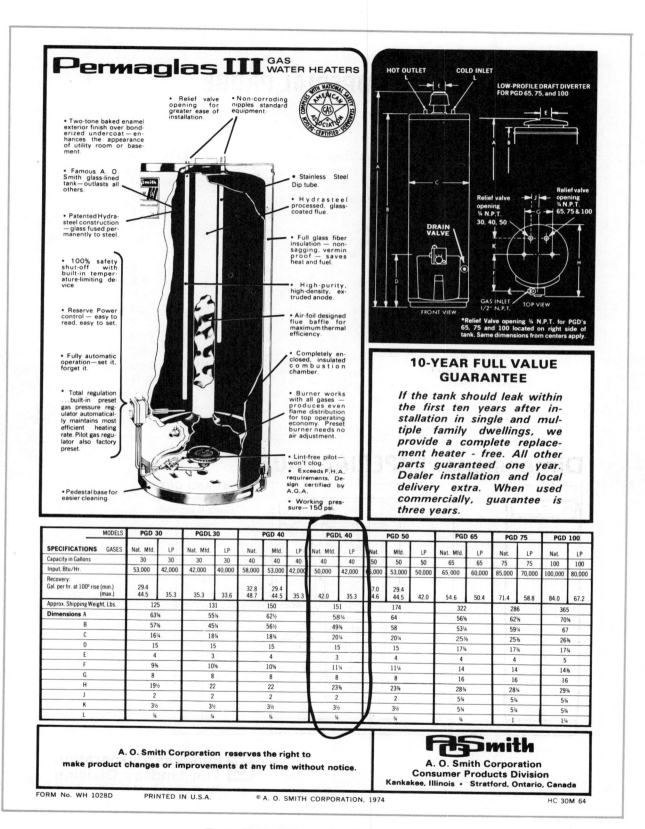

Figure 12-17. Water heater rough-in sheet.

Specifications: IMPERIAL PROBE

Rated Capacity (Grains) @ Salt Usage (lbs.) .... 21,000 @ 13.5
18,000 @ 7

Type of Ion Exchange Material .... High Capacity Resin
Amount of Ion Exchange Material (Cu. Ft.).................................. .95
Water Pressure Limits (PSI)... 20-125
Maximum Water Temperature (°F) 120
Minimum Pump Capacity (Gals. Per Hour) 180
Service Flow Rate (Gallons Per Minute @
      15 PSI Pressure Loss) 9
Backwash Flow (Gals Per Min.) 1.8
Brine Flow (Gals Per Min.) 5
Brine Rinse Flow (Gal. Per Min.)....................................... .33
Backwash Flow (Gal. Per Min.) 1.8
Fast Rinse Flow (Gal. Per Min.) 1.6
Salt Storage Capacity (Lbs.) 200
Electrical Rating .... 115 Volt, 60 Cycle
Water Temp. .... 100°F

Dimensions: IMPERIAL PROBE

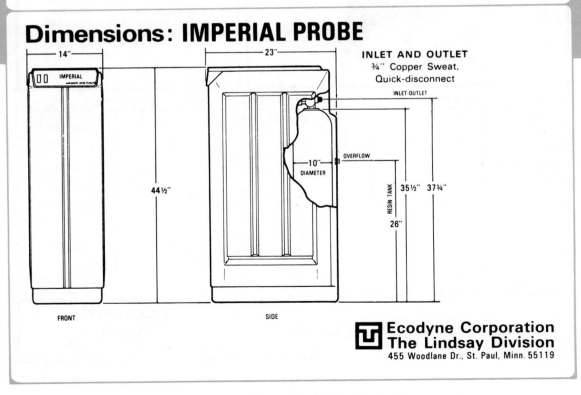

Figure 12-18. Water softener rough-in sheet.

◯ INSTALLATION OF THE ROUGH PLUMBING

As mentioned in Chapter 10, the installation of the rough plumbing, or roughing-in, consists of the installation of all parts of the plumbing system that can be completed prior to the installation of fixtures. This includes the drainage, vent, and water supply piping, as well as any necessary fixture supports. The rough plumbing in a single-family dwelling of the type illustrated in this chapter is usually installed after all the framing of the house is completed and the roof is covered.

Before starting the rough plumbing, the plumber should study Sheets 2, 3, and 6 of the *House Plans* to locate the various plumbing fixtures and appliances. The specifications on Figure 12-1, page 3 should also be checked for the type(s) of piping material to be used. The specifications in this example specify the use of hubless (no-hub) cast iron for the building drainage piping.

Sketch of Waste and Vent Piping and List of Material

After studying the plans and specifications, it would be well for the plumber to sketch an isometric drawing of the sanitary waste and vent piping, to give an idea of how this piping will appear when it is completed and also to aid him in ordering the proper material for the installation. Figure 12-19 is such an isometric drawing of the sanitary drainage and vent piping for the example presented in this chapter, with the pipe sized as it would be installed with no-hub pipe and fittings. You will notice the absence of $1\frac{1}{4}$-inch size pipe, but you must remember that no-hub pipe and fittings are not manufactured in this size. Figure 12-20 is a reproduction of Figure 12-19, with all the fittings to be used marked in their proper places on the drawing. Figure 12-21 is a list of the material required to pipe the sanitary drainage and vent piping for this example.

As the plumber becomes more experienced in housing work, he will eventually form the habit of visualizing this part of the plumbing installation without having to make isometric drawings of this type.

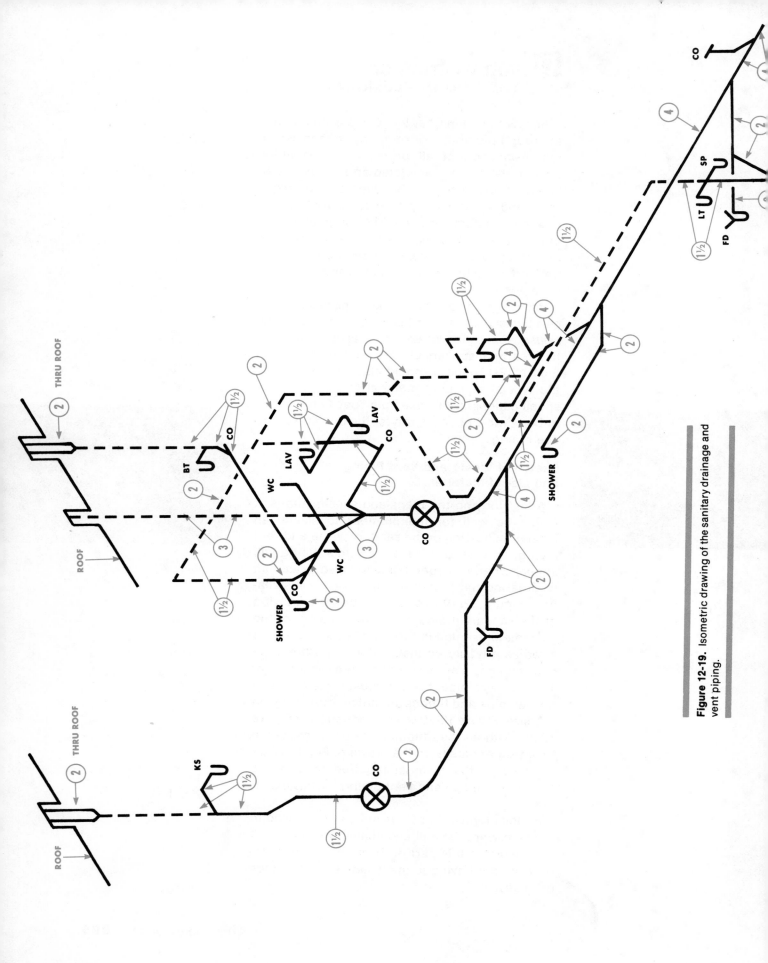

Figure 12-19. Isometric drawing of the sanitary drainage and vent piping.

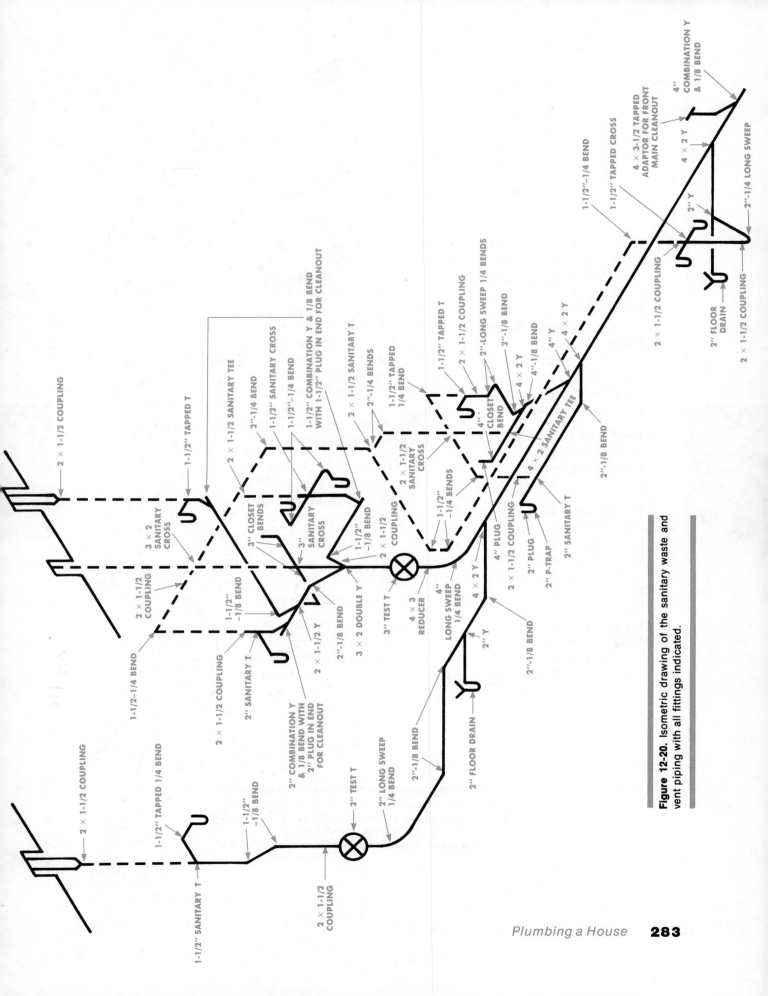

Figure 12-20. Isometric drawing of the sanitary waste and vent piping with all fittings indicated.

1—4″ combination Y and ⅛ bend
1—4 × 3½ tapped adaptor
    (for front main cleanout)
1—4″ Y
4—4 × 2 Ys
1—4″ long sweep ¼ bend
1—4″ ⅛ bend
1—4 × 2 sanitary T
1—4 × 4 × 16 × 16 closet bend
1—4″ blind plug
1—4 × 3 reducer

1—3″ test T
1—3 × 2 double Y
1—3″ sanitary cross
2—3 × 3 × 16 × 16 closet bends

2—2″ floor drains
2—2″ P-traps
2—2″ Ys
4—2″ long sweep ¼ bends
3—2″ ¼ bends
5—2″ ⅛ bends
2—2″ Ts
2—2 × 1½ Ts
1—2 × 1½ cross
1—2 × 1½ Y
1—2″ combination Y and ⅛ bend

2—2″ blind plugs
1—2″ test T
1—1½″ tapped cross
2—1½″ tapped Ts
3—1½″ tapped ¼ bends
1—1½″ T
1—1½″ cross
2—1½″ combination Y and ⅛ bends
7—1½″ ¼ bends
3—1½″ ⅛ bends

The following list of no-hub couplings is only an approximation:

19—4″ no-hub couplings
10—3″ no-hub couplings
55—2″ no-hub couplings
45—1½″ no-hub couplings
 7—2 × 1½″ no-hub couplings

The following list of pipe is also an approximation as no allowance was made for fittings when scaling the drawings:

20 feet 4″ no-hub pipe
20 feet 3″ no-hub pipe
60 feet 2″ no-hub pipe
80 feet 1½″ no-hub pipe

Figure 12-21. Material required to pipe the sanitary waste and vent pipe system illustrated in Figure 12-19, as taken from Figure 12-20. (All fittings are no-hub soil pipe.)

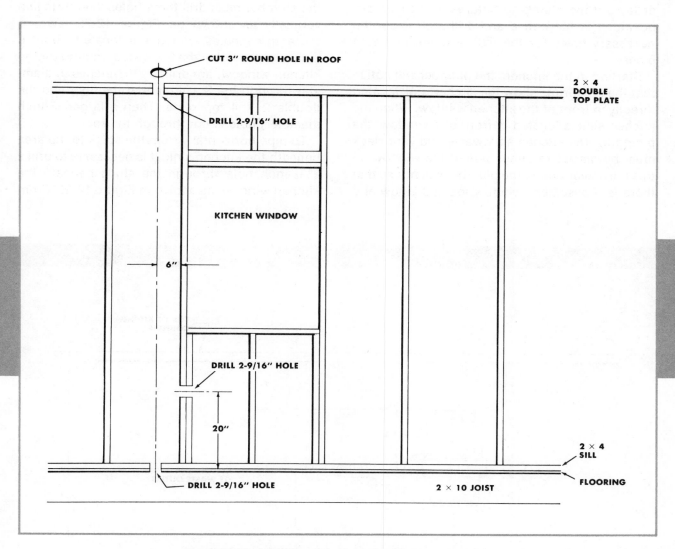

Figure 12-22. Location of the holes for kitchen sink waste and vent piping.

Location of Fixtures on the First Floor and Drilling the Necessary Holes

After sketching the rough plumbing and obtaining the material, the plumber is ready to proceed with the next step. He must now locate or lay out the plumbing fixtures in their proper positions in the house and drill and/or cut the necessary holes for the fixture waste and vent pipes.

Starting in the kitchen, the plumber will notice that the blueprints show the kitchen sink located directly in front of the kitchen window. When the kitchen sink is located in front of a window, the piping for the kitchen sink waste and vent stack must be placed to either side of the window. In this particular home, the plumber will notice that there is a basement window located below and to the right of the kitchen sink; to avoid a conflict with this window, the kitchen sink waste stack will have to be placed to the left of the kitchen window. To provide the proper opening for the waste and vent stack serving the kitchen sink, the plumber must drill three holes directly in line vertically as shown on Figure 12-22. Two of these three holes, which are located about 6 inches to the left of the 2 × 4 stud framing for the kitchen window, are drilled 2$^9/_{16}$-inches in diameter through the 2 × 4 plate at the floor and the double 2 × 4 top plate. Then a larger (3-inch diameter) hole is cut through the roof.

To pipe horizontally from the stack to the area beneath the kitchen sink, it is necessary to drill a 2$^9/_{16}$-inch hole through the studs beneath the kitchen window, as shown in Figure 12-22. From

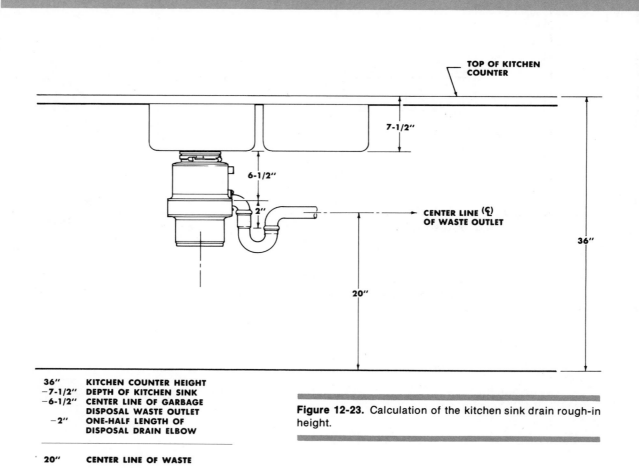

```
 36"       KITCHEN COUNTER HEIGHT
-7-1/2"    DEPTH OF KITCHEN SINK
-6-1/2"    CENTER LINE OF GARBAGE
           DISPOSAL WASTE OUTLET
  -2"      ONE-HALF LENGTH OF
           DISPOSAL DRAIN ELBOW
─────────────────────────────────
 20"       CENTER LINE OF WASTE
```

Figure 12-23. Calculation of the kitchen sink drain rough-in height.

the data in Figure 12-23, the location of this hole can be computed as follows:

Kitchen counter height: 36 inches

Subtract the following measurements:
- 7½"   Depth of kitchen sink
- 6½"   Center line of garbage disposal waste outlet
- 2"   One-half length of disposal drain elbow
_____
−16"

When this 16 inches is subtracted from the 36-inch height of the kitchen sink, it indicates that the hole should be centered at 20 inches above the floor.

Moving to the bathroom area of the house, the plumber should refer to Sheet 9 of the house plan blueprints, which is the detail sheet for the full and three-quarter bathrooms.

A floor plan of the holes required for the waste and vent pipes for these two bathrooms is illustrated in Figure 12-24.

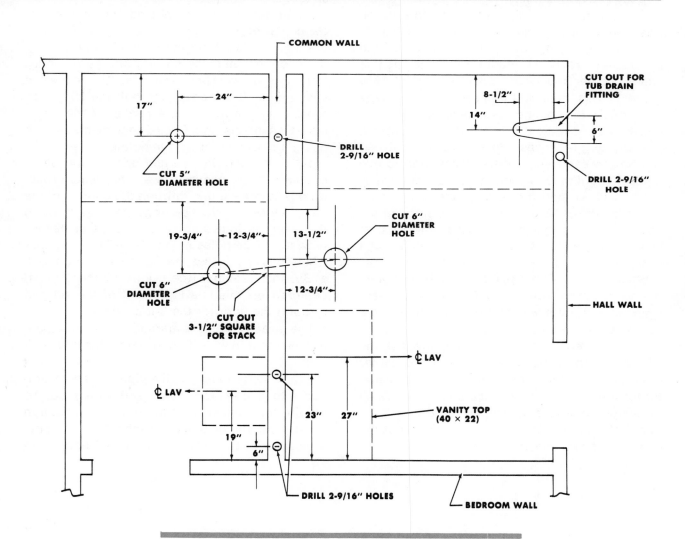

Figure 12-24. Location of the holes for bathroom waste and vent pipes.

Starting with the bathtub waste, lay out on the floor the location of the bathtub drain opening, as shown on Figure 12-6 (the bathtub rough-in sheet), 14 inches from the back wall and $8^1/_2$ inches from the hall wall, and drill a $2^9/_{16}$-inch hole. From this hole to the end wall, lay out a triangle-shaped opening through the stud wall, as shown on Figure 12-24, and cut this opening to provide clearance for the bathtub waste fitting. Drill a $2^9/_{16}$-inch hole through the sill alongside the triangle-shaped opening for the bathtub vent. Since the bathtub vent continues through the roof, it is also necessary to drill through the top plate and cut a hole for the vent in the roof directly above this hole.

Lay out the 40- by 22-inch vanity top on the floor. Measure the space left between the vanity top and the short stub wall at the end of the bathtub—there is 27 inches of space. In this space, center the water closet waste (at $13^1/_2$ inches). This hole will be centered out $12^3/_4$ inches from the common wall, because the water closet rough-in sheet (Figure 12-9) indicates a 12-inch rough-in measurement from the finished wall to the center line of the waste, and to this measurement must be added $^1/_2$ inch for plasterboard and $^1/_4$ inch for ceramic tile thickness—giving a total of $12^3/_4$ inches. After laying out the center lines of this hole for the water closet waste, use the closet collar as a template to cut a 6-inch diameter hole.

While in this bathroom, mark on the floor plate of the common wall the 27-inch measurement from the bedroom wall to the center line of the vanity lavatory waste.

In the three-quarter bathroom, lay out the 34-inch width of the shower base. In the center of this base, cut a hole about 4 inches in diameter for the shower drain. In the common wall sill (directly behind the shower base), drill a $2^9/_{16}$-inch hole, as indicated, at the center line of the shower base for the shower bath vent.

A space of 68 inches is left along the common wall in which to install the water closet and lavatory. As the water closet tank is $20^1/_2$ inches wide and the lavatory is 19 inches wide (a total of $39^1/_2$ inches), this leaves $28^1/_2$ inches of space ($68 - 39^1/_2 = 28^1/_2$) to be divided out in spacing between the fixtures. The usual procedure is to divide this space equally between the fixtures. Since $28^1/_2$ inches divided by 3 equals $9^1/_2$ inches, there would be a $9^1/_2$-inch space between the water closet and the lavatory, a $9^1/_2$-inch space between the water closet and the shower, and another $9^1/_2$-inch space between the lavatory and the wall. Figure 12-25 illustrates this fixture spacing.

Using these fixture spacing measurements, the center of the water closet opening is marked out $12^3/_4$ inches from the common wall (as figured for the other bathroom) and $19^3/_4$ inches from the shower base—that is, $9^1/_2$ inches of space plus $10^1/_4$ inches (one-half the $20^1/_2$-inch toilet tank width)—for the center of the water closet waste opening. Cut a 6-inch diameter hole using a closet collar as a template.

To determine the center line of the stack opening in the common wall, the usual procedure is to draw a line through the centers of the water closet waste openings passing through this wall, as indicated by the series of dashes between these holes in Figure 12-24. In the area where this line passes through the common wall, a $3^1/_2$-inch square (the width of the 2 × 4) centered on this line is cut out of the sill and through the floor for the 3-inch stack.

The lavatory waste is marked on the floor plate of the common wall 19 inches from the bedroom wall.

Figure 12-19 indicates the piping of the lavatories into a common waste and vent pipe. To obtain the proper location for this common waste pipe, the distance between the two marks previously put on the common wall floor plate for the lavatory waste center lines is divided in half, and a 2⁹/₁₆-inch hole is drilled, as indicated, at 23 inches from the bedroom wall.

Another 2⁹/₁₆-inch hole is drilled in the common wall floor plate for the vent pipe for the basement fixtures. This hole—6 inches from the bedroom wall—is shown on Figure 12-24.

After cutting the holes for the fixture waste and vent pipes in the floor, the plumber proceeds to the holes for the stack and horizontal vent piping. For the vent portion of the 3-inch stack, the top plate is cut off directly above the previously located stack opening in the floor plate, and a hole about 6 inches square is also cut through the roof above so the stack can be piped through the roof.

The studs on the common wall are then marked and drilled with 2⁹/₁₆-inch holes for the horizontal branch vent pipe from the shower vent hole to the stack and from the basement fixture vent hole to the stack. These holes are centered at 37 inches above the floor, as calculated by adding 6 inches to the 31-inch height of

the wall-hung lavatory rim. (The apprentice should remember from Chapter 6 that horizontal branch vent piping must be piped at least 6 inches above the flood level rim of the highest fixture connected to the vent.)

Location of Basement Plumbing Fixtures

To locate the plumbing fixtures in the basement of the house, the plumber will need to refer to Sheet 2, the Basement Plan, of the blueprints. You will notice that on this sheet the laundry tray, floor drains, and basement three-quarter bathroom rough-in are not marked with any locating dimensions. This is quite often true on blueprints for residential work. When the blueprints do not have any dimensions, the plumber must measure or *scale* the location of the fixtures from the blueprint. (*Note:* Since the blueprints for the example in this chapter have been reduced to fit the book page, they cannot be scaled.) Figure 12-26, a plan view of the building drain, is provided for the apprentice's information, with the necessary dimensions marked on it.

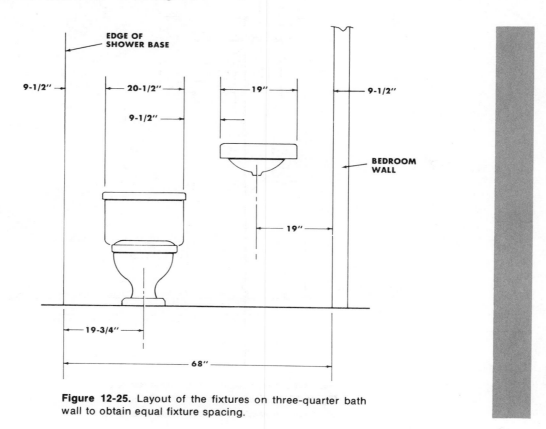

Figure 12-25. Layout of the fixtures on three-quarter bath wall to obtain equal fixture spacing.

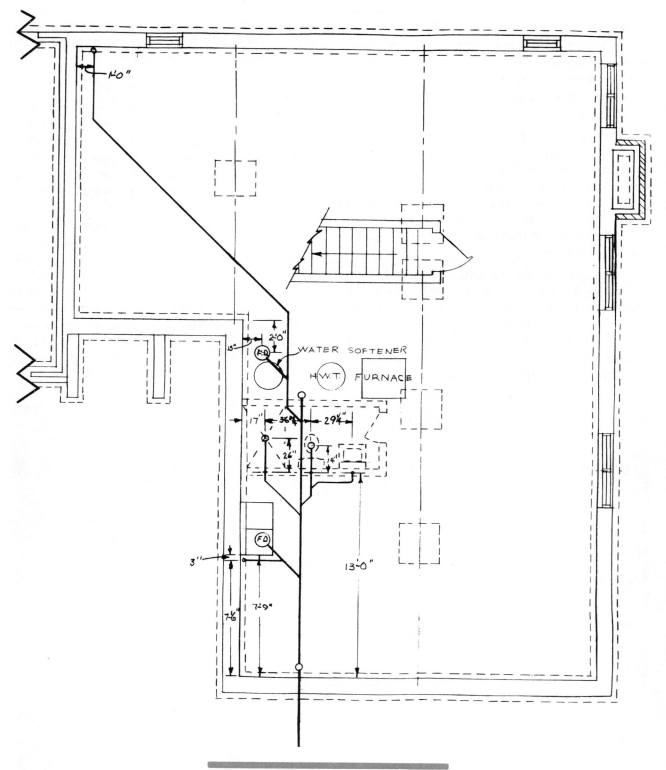

Figure 12-26. A plan view of the building drain.

Using the dimensions provided on Figure 12-26 (which would normally be scaled from Sheet 2), the plumber would proceed to lay out the location of the basement plumbing fixtures. The laundry tray waste would be marked on the side wall 7 feet 6 inches from the front wall, as indicated, and a stake driven into the dirt under the laundry tray location to indicate a floor drain.

The center line of the future three-quarter bathroom plumbing wall would be located on the side wall, 13 feet from the front wall. The openings for the shower waste and vent, the water closet waste and vent, and the lavatory waste would all be located with stakes driven into the dirt at their appropriate dimensions, as taken from Figure 12-26.

The 3-inch soil stack would be located by dropping a plumb bob through the opening cut in the common bathroom wall on the first floor. A stake would then be driven at this location.

The floor drain serving the water softener would also be marked with another stake driven into the dirt.

Finally, the location of the opening for the kitchen sink waste stack would be marked on the back wall of the basement one foot from the garage wall.

Layout of the Building Drain Trench

Having laid out the basement plumbing fixture and stack openings, the plumber proceeds to lay out the trench for the building drain. The usual method is to center the trench at the front wall on the stack opening and scratch a line in the dirt from the front wall to the stack. The trenches for the other fixtures draining into the building

drain are then marked in the dirt to provide for the building drain piping shown in Figure 12-26.

Before starting to dig the trench, the plumber must know the invert elevation of the building drain where it leaves the front of the house. (*Invert elevation,* is the term applied to the lowest portion of the inside of a horizontal pipe.) This depth may be calculated by scaling the longest run of pipe on the building drain (from Figure 12-26) and multiplying this measurement by $1/4$ inch per foot for grade. To this measurement must be added 4 inches for the thickness of the basement concrete floor and 4 more inches for the diameter of the building drain. For example, on Figure 12-19, the longest run of building drain piping is to the kitchen sink waste stack. The distance from the front main cleanout to the kitchen sink stack scales approximately 48 feet along the length of the pipe, as drawn in Figure 12-26. The depth of the trench would then be: 48 feet × $1/4$ inch per foot of grade equals 12 inches, *plus* 4 inches for the basement floor thickness, *plus* 4 inches for the diameter of the building drain, to *total* 20 inches, as illustrated on Figure 12-27.

With this measurement, the plumber may dig a trench 20 inches deep at the front of the building to the stack and kitchen sink openings, grading this trench approximately $1/4$ inch per foot. The trenches required for the drains from the laundry tray, floor drains, and basement three-quarter bathroom are also dug at this time. It is also necessary to tunnel several feet underneath the footing at the front of the house so that a piece of 4-inch soil pipe may be extended out for eventual connection of the building drain to the building sewer.

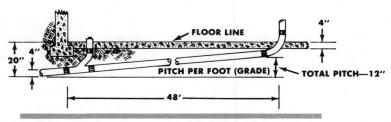

Figure 12-27. Invert elevation showing the depth of the building drain outlet.

Installation of the Building Drain

Once the digging of the trench is completed, the plumber is ready to install the building drain piping. Referring to Figures 12-20 and 12-26, the installation of the building drain would proceed as described in the following paragraphs.

(Since the purpose of this section is to familiarize the apprentice with the method of installation, no actual piping measurements will be given. Instead, the layout of the fittings will be described as they would be installed with no-hub soil pipe and fittings.)

A piece of 4-inch no-hub soil pipe is joined to a 4-inch no-hub combination wye and 1/8 bend and pushed out beneath the footing at the front of the house for future connection to the building sewer. The top opening of the combination fitting is extended up to a $4 \times 3^{1}/_{2}$ tapped adapter for the front main cleanout. (The top of this adaptor must be at least 2 inches above the finished basement floor.) From the 4-inch combination wye and 1/8 bend, the 4-inch building drain pipe would extend to a 4×2 no-hub wye for the laundry tray and floor drain. Proceeding from this 4×2 wye, the drain is continued 4-inch size to another 4×2 wye for the future basement shower drain. Directly in front of this second 4×2 wye, a 4-inch no-hub wye is placed for the future basement water closet and lavatory wastes.

The building drain then continues 4-inch size to a third 4×2 no-hub wye for the kitchen sink waste and the floor drain provided for the water softener backwash. Finally, the 4-inch building drain is extended to the 4-inch long sweep 1/4 bend to receive the 3-inch soil stack.

Going back to the 4×2 wye placed for the laundry tray and floor drain wastes, the 2-inch side opening of the 4×2 wye is extended to a 2-inch wye. The end of this 2-inch wye is extended to the 2-inch floor drain, which will be beneath the laundry tray. The side opening of the 2-inch wye is extended to a 2-inch long sweep 1/4 bend, which will receive the wastes from the laundry tray and the standpipe for the automatic clothes washer.

Moving up the trench to the 4×2 wye for the future basement shower drain: the 2-inch side opening of the 4×2 wye is extended to a 2-inch 1/8 bend positioned so that the end of the 1/8 bend is parallel to the side wall of the basement, directly in line with the center of the shower drain. From this 2-inch 1/8 bend, the 2-inch pipe continues to a 2-inch sanitary tee placed on its back (with the side opening of the tee looking straight up at the center line of the future basement wall) to vent the shower drain trap. The 2-inch pipe then continues to a 2-inch P-trap for the shower drain. This P-trap is extended above the basement floor and plugged with a 2-inch no-hub blind plug.

At the 4-inch wye placed for the future basement water closet and lavatory drains, a piece of 4-inch no-hub pipe is extended to a 4-inch 1/8 bend, which is placed directly in line with the basement water closet opening. A 4×2 wye for the lavatory drain is joined to this 1/8 bend. From the 4×2 wye, the 4-inch water closet drain continues to a 4×2 sanitary tee placed on its back, with the 2-inch side opening of the tee looking straight up at the center line of the future basement wall, to provide a vent for the water closet. The $4 \times 4 \times 16 \times 16$ closet bend is then cut to fit the proper rough-in measurement from the future basement wall and joined to the 4×2 sanitary tee. The top of this closet bend is plugged with a 4-inch no-hub blind plug.

Moving back to the 4×2 wye placed for the future basement lavatory drain: a 2-inch 1/8 bend is joined to the 4×2 wye so that the 2-inch pipe will run parallel to the future basement wall. From this 1/8 bend, a 2-inch pipe is extended to a 2-inch long sweep 1/4 bend placed in line with the center of the future basement lavatory waste. Another 2-inch long sweep 1/4 bend, looking up in the center of the future basement wall, is joined to the first 1/4 bend to provide the lavatory waste opening in the building drain.

Working up the trench to the 4×2 wye left for the kitchen sink and the floor drain by the water softener, a 2-inch pipe is extended to a 2-inch 1/8 bend. From the 2-inch 1/8 bend, the 2-inch drain continues to a 2-inch wye to provide an opening for the 2-inch floor drain for the water softener backwash. This floor drain is then joined to the side opening of the 2-inch wye.

The 2-inch drain then continues to a second 2-inch ¹⁄₈ bend. From this second ¹⁄₈ bend, the 2-inch drain continues to a third 2-inch ¹⁄₈ bend located one foot from the garage wall. The 2-inch drain then is extended to a 2-inch long sweep ¹⁄₄ bend at the back wall of the basement to receive the kitchen sink waste. This completes the building drain piping.

Installation of the Kitchen Sink Stack

Having completed the building drain piping, the plumber proceeds to pipe the kitchen sink waste and vent stack, as detailed in Figure 12-28. From the 2-inch long sweep ¹⁄₄ bend, a piece of 2-inch no-hub pipe is extended above the basement floor to a 2-inch test tee for a stack base cleanout. (*Note:* a 2-inch test tee is used because 1¹⁄₂-inch no-hub test tees are not available.)

A 2×1¹⁄₂ no-hub coupling is used to reduce the top of the test tee to the 1¹⁄₂-inch size stack required for a kitchen sink drain. A piece of 1¹⁄₂-inch pipe is extended from the top of the test tee to the top of the concrete block basement wall. At the top of the wall, two 1¹⁄₂-inch ¹⁄₈ bends are used to offset back over the 12-inch concrete block wall to the 2×4 stud wall in the kitchen. The 1¹⁄₂-inch pipe then continues to the 1¹⁄₂-inch sanitary tee for the kitchen sink drain. The top of this tee is extended with 1¹⁄₂-inch pipe (to vent the kitchen sink trap) to a point below the roof, where it is increased to 2-inch size with a 2×1¹⁄₂ inch no-hub coupling. The vent then continues through the roof 2-inch size. (*Note:* 1¹⁄₂-inch pipe is used to vent the kitchen sink in this example because 1¹⁄₄-inch no-hub pipe and fittings are not available.)

At the 1¹⁄₂-inch sanitary tee left for the kitchen sink drain, a piece of 1¹⁄₂-inch pipe is added to extend the drain over to a 1¹⁄₂-inch tapped ¹⁄₄ bend for the actual kitchen sink drain connection.

Installation of the First-Floor Bathrooms

After the kitchen sink stack, the plumber installs the waste and vent piping for the first-floor bathrooms, as detailed in Figure 12-29. The first pipe to be installed in the bathroom area is the

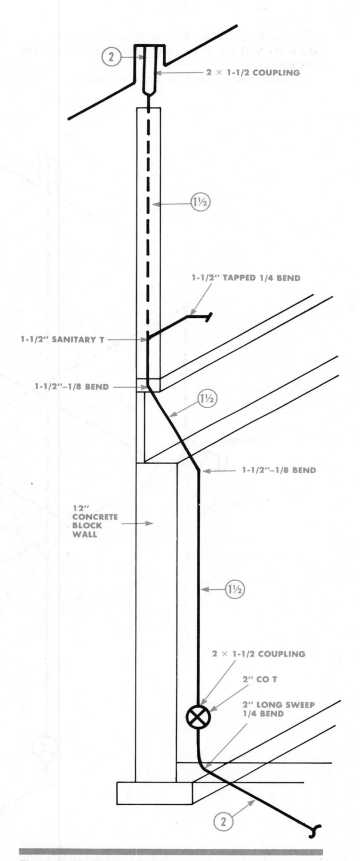

Figure 12-28. Detail of the kitchen sink waste and vent stack.

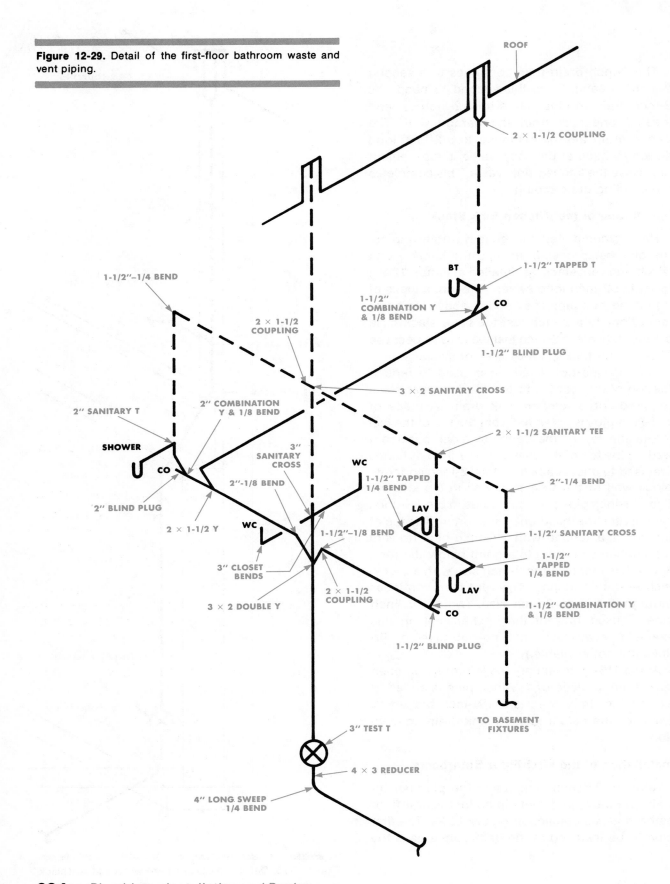

Figure 12-29. Detail of the first-floor bathroom waste and vent piping.

ROOF

2 × 1-1/2 COUPLING

1-1/2"–1/4 BEND

BT

1-1/2" TAPPED T

1-1/2" COMBINATION Y & 1/8 BEND

CO

2 × 1-1/2 COUPLING

1-1/2" BLIND PLUG

3 × 2 SANITARY CROSS

2" COMBINATION Y & 1/8 BEND

2 × 1-1/2 SANITARY TEE

2" SANITARY T

SHOWER

3" SANITARY CROSS

WC

2"-1/4 BEND

CO

2"-1/8 BEND

1-1/2" TAPPED 1/4 BEND

LAV

2" BLIND PLUG

1-1/2" SANITARY CROSS

2 × 1-1/2 Y

WC

1-1/2"–1/8 BEND

1-1/2" TAPPED 1/4 BEND

3" CLOSET BENDS

LAV

3 × 2 DOUBLE Y

2 × 1-1/2 COUPLING

CO

1-1/2" COMBINATION Y & 1/8 BEND

1-1/2" BLIND PLUG

TO BASEMENT FIXTURES

3" TEST T

4 × 3 REDUCER

4" LONG SWEEP 1/4 BEND

3-inch main soil stack. A 4×3 no-hub reducer is joined to the previously installed 4-inch long sweep ¼ bend, and a 3-inch test tee is joined to the reducer to provide a stack base cleanout. A piece of 3-inch no-hub soil pipe is extended from the test tee up to a 3×2 double wye for the drains from the shower bath, bathtub, and lavatories. A 3-inch sanitary cross is joined to the top of the 3×2 double wye. The two 3×3×16×16 closet bends are cut to length and joined to the side openings of the sanitary cross to provide the water closet waste openings. A piece of 3-inch pipe is extended from the top of the 3-inch sanitary cross to a 3×2 sanitary cross centered at 37 inches above the first floor. This 3×2 sanitary cross will be the vent connection for the shower bath and lavatories. From the top of the 3×2 sanitary cross, the stack is extended full 3-inch size through the roof.

Lavatory Drains. Dropping back below the floor and to the right of the stack, a 2×1½ no-hub coupling is put on the right-hand 2-inch opening in the 3×2 double wye. A 1½-inch ⅛ bend is then installed in this coupling. From the 1½-inch ⅛ bend, a 1½-inch pipe is extended to a 1½-inch combination wye and ⅛ bend with the side opening looking up directly beneath the hole drilled for the common lavatory drain. The end of this combination fitting is capped with a 1½-inch blind plug for a cleanout. The top of the combination fitting is extended as 1½-inch size above the floor to a 1½-inch sanitary cross for the common lavatory waste. This cross is centered at 17 inches above the floor. (See Figure 12-11 for the lavatory waste rough-in height.) A 1½-inch tapped ¼ bend is attached to each side opening at the 1½-inch sanitary cross. The ¼ bend on the left side faces into the full bathroom, while the ¼ bend on the right faces into the three-quarter bathroom.

Shower Drain. At the left of the stack, a 2-inch ⅛ bend is joined to the left side opening of the 3×2 double wye. A 2-inch pipe is extended from this ⅛ bend to a 2×1½ wye for the bathtub drain. A 2-inch combination wye and ⅛ bend, with the side opening looking up directly beneath the hole drilled for the shower bath vent, is joined to the end of the 2×1½ wye. The end of

this combination fitting is capped with a 2-inch no-hub blind plug for a cleanout. A 2-inch sanitary tee is set on the top opening of the combination fitting. The side opening of this tee is extended to a 2-inch P-trap for the shower drain, and the inlet of this trap is extended above the floor.

To vent the shower, the top of the 2-inch sanitary tee is reduced with a 2×1½ no-hub coupling, and a 1½-inch pipe is extended through the floor to a 1½-inch ¼ bend centered at 37 inches above the floor. This ¼ bend is then extended to the left side of the 3×2 sanitary cross with a piece of 1½-inch pipe, and joined to the cross with a 2×1½ no-hub coupling.

Bathtub Drain and Vent. At the 2×1½ wye left for the bathtub drain, the 1½-inch opening is extended to a 1½-inch ⅛ bend, which is centered on the previously drilled hole for the bathtub vent. From the ⅛ bend, a 1½-inch pipe is extended to a 1½-inch combination wye and ⅛ bend lying on its back with the side opening of the combination fitting looking up directly beneath the tub vent hole. The end of the combination fitting is capped with a 1½-inch no-hub blind plug to provide a cleanout. A 1½-inch tapped tee is joined to the top of the combination fitting for the bathtub waste connection. The top of this tee is extended with 1½-inch pipe to a point below the roof, where it is increased to 2-inch size with a 2×1½ no-hub coupling and continues as 2-inch size through the roof.

Basement Fixture Vent. The right side opening of the 3×2 sanitary cross provides the vent opening for the first-floor lavatories and the basement plumbing fixtures. A 2-inch pipe is joined to the right side of the cross. This pipe is extended to a 2×1½ sanitary tee, with the side opening of the tee looking down directly above the previously installed 1½-inch sanitary cross for the common lavatory waste. A piece of 1½-inch pipe joins these two fittings to complete the lavatory vent. From the end of the 2×1½ sanitary tee, a 2-inch pipe is extended to a 2-inch ¼ bend looking down toward the hole cut for the basement fixture vent. A piece of 2-inch pipe is then joined to the ¼ bend and extended below the floor into the basement ceiling.

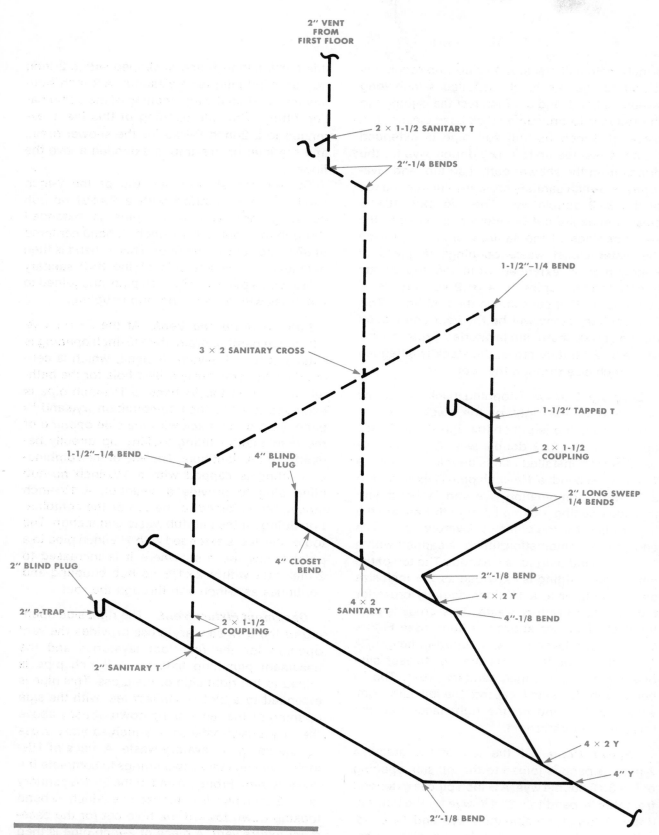

2" VENT
FROM
FIRST FLOOR

2 × 1-1/2 SANITARY T

2"-1/4 BENDS

1-1/2"-1/4 BEND

3 × 2 SANITARY CROSS

1-1/2" TAPPED T

2 × 1-1/2
COUPLING

1-1/2"-1/4 BEND

4" BLIND
PLUG

2" LONG SWEEP
1/4 BENDS

2" BLIND PLUG

4" CLOSET
BEND

2"-1/8 BEND

4 × 2 Y

2" P-TRAP

4 × 2
SANITARY T

4"-1/8 BEND

2 × 1-1/2
COUPLING

2" SANITARY T

4 × 2 Y

4" Y

2"-1/8 BEND

Figure 12-30. Detail of the basement three-quarter bathroom
waste and vent piping.

Basement Three-Quarter Bathroom Vents

On completion of the first-floor waste and vent piping, the plumber may proceed to install the vent piping for the future three-quarter bathroom in the basement, as detailed in Figure 12-30. Starting at the 4×2 sanitary tee left for the water closet vent, a 2-inch pipe is extended up to a 2×1½ sanitary cross centered at 37 inches above the basement floor. This cross will provide the vent openings for the future basement shower bath and lavatory. From the top of the cross, a 2-inch pipe is extended up to a 2-inch ¼ bend near the ceiling. This ¼ bend is joined to a second ¼ bend, which looks up directly in line with the 2-inch pipe previously extended through the first floor for the basement fixture vents. The 2-inch pipe and the 2-inch ¼ bend are joined by placing a 2×1½ sanitary tee between them in the ceiling joist space. The side opening of this tee faces the side wall of the basement to provide a vent opening for the laundry tray.

At the 2×1½ sanitary cross, the left side opening is extended to a 1½-inch ¼ bend looking down at the 2-inch sanitary tee previously installed for the shower bath vent. A piece of 2-inch pipe is extended up above the floor from the sanitary tee. (*Note:* this is necessary because most plumbing codes do not allow any pipe smaller than 2-inch size to be buried below a concrete floor.) A 2×1½ no-hub coupling is placed on this 2-inch pipe, and a piece of 1½-inch pipe is used to join the coupling and the 1½-inch ¼ bend.

The right side opening of the 2×1½ sanitary cross is extended to a 1½-inch ¼ bend looking down at the 2-inch long sweep ¼ bend previously installed for the lavatory waste. The 2-inch long sweep ¼ bend must also be extended 2-inch size above the basement floor line, at which point it may be reduced to 1½-inch size with a 2×1½ no-hub coupling. From this coupling, a piece of 1½-inch pipe is extended up to a 1½-inch tapped tee placed at 17 inches to center above the basement floor for the lavatory drain connection. (This 17-inch measurement

was taken from Figure 12-11—the rough-in sheet for this lavatory.) The top of the tapped tee is then extended with a piece of 1½-inch pipe to join the 1½-inch ¼ bend to complete the lavatory vent.

Laundry Tray Waste and Vent

The only fixture remaining to be piped is the laundry tray, as detailed in Figure 12-31. The 2-inch long sweep ¼ bend previously installed for the laundry tray waste is extended with a piece of 2-inch pipe above the basement floor line, at which point it reduces to 1½-inch size with a 2×1½ no-hub coupling. A piece of 1½-inch pipe is extended up to a 1½-inch tapped cross, centered at 18 inches above the basement floor.

This location is calculated in the following way (see Figure 12-15, "Laundry Tray Rough-in Sheet"):

Laundry tray height: 35¼ inches
Subtract the following:

−15³/₁₆″	Depth of laundry tub
− 2″	One-half length of a 1½ × 4 flanged tailpiece
−17³/₁₆″	

When this figure (rounded to 17¼″) is subtracted from the height of the laundry tray, it gives 18 inches for the center line of the waste. This tapped cross will provide waste openings for both the laundry tray and the standpipe for the automatic clothes washer.

From the top of the tapped cross, a piece of 1½-inch pipe is extended up to a 1½-inch ¼ bend near the ceiling. A piece of 1½-inch pipe extends across the basement ceiling from this ¼ bend toward the basement three-quarter bathroom. At the center of the 2×1½ sanitary tee left for the laundry tray vent, another 1½-inch ¼ bend is joined to the 1½-inch pipe. This ¼ bend faces up to another ¼ bend, which faces the 2×1½ tee. A piece of 1½-inch pipe joins the ¼ bend and the tee to complete the laundry tray vent and the waste and vent piping for the entire house.

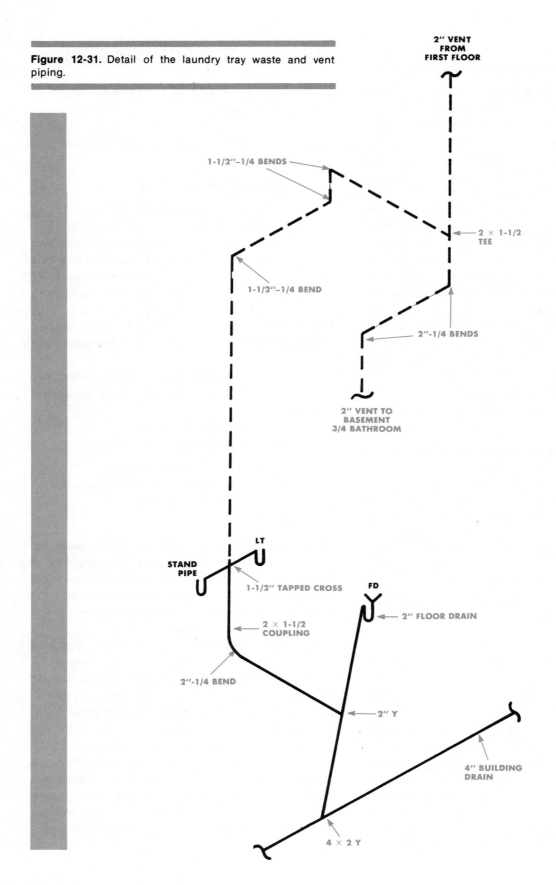

Figure 12-31. Detail of the laundry tray waste and vent piping.

2″ VENT FROM FIRST FLOOR

1-1/2″–1/4 BENDS

2 × 1-1/2 TEE

1-1/2″–1/4 BEND

2″-1/4 BENDS

2″ VENT TO BASEMENT 3/4 BATHROOM

LT

STAND PIPE

1-1/2″ TAPPED CROSS

FD

2″ FLOOR DRAIN

2 × 1-1/2 COUPLING

2″-1/4 BEND

2″ Y

4″ BUILDING DRAIN

4 × 2 Y

Test and Inspection of the Rough Plumbing

When sanitary drainage and vent piping has been completed, the plumber is ready to test the installation and have it inspected. The methods for testing sanitary drainage and vent piping were discussed in Chapter 11, and the apprentice should refer back to that chapter for the appropriate testing method required by the local plumbing code.

After the test and inspection, the plumber must backfill the trenches. During the backfilling, large rocks and other debris must be kept out of the fill material being put back into the trench. This will prevent damage to the pipe and fittings. When the backfilling is finished, recheck and level the floor drains and the front main cleanout to see that they have not been knocked out of alignment.

Installation of Bathtub and Shower Base

The bathtub and shower base (because they are *built-in* plumbing fixtures) must be installed before the plasterboard (sheet rock) and ceramic tile are applied to the rough walls. The plumber sets these fixtures after he completes the installation of the sanitary drainage and vent piping.

Shower Base. The shower base is set in its location in the three-quarter bathroom. The rough floor must be cleaned before the base is set into place so that it rests evenly on the floor. The connection to the 2-inch pipe inlet of the P-trap is then sealed with a caulked lead and oakum joint.

Bathtub. Before the bathtub is set, it is necessary to nail a piece of 2×4 along the back wall of the tub space approximately 13½ inches above the floor to support the back edge of the bathtub. After this board is installed, the bathtub may be set into place. The waste and overflow fitting is then attached to the bathtub and the P-trap connected to complete the bathtub drain.

Fixture Backing. Since plumbing fixtures cannot be supported by plasterboard walls, backing or reinforcement is required for wall-hung fixtures (such as the lavatory to be installed in the three-quarter bath of this example home). The backing consists of a 1×8 board nailed between the studs on the common bathroom wall. The 1×8 should run at least the full width of the lavatory and be centered 29¾ inches above the floor, as shown in Figure 12-11—the rough-in sheet for this lavatory.

Backing is also required to support the rough water piping. The installation of this backing will be elaborated on in the next sections of this chapter.

Rough Water Piping. The concealed water supply piping is also installed as a part of the rough plumbing. For the example presented in this chapter, the plumber would be required to pipe the shower water supply fitting and the bathtub water supply fitting, and to provide water piping for the water closets and lavatories in the first-floor bathrooms.

Shower Fitting. The shower water supply fitting illustrated in Figure 12-14 is assembled with ½-inch copper tubing and fittings, as illustrated in Figure 12-32. Half-inch copper 90° street elbows and pieces of ½-inch copper tubing approximately 60 inches long are soldered to the shower fitting body. These pieces of copper tubing are long enough to pass down through the first floor to the basement ceiling below the joists, for connection to the water piping in the basement when it is installed. A piece of ½-inch copper tubing approximately 24 inches long is extended from the top of the shower fitting up to a ½-inch copper × ½-inch female iron pipe (FIP) drop ell. This ell is the future connection for the shower arm.

A piece of 1×4 board is nailed in the wall centered at 6 feet above the floor to provide backing for the drop ell. Another piece of 1×4 board backing is required at the center of the shower fitting body to provide a firm support for anchoring the shower fitting.

After the backing is installed, two ¾-inch diameter holes may be drilled through the 2×4 wall plate to permit the 60-inch pieces of copper tubing to pass through the floor into the basement. The shower fitting assembly may then be inserted into these holes and anchored to the backing.

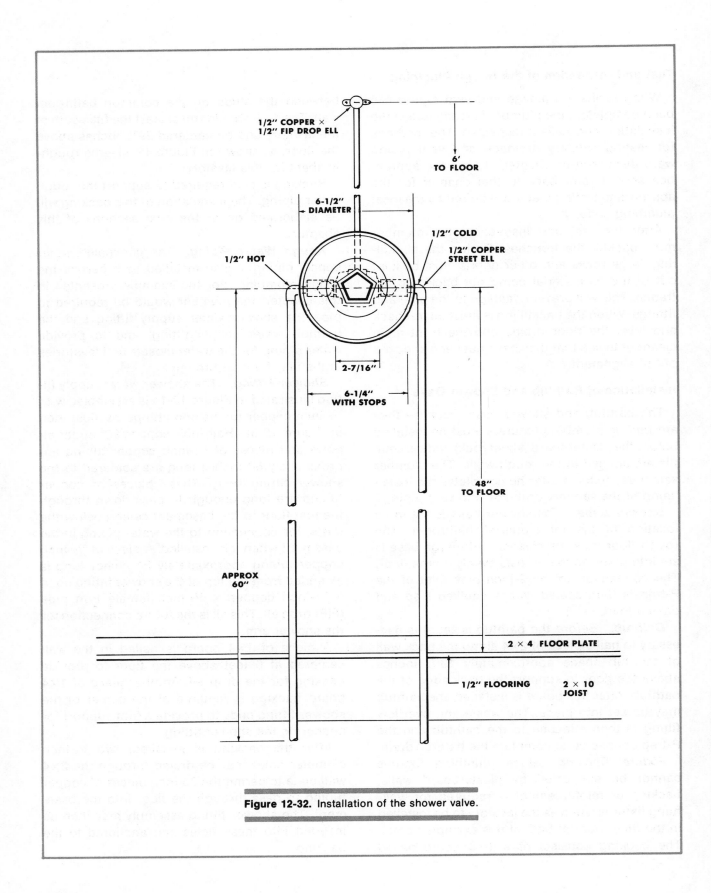

1/2" COPPER ×
1/2" FIP DROP ELL

6'
TO FLOOR

6-1/2"
DIAMETER

1/2" COLD

1/2" COPPER
STREET ELL

1/2" HOT

2-7/16"

6-1/4"
WITH STOPS

48"
TO FLOOR

APPROX
60"

2 × 4 FLOOR PLATE

1/2" FLOORING

2 × 10
JOIST

Figure 12-32. Installation of the shower valve.

Bathtub Fitting. The bathtub water supply fitting illustrated in Figure 12-8 is assembled with 1/2-inch copper tubing and fittings, as illustrated in Figure 12-33. As with the shower fitting, 1/2-inch copper 90° street ells with pieces of 1/2-copper tubing approximately 40 inches long are soldered to the bath fitting body to extend through the floor into the basement. Another piece of 1/2-inch copper tubing, approximately 10 inches long, is soldered to the bottom outlet of the bath fitting. Onto this 10-inch piece of copper tubing, a 1/2-inch copper 90° drop elbow and a piece of copper approximately 6 inches long are soldered to extend through the wall and provide the connection for the bathtub fitting spout. A piece of 1×4 board backing is placed in the wall, centered 4 inches above the bathtub to provide support for the drop ell. Another piece of backing is required at the center of the valve body to provide a firm support.

After the backing is installed, two 3/4-inch diameter holes may be drilled through the 2×4 wall plate to permit the 40-inch pieces of copper tubing to pass through into the basement. The bath fitting assembly may then be inserted into these holes and anchored to the backing.

Lavatory Water Piping. To provide the cold and hot water supply to the lavatories in each bathroom, four water pipe drops are constructed of 1/2-inch copper tubing and drop ells, as illustrated in Figure 12-34. These drops are made 30 inches long to pass through the floor. They have 6-inch pieces of copper to extend through the wall. Backing is provided 21 inches above the floor, and 3/4-inch diameter holes are drilled through the 2×4 floor plate, 4 inches to each

side of the previously marked lavatory center lines in both the three-quarter and full bathrooms. The water pipe drops are then inserted into these holes and fastened to the backing 21 inches above the floor.

Water Closet Water Piping. Two 24-inch long water pipe drops of the type illustrated in Figure 12-34 are constructed to provide water supply for the water closets. The backing is centered 10 1/2 inches above the floor and 6 inches to the left of the center of each water closet, as viewed in Figure 12-9—the water closet rough-in sheet. A 3/4-inch diameter hole is drilled through the 2×4 floor plate 6 inches to the left of the center of each water closet. The water pipe drops may then be inserted into these holes and fastened to the backing.

Installation of Roof Jackets

The plumber furnishes the roof jackets, which are installed around the vent pipe terminals passing through the roof to make the roof watertight. The installation of this jacket is usually the work of the roofer. However, the plumber places the necessary jackets over the vent terminals to seal these openings temporarily until the roof is shingled.

In the example home in this chapter, the following roof jackets are required: two 2-inch roof jackets for the kitchen sink and bathtub vent terminals; and one 3-inch roof jacket for the 3-inch stack vent terminal.

Note that the shower bath base, bathtub, bathtub drain fitting and P-trap, water supply piping, and roof jackets installed in the previous sections of this chapter are all part of the material necessary to install the rough plumbing.

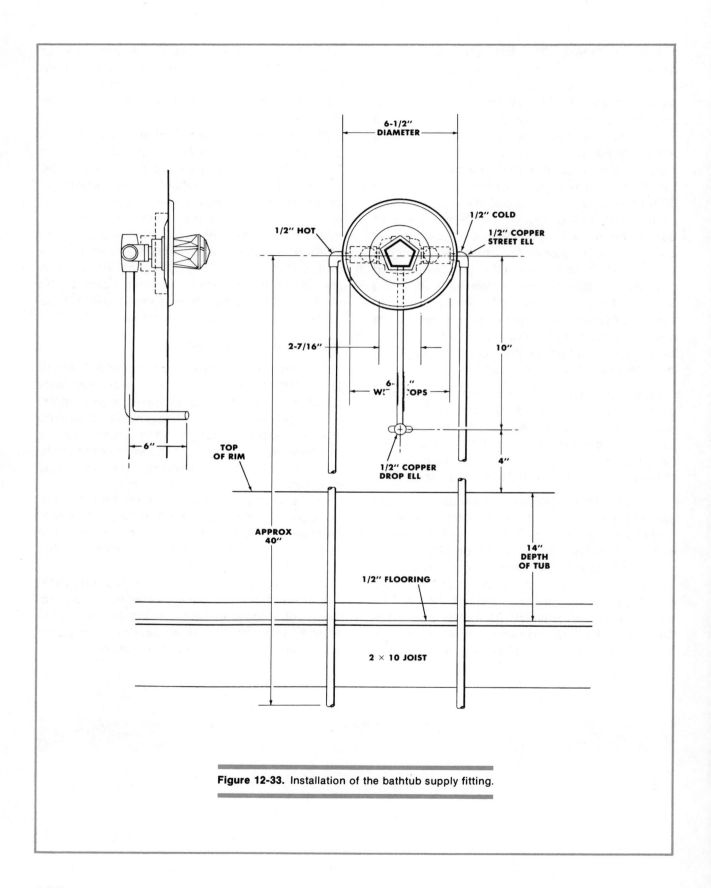

6-1/2" DIAMETER

1/2" COLD

1/2" HOT

1/2" COPPER STREET ELL

2-7/16"

10"

6-" W: OPS

1/2" COPPER DROP ELL

4"

TOP OF RIM

APPROX 40"

14" DEPTH OF TUB

1/2" FLOORING

2 × 10 JOIST

6"

Figure 12-33. Installation of the bathtub supply fitting.

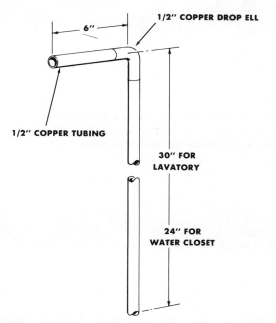

Figure 12-34. Lavatory and water closet ½-inch copper water pipe drops.

INSTALLATION OF THE BUILDING SEWER AND WATER SERVICE

When developing suburban housing, it is common practice for the developer to install the sanitary sewer, storm sewer, and water main pipes in the streets before the streets are paved or any homes built in the development. In colder areas of the country, the water main is buried beneath the street at a sufficient depth to prevent it from freezing. The sanitary sewer main is buried at a sufficient depth to allow all the homes to have a gravity flow building drain beneath the basement floor.

When the mains in the street are being installed, connections for both the sanitary sewer and the water service are piped to each home lot past the curb line. The water service is piped to the stop box and the sanitary sewer is piped to a point near the stop box. The locations of these connections are marked on the lot with a stake, and the location and depth of the connections are marked on master plans, which are kept at the municipal building department office.

The locations of both the sanitary sewer connection and the water service stop box for the example home in this chapter are indicated on Sheet 1, the *Plot Plan*.

Installation of the Building Sewer

A number of factors must be determined before the actual installation of the building sewer can begin—namely, the location and depth of the building drain outlet, the depth of the sanitary sewer connection at the lot line, and the grade or pitch of the building sewer.

The depth of the building drain outlet for the example presented in this chapter has already been calculated as being 20 inches below the basement floor, and the building drain is assumed to be installed. However, if the building drain had not been installed, the plumber would have to calculate the proper depth and location for piping the building sewer to the building drain outlet. The method for determining the depth and location of the building drain outlet was described in a previous section of this chapter, "Layout of the Building Drain Trench."

The depth of the building sewer at the front lot line must next be established. This depth is usually marked on a master sewer plan, which is kept at the municipal building department office. The plumbing contractor would normally be given this information (as well as the locations of the sewer and water connections to the lot line) when he purchased the permits for the installation of the building sewer and water service. (These locations would be needed if the marking stake had been removed.)

The depth of the building drain outlet and the depth of the sanitary sewer connection at the front lot line must be known to determine the grade of the building sewer. The total pitch (grade) of the building sewer may be found as the difference between the depth of the building drain outlet and the depth of the sanitary sewer connection at the front lot line. For example, if the basement floor is 6 feet 6 inches below the street grade and the building drain outlet is 1 foot 8 inches below the basement floor, the building drain outlet will then be 8 feet 2 inches (98 inches) below the street grade. If the sanitary

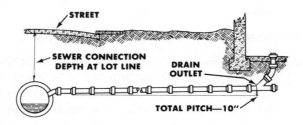

Figure 12-35. Total pitch of the building sewer.

sewer connection is 9 feet (108 inches) below the street grade, then the total pitch of the building sewer would be 108 inches minus 98 inches, or 10 inches of total fall, as illustrated in Figure 12-35.

To arrive at the pitch per foot, the building sewer location must be given and its length determined. Referring to Sheet 1, the *Plot Plan,* the plumber would measure from the front of the house to the marked sewer connection and find the length to be approximately 40 feet. The total pitch is divided by this figure. In this example, the 10 inches of pitch determined above, divided by the 40 feet of building sewer length, would equal ¹/₄ of an inch of pitch to each foot.

The pitch of a building sewer is usually not less than ¹/₄ inch for each foot of length. However, should it be necessary because of certain job conditions, less pitch per foot may be allowed.

Sometimes a building sewer is found to have more pitch than ¹/₄ of an inch per foot. This is particularly true when a house sits high on the lot above the street. In Chapter 6, it was stated that it is undesirable to grade horizontal drainage piping at more than ¹/₄ of an inch per foot because only the liquid tends to flow, leaving the solids in the waste pipe. If the pitch would be more than ¹/₄ of an inch per foot, the general procedure is to grade the building sewer at ¹/₄ of an inch per foot backwards from the house to

the sewer connection at the lot line, and then to make the final connection to the sewer with the use of two ¹/₈ bends, as illustrated in Figure 12-36.

To help you better understand the many points that must be clearly understood before a plumber can proceed with the actual installation of a building sewer, a cross-section drawing of the building sewer is shown in Figure 12-37. This drawing shows the sewer connection at the lot line and the building sewer, and gives the numerical elevations necessary to complete the installation. Reference to this drawing will be made frequently in the following paragraphs to explain the construction details more clearly.

The blueprints, in most instances, indicate grades by means of whole numbers or whole numbers and decimals that are related to a grade referred to by building engineers as a *bench mark.* The bench mark, which is usually a small copper plate mounted on a concrete pier, indicates the height of the plate above sea level or a previously established grade. All land within a given area of this marker can be surveyed and may be found to be above or below the original mark. City engineers and architects signify the depth of public sewers, street grades, water main grades, sidewalk levels, and other public utilities, as well as building foundations, floor levels, landscape contours, and many other

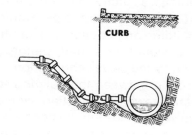

CURB

Figure 12-36. Use of fittings to provide additional grade on the building sewer.

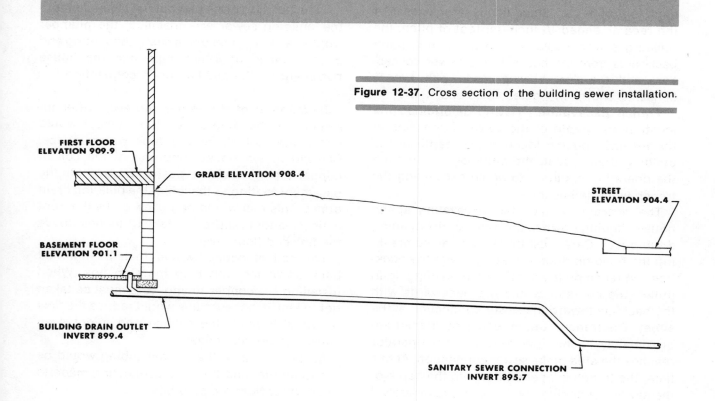

Figure 12-37. Cross section of the building sewer installation.

FIRST FLOOR
ELEVATION 909.9

GRADE ELEVATION 908.4

STREET
ELEVATION 904.4

BASEMENT FLOOR
ELEVATION 901.1

BUILDING DRAIN OUTLET
INVERT 899.4

SANITARY SEWER CONNECTION
INVERT 895.7

phases of building construction, *as either above or below the fixed bench mark.*

In the municipality in which this example house was built, there is a bench mark of 900 feet above sea level. The elevation of the first floor of the house is established as 909.9 feet, the street as 904.4 feet, the basement floor as 901.1 feet, and the invert elevation of the sewer connection at the lot line as 895.7 feet. Most of these figures were taken from Sheet 1, the *Plot Plan.*

The builder of this home thus knows that the level of the first floor will be 9.9 feet above the bench mark when the home is completed. Likewise, the sewer connection at the lot line (elevation 895.7 feet) is 5.4 feet under the level of the basement floor which has an elevation of 901.1 feet (901.1 feet − 895.7 feet = 5.4 feet).

(These elevations may confuse an uninformed reader, because they are given in decimals—tenths of a foot—rather than in feet and inches. This is surveyor's practice, and the plumber must convert them.)

The invert elevation of the sewer connection at the lot line is 895.7 feet. The difference between this and the building drain invert of 899.4 feet is 3.7 feet, or approximately 3 feet 8 inches. The total pitch of the building sewer divided by its length gives the pitch of the sewer per foot. For this example, 44 inches (3 feet 8 inches) divided by 40 feet of length equals 1.1 inch of pitch for each foot of sewer. Since this is far in excess of the recommended $1/4$ inch per foot of pitch, the building sewer would be run at $1/4$ inch pitch backwards from the house to the sewer connection, and the final hookup made with two $1/8$ bends, as shown on Figures 12-36 and 12-37.

Trench Excavation. Having determined the location and depth of the sewer connection at the lot line and the location and depth of the building drain outlet, the plumber can instruct the operating engineer to begin excavating the trench with the backhoe.

The operator would start excavating at the house, digging down to the depth of the building drain outlet. Care must be taken to avoid breaking the building drain outlet pipe with the backhoe. On reaching the depth of the building drain outlet, the operator would dig backwards with the backhoe toward the sewer connection at the street. The trench bottom would be graded approximately $1/4$ inch per foot until the operator reaches the area of the sewer connection. At this time, the trench is opened down to the depth of the sewer connection, which is approximately 3 feet below the depth of the main part of the trench.

The water service is usually installed in the same trench with the building sewer to save excavation expense. At this time, the trench would be opened wide enough to expose the *water service stop box.* This would complete the excavation of the trench for both the building sewer and water service.

Installation of the Building Sewer. Once the trench has been excavated to expose the build-

ing drain outlet and the sewer connection at the lot line, the plumber may proceed to connect these two openings, as pictured on Figure 12-37.

A length of 4-inch no-hub pipe would be joined to the building drain outlet, and additional lengths of pipe would be added, piping back down the trench to the sewer connection.

The final hookup between the building sewer and the sanitary sewer connection would be made with two 4-inch no-hub $1/8$ bends with a piece of pipe between them to make up the 3-foot difference in elevation discussed previously.

If the building drain had not been piped before the building sewer was installed, the plumber would have to tunnel underneath the footing and push a piece of 4-inch pipe into the house basement for the building drain connection.

Installation of the Water Service. After installing the building sewer, the plumber would install the $3/4$-inch copper tube water service. Starting at the house, one end of the coil of copper tubing would be pushed underneath the basement footing to the left of the building drain outlet. This end would be pulled up by the front main cleanout, approximately 12 inches above the finished floor level.

The coil of copper would then be unrolled back down the trench to the stop box. When installing the copper tubing, care must be taken not to kink it, because this would reduce the flow of water through the service and might even cause the service to leak.

At the stop box, the copper tubing would be cut to length, and the flare connection made to the curb cock in the stop box.

A $3/4$-inch copper flared gate valve would then be installed on the tubing end inside the basement near the front main cleanout.

At this time, the building sewer and water service would be ready for the appropriate tests and inspection as detailed in Chapter 11.

The trench can be filled with dirt after the proper tests and inspection have been made. To prevent damage to the sewer and water pipes, take care not to permit heavy rocks or large pieces of dirt to fall directly on them.

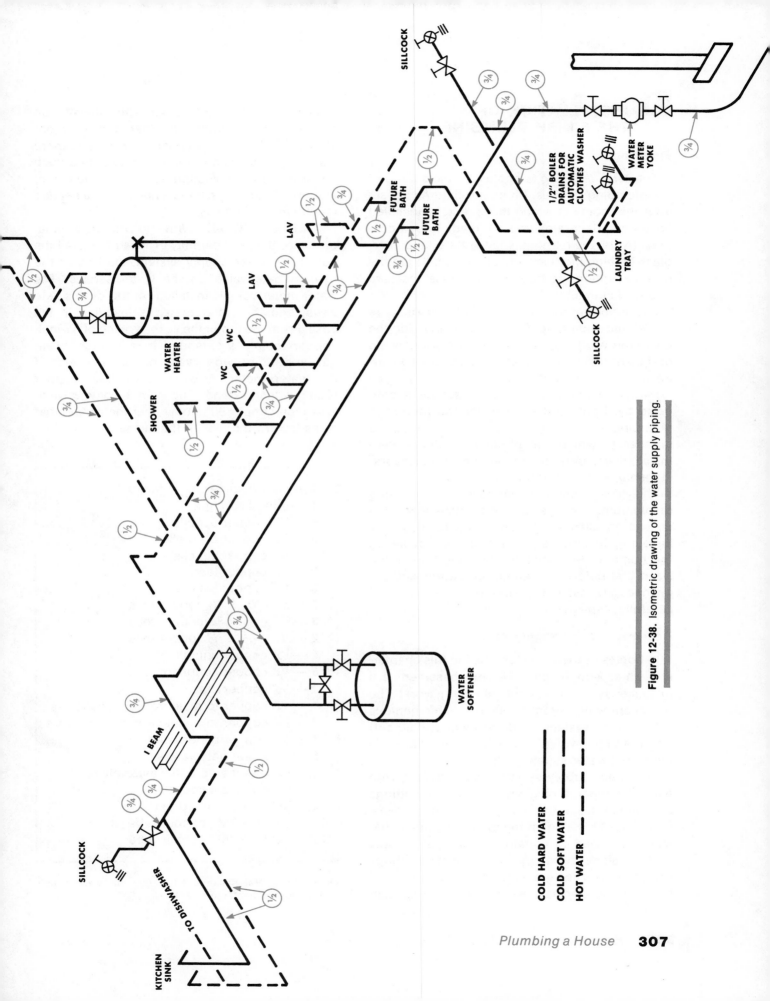

Figure 12-38. Isometric drawing of the water supply piping.

SILLCOCK

3/4
3/4
3/4
3/4
3/4

WATER METER YOKE

1/2" BOILER DRAINS FOR AUTOMATIC CLOTHES WASHER

3/4

1/2
1/2
1/2
1/2

FUTURE BATH

FUTURE BATH

LAV

3/4
1/2

LAV

3/4
1/2

3/4

WC

1/2

WC

1/2

3/4

1/2
3/4

3/4

1/2

SHOWER

WATER HEATER

3/4

3/4

1/2

LAUNDRY TRAY

SILLCOCK

1/2

3/4

WATER SOFTENER

3/4

3/4

I BEAM

1/2

3/4
3/4

SILLCOCK

TO DISHWASHER

1/2

KITCHEN SINK

COLD HARD WATER ————
COLD SOFT WATER ———
HOT WATER — — —

Plumbing a House **307**

INSTALLATION OF THE FINISH PLUMBING

Finishing

The term *finishing* is applied to the setting of the plumbing fixtures. Plumbing fixtures are set after the rooms in which they are to be installed have been decorated and are ready to be occupied. (*Note:* the bathtub, shower base, and fiberglass bathing modules and shower stalls are the only exceptions to this rule because they are built-in fixtures.)

It is important that the plumber take extra care to set the plumbing fixtures properly, for the customer will judge the quality of workmanship of the entire plumbing installation by the appearance of the plumbing fixtures.

Many plumbing installations have been criticized by the customer, even though the rough plumbing installation was perfect, because the manner in which the plumbing fixtures were installed was poor in comparison with the workmanship of the other trades.

The general methods employed in the setting of plumbing fixtures and appliances were discussed in Chapter 10 and may be used for installing the fixtures and appliances specified in the home in this example. It should be emphasized that before the plumbing fixtures and appliances are set into place, each one must be carefully inspected for defects.

Basement Water Supply System

In homes of the type illustrated in this chapter, the water supply pipes in the basement are installed as a part of the finishing work. The basement water supply system for this home is a three-pipe system: cold hard water, cold soft water, and hot water. Figure 12-38 illustrates this basement water supply system.

The water supply system is constructed with type M copper tubing and solder-joint fittings suspended from the basement ceiling joists. Figure 12-39 is a list of the pipe and fittings that would be used to construct this water supply system. Since these copper solder-joint fittings are small and inexpensive, it is recommended that the plumber have extra fittings on hand in case it is necessary to change the installation slightly from the way it is illustrated in Figure 12-38. Conditions that cannot be forseen before starting the installation might necessitate such an alteration. Neglecting to have a sufficient amount of fittings on hand when they are needed results in costly delay.

Cold Hard Water. With the proper material on hand, the plumber may proceed to install the basement water supply system, starting at the 3/4-inch gate valve on the water service. The water meter yoke is installed on top of this gate valve, and the water meter is installed in the yoke. From the top of the yoke, a piece of 3/4-inch copper tubing is extended to a 3/4-inch sweat gate valve to provide a valve on the outlet side of the water meter. A piece of 3/4-inch copper tubing is then extended up from the gate valve to a 3/4-inch copper 90° elbow near the ceiling that faces toward the rear of the house.

1	— pound solder
1	— pound can flux
26	— 1/2″ copper 90° ells
2	— 1/2″ copper drop ells
14	— 3/4″ copper 90° ells
2	— 1/2″ copper tees
6	— 3/4″ copper tees
9	— 3/4 × 3/4 × 1/2 copper tees
3	— 3/4 × 1/2 × 3/4 copper tees
2	— 3/4 × 1/2 × 1/2 copper tees
4	— 1/2″ copper couplings
6	— 3/4″ copper couplings
2	— 1/2″ copper caps
1	— 3/4″ copper to male iron pipe adaptors
2	— 3/4″ copper to female iron pipe adaptors
1	— 3/4″ copper meter yoke
5	— 3/4″ sweat gate valves
3	— 3/4″ sweat stop and waste valves
3	— 3/4″ sweat sill cocks
2	— 1/2″ sweat boiler drains
80′	— 1/2″ type "M" copper tube
120′	— 3/4″ type "M" copper tube

Figure 12-39. Material required to pipe the water supply system illustrated in Figure 12-38.

A short piece of ³/₄-inch copper tubing joins this elbow with a ³/₄-inch tee. The side opening of the tee faces up into the second ceiling joist space (from the front of the house), to provide the cold water supply for the two sill cocks at the front of the house. The side opening of this tee is extended up to the side opening of another ³/₄-inch tee located up in the joist space. The end, or *run,* openings of this second tee are extended with ³/₄-inch tubing to ³/₄-inch stop and waste valves to control the sill cocks. From the stop and waste valves, pieces of ³/₄-inch copper tubing are extended through the side walls of the house to the ³/₄-inch sweat sill cocks located on the left and right sides of the house.

Moving back to the first ³/₄-inch tee on the cold hard water main, a length of ³/₄-inch copper tubing is extended, below the ceiling joists, toward the rear of the house. A ³/₄-inch tee is provided in the line, with the side opening facing down, to supply cold hard water to the water softener. From the end of this tee, the cold hard water pipe is extended as ³/₄-inch size to an offset constructed with four ³/₄-inch 90° elbows and short pieces of ³/₄-inch tubing to convey the water up, over, and back down on the kitchen side of the steel support I-beam.

On the kitchen side of the I-beam, a piece of ³/₄-inch copper tubing extends back toward the rear of the house below the ceiling joists. Near the back wall of the house, a ³/₄ × ³/₄ × ¹/₂ copper tee (commonly called ³/₄ × ¹/₂) is placed facing the kitchen sink to provide cold hard water to the kitchen sink faucet. The end of this ³/₄ × ¹/₂ tee is extended to a ³/₄-inch stop and waste valve. From the end of the valve, a piece of ³/₄-inch copper tubing is extended to two ³/₄-inch 90° ells to offset the pipe up into the joist space. From the second ell, a piece of ³/₄-inch copper tubing is extended through the back wall of the house to a ³/₄-inch sweat sill cock.

The side opening of the ³/₄-× ¹/₂-inch tee is then extended with ¹/₂-inch copper tubing across the basement ceiling to a point beneath the kitchen sink. At this point, the pipe turns with a ¹/₂-inch 90° ell and then is piped to another ¹/₂-inch 90° ell. This second ell is extended with ¹/₂-inch pipe up through the bottom of the kitch-

en sink cabinet near the back wall, where it may be connected to the kitchen sink faucet.

Since the kitchen sink in this home is located on an outside wall, the cold and hot water supply pipes to the kitchen sink faucet are piped up from the basement through the bottom of the kitchen sink cabinet rather than through the wall of the house to eliminate the possibility of freezing these pipes.

At the ³/₄-inch tee left to supply cold hard water to the water softener, a short piece of ³/₄-inch copper tubing is extended to a ³/₄-inch 90° ell and then across the ceiling to another ³/₄-inch 90° ell, which turns down to the water softener inlet. From this ell, ³/₄-inch copper tubing is extended down to a ³/₄-inch tee to provide a by-pass for the water softener. From the bottom of this tee, ³/₄-inch copper tubing is extended to a ³/₄-inch gate valve to control the water entering the softener and then into the inlet of the water softener, which completes the installation of the cold hard water piping.

Cold Soft Water. The cold soft water piping starts at the outlet of the water softener. A piece of ³/₄-inch copper tubing is extended up to a ³/₄-inch gate valve and then to a ³/₄-inch tee for the other side of the softener by-pass. This tee is then connected to the other by-pass tee (previously placed on the cold hard water pipe) with two short pieces of ³/₄-inch tubing and a gate valve between them to complete the softener by-pass.

From the by-pass tee, a piece of ³/₄-inch copper tubing is extended up to a ³/₄-inch 90° ell placed near the ceiling, which faces the water heater. A piece of ³/₄-inch copper tubing extends from this ell toward the water heater. As this pipe passes the previously installed cold hard water main, a ³/₄-inch tee is left with the side opening facing up approximately 4 inches to the left of the cold hard water main.

The ³/₄-inch cold soft water pipe then continues to the water heater. At the heater, a ³/₄ × ¹/₂ × ³/₄ copper tee is placed to provide a ¹/₂-inch cold water supply to the bathtub. The ³/₄-inch side opening of this tee is piped down through a ³/₄-inch gate valve and to a ³/₄-inch copper to female iron pipe adaptor, which is threaded onto the $3/4$-inch cold water inlet nipple on the water heater to provide cold water to the water heater.

The end of the $3/4 \times 1/2 \times 3/4$ tee is then extended as $1/2$-inch size to the $1/2$-inch copper tubing that was attached to the bathtub water supply fitting installed with the rough plumbing. These two $1/2$-inch pipes are then joined with a $1/2$-inch $90°$ ell to complete the cold water supply to the bathtub.

Returning to the $3/4$-inch tee placed to the left of the cold hard water main, a short piece of $3/4$-inch copper tubing is extended up to a $3/4$-inch copper 90 degree ell at the same elevation as the cold hard water pipe. From this ell, a piece of $3/4$-inch tubing is extended to a $3/4 \times 1/2$ tee placed in line with the $1/2$-inch copper cold water supply to the shower valve. The shower valve pipe is then connected to this tee with two short pieces of $1/2$-copper tubing and two $1/2$-inch copper $90°$ ells.

From this $3/4 \times 1/2$ tee, the cold soft water is piped $3/4$-inch size to four $3/4 \times 1/2$ tees to supply cold water for the water closet in the three-quarter bath, the water closet in the full bath, the vanity lavatory in the full bath, and the wall-hung lavatory in the three-quarter bath. Each of these tees is then connected with two short pieces of $1/2$-inch copper tubing and two $1/2$-inch copper $90°$ ells to the $1/2$-inch copper water pipe drops installed for these fixtures at the time the rough plumbing was installed.

A piece of $3/4$-inch copper tubing is extended from the last $3/4 \times 1/2$ copper tee to a $3/4 \times 1/2 \times 1/2$ tee located over the plumbing wall (the wall with the vent piping) in the future basement three-quarter bath. The side opening of this tee faces down into this future plumbing wall and is capped with a short piece of $1/2$-inch copper tubing and a $1/2$-inch copper cap.

The end opening of the $3/4 \times 1/2 \times 1/2$ tee is extended with $1/2$-inch copper tubing across the ceiling to the laundry tray. At the laundry tray, the pipe is offset down and across to the laundry tray faucet with three $1/2$-inch copper $90°$ ells and pieces of $1/2$-inch copper tubing. Just above the laundry tray faucet, a $1/2$-inch copper tee is placed with the side opening facing the front of the house. This opening is extended with $1/2$-inch

copper tubing across the back edge of the laundry tray to a point approximately 12 inches past the edge of the laundry tray, and connected with a $1/2$-inch copper drop ell to a $1/2$-inch sweat boiler drain. The drop ell is fastened to the wall to support the boiler drain. This boiler drain will be directly behind the automatic clothes washer to supply it with cold water.

This connection completes the cold soft water piping.

Hot Water. The installation of the hot water piping starts at the $3/4$-inch hot water outlet nipple of the water heater. A $3/4$-inch copper to female iron pipe adaptor is threaded onto the $3/4$-inch iron pipe size outlet nipple. From this adaptor, a piece of $3/4$-inch copper tubing is extended up to a $3/4$-inch copper $90°$ ell near the ceiling. A piece of $3/4$-inch tubing extends from this ell over the $1/2$-inch cold water supply to the bathtub to a $3/4 \times 1/2 \times 3/4$ copper tee, the $1/2$-inch opening of which faces the $1/2$-inch hot water tubing on the bathtub water supply fitting. This $1/2$-inch opening of the tee is then joined to the $1/2$-inch hot water tubing on the bath fitting with a piece of $1/2$-inch copper tubing and a $1/2$-inch copper $90°$ ell.

The $3/4$-inch opening of the $3/4 \times 1/2 \times 3/4$ tee is then extended with $3/4$-inch copper tubing back toward the cold hard and cold soft water mains. At a point 4 inches to the right of the cold soft water main, another $3/4 \times 1/2 \times 3/4$ tee is placed, with the $3/4$-inch end opening facing the front of the house and the $1/2$-inch end opening facing the rear of the house. The $1/2$-inch opening is extended toward the rear of the house with $1/2$-inch copper tubing. At a point approximately 4 inches toward the front of the house from the $3/4$-inch cold hard water pipe supplying the kitchen sink and rear sill cock, the $1/2$-inch hot water pipe is offset up, over, and back down on the other side of the I-beam. On the other side of the I-beam, a piece of $1/2$-inch copper tubing is extended toward the rear of the house, 4 inches to the left of the cold hard water pipe.

The $1/2$-inch hot water pipe then follows the previously installed cold hard water supply up to the kitchen sink faucet, running parallel and approximately 4 inches to the left of the cold water pipe. After the $\frac{1}{2}$-inch hot water line passes through the bottom of the kitchen cabinet, a $\frac{1}{2}$-inch copper tee is left in the line to provide a hot water supply to the automatic dishwasher. The hot water may then be connected to the kitchen sink faucet.

Back at the $\frac{3}{4} \times \frac{1}{2} \times \frac{3}{4}$ tee on the hot water pipe facing the front of the house, a piece of $\frac{3}{4}$-inch copper tubing is extended to a $\frac{3}{4} \times \frac{1}{2}$ copper tee with the $\frac{1}{2}$-inch side opening of the tee directly beneath the $\frac{1}{2}$-inch hot water supply tubing connected to the shower valve. This tubing and the tee are then connected.

From this tee, the $\frac{3}{4}$-inch hot water pipe is extended to two $\frac{3}{4} \times \frac{1}{2}$ tees to supply hot water to the vanity lavatory in the full bathroom and the wall-hung lavatory in the three-quarter bathroom. These two tees are also connected to the hot water supply pipes for these fixtures which were installed as a part of the rough plumbing.

A piece of $\frac{3}{4}$-inch copper tubing is extended from the last $\frac{3}{4} \times \frac{1}{2}$ copper tee to a $\frac{3}{4} \times \frac{1}{2} \times \frac{1}{2}$ tee located over the plumbing wall in the future basement $\frac{3}{4}$ bathroom. The side opening of this tee, which faces down into the future wall, is capped with a short piece of $\frac{1}{2}$-inch copper tubing and a $\frac{1}{2}$-inch copper cap.

The end opening of this $\frac{3}{4} \times \frac{1}{2} \times \frac{1}{2}$ tee is extended with $\frac{1}{2}$-inch copper tubing to run parallel the $\frac{1}{2}$-inch cold water supply to the laundry tray faucet. Above the laundry tray faucet, a $\frac{1}{2}$-inch tee is also left in the hot water pipe. The side opening of this tee is extended with $\frac{1}{2}$-inch copper tubing across the back of the laundry tray to a $\frac{1}{2}$-inch drop ell and a $\frac{1}{2}$-inch boiler drain to provide a hot water connection for the automatic clothes washer.

This completes the installation of the hot water piping.

Water Pipe Test

After the entire basement water supply system has been completed and all the plumbing fixtures and appliances have been installed, the system should be filled with water to test it for leaks. Any leaks that are found must be repaired. It is usually not necessary to test water piping in homes of the type illustrated in this chapter with water pressure that exceeds that of the city water main.

Final Air Test

Upon completion of the installation of the finish plumbing, the house is ready for the finished plumbing test. The methods for applying finished plumbing tests were described in Chapter 11, and the apprentice should refer back to that chapter for the testing method required by the local plumbing code.

After the finished plumbing test, the plumber must remove all testing plugs placed in the roof vent terminals, or the effectiveness of the entire vent system will be seriously impaired.

Final Cleanup

The plumber should clean all the plumbing fixtures and appliances so that they will be ready for use by the homeowner. In addition, he must also remove all the packing cartons, paper, and other debris that accumulated during the finishing operations.

The plumber must leave for the homeowner all operating instructions and warranty papers that apply to the plumbing fixtures and appliances so that the homeowner will have these papers for the proper use of the plumbing fixtures and appliances.

1. How does the plumber determine the layout of piping in a building? From what sources does he get measurements for roughing-in pipes?

2. Who provides the lists of pipe, fittings, and fixtures?

3. What parts of the plumbing system are included in a rough-in? What pipe is usually installed first?

4. From what sources does the plumber get the data he uses to mark out locations and pipe measurements for fixtures?

5. From what point does the plumber begin to lay the building drain?

6. How high above the floor must the main cleanout shown in Figure 12-19 be extended? Why?

7. How many other cleanouts are provided in the drainage system shown in Figure 12-19?

8. How is the depth of the building sewer trench determined?

9. Obtain a set of blueprints for a small home and draw an isometric drawing of the sanitary drainage and vent piping (similar to Figure 12-19). List the material required to install this system.

10. Draw an isometric sketch of the water supply piping in this home (similar to Figure 12-38). List the material required to install this system.

House Plans for Chapter 12

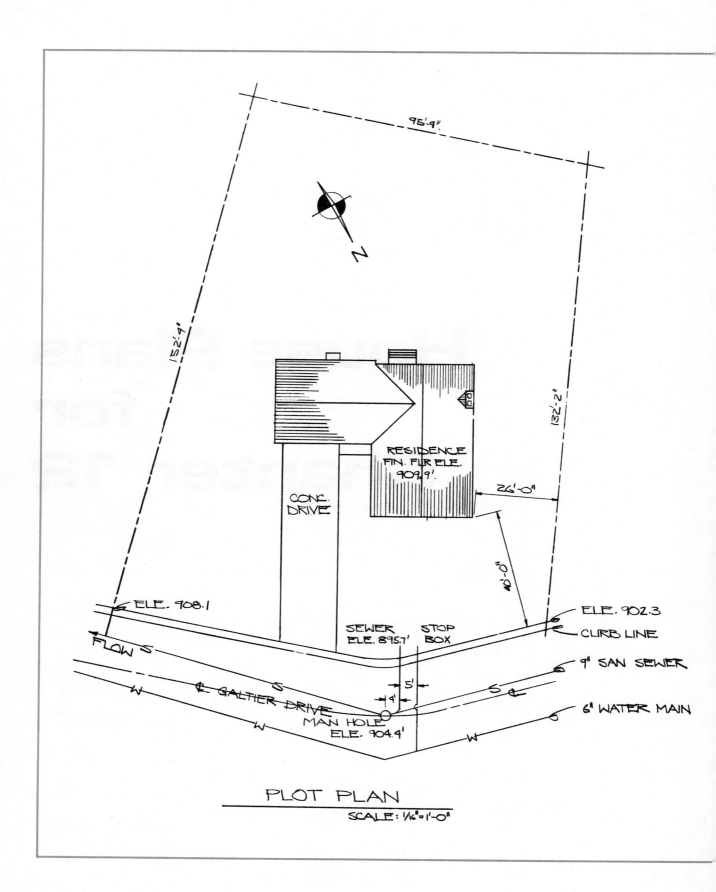

PLOT PLAN

SCALE: 1/16" = 1'-0"

INDEX TO DRAWINGS

1   PLOT PLAN
2   BASEMENT PLAN
3   FLOOR PLAN
4   WALL SECTION - STAIR
    SECTION
5   KITCHEN ELEVATIONS
6   BATHROOM ELEVATIONS
7   ELEVATIONS
8   ELEVATIONS

RESIDENCE FOR LESLIE V. RIPKA

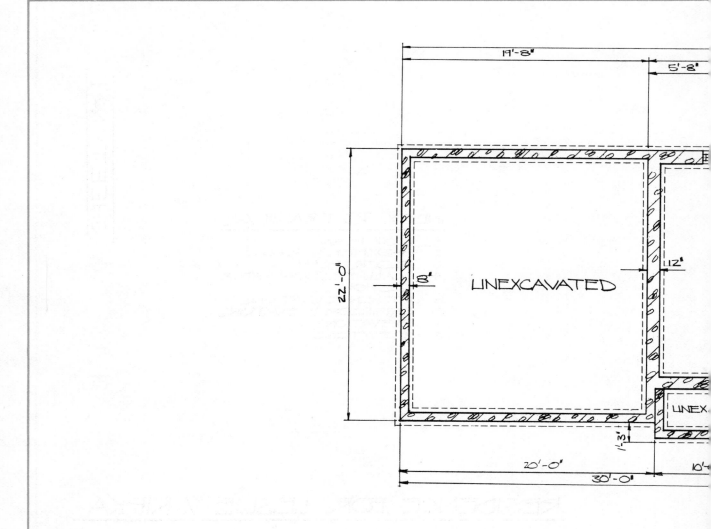

19'-8" 5'-8"

22'-0"

8" 12"

UNEXCAVATED

UNEX

1'-3"

20'-0" 10'-

30'-0"

BASEMENT PLAN
SCALE: 1/4"=1'-0"

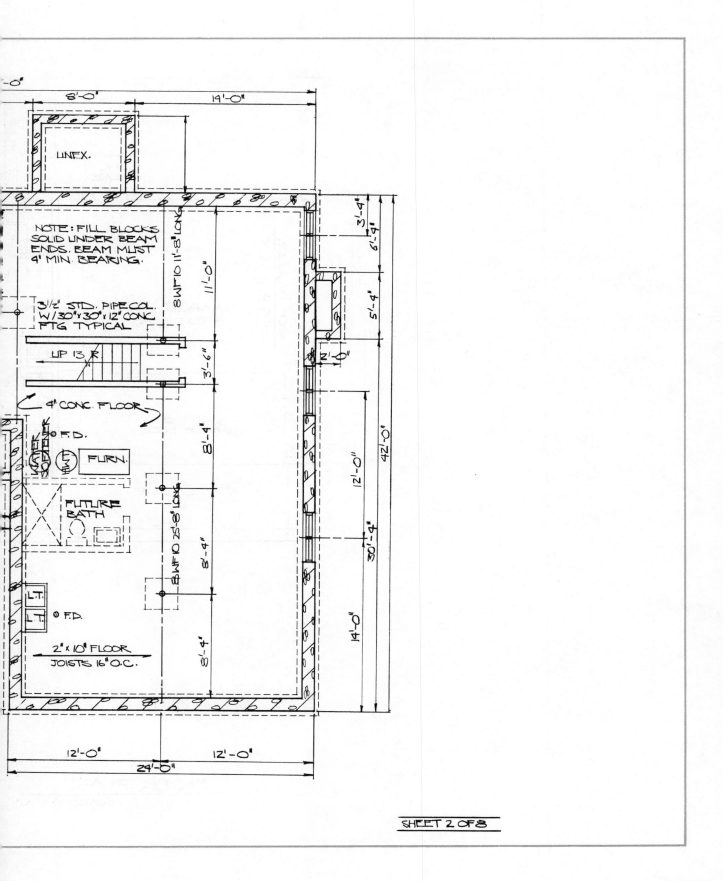

UNEX.

NOTE: FILL BLOCKS
SOLID UNDER BEAM
ENDS. BEAM MUST
9' MIN. BEARING.

3½" STD. PIPE COL.
W/30"x 30"x 12" CONC.
FTG TYPICAL

UP 13 R

4" CONC. FLOOR

F.D.

WATER
SOFTENER

FURN.

FUTURE
BATH

L.T.

L.T.

F.D.

2"x 10" FLOOR
JOISTS 16" O.C.

8'-0"   19'-0"

0"

3'-4"

6'-4"

5'-4"

2'-0"

8WF10 11'-8" LONG

11'-0"

3'-6"

8'-4"

8WF10 25-8" LONG

8'-4"

12'-0"

42'-0"

30'-4"

14'-0"

8'-4"

12'-0"   12'-0"

24'-0"

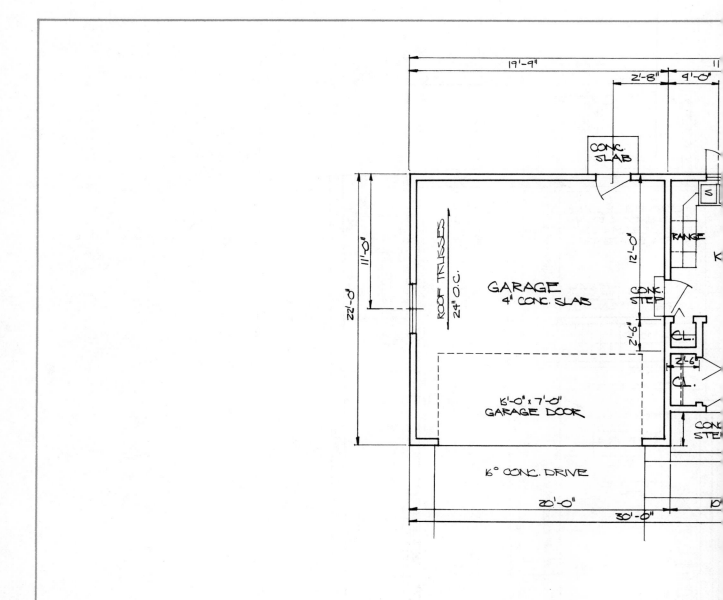

GARAGE
4" CONC. SLAB

ROOF TRUSSES
24" O.C.

CONC.
SLAB

RANGE

CONC.
STEP

CL.

CL.

CONC.
STEP

6'-0" x 7'-0"
GARAGE DOOR

6" CONC. DRIVE

19'-9"

2'-8"

9'-0"

11'-0"

22'-0"

12'-0"

2'-9"

2'-6"

20'-0"

30'-0"

FLOOR PLAN

SCALE: 1/4" = 1'-0"

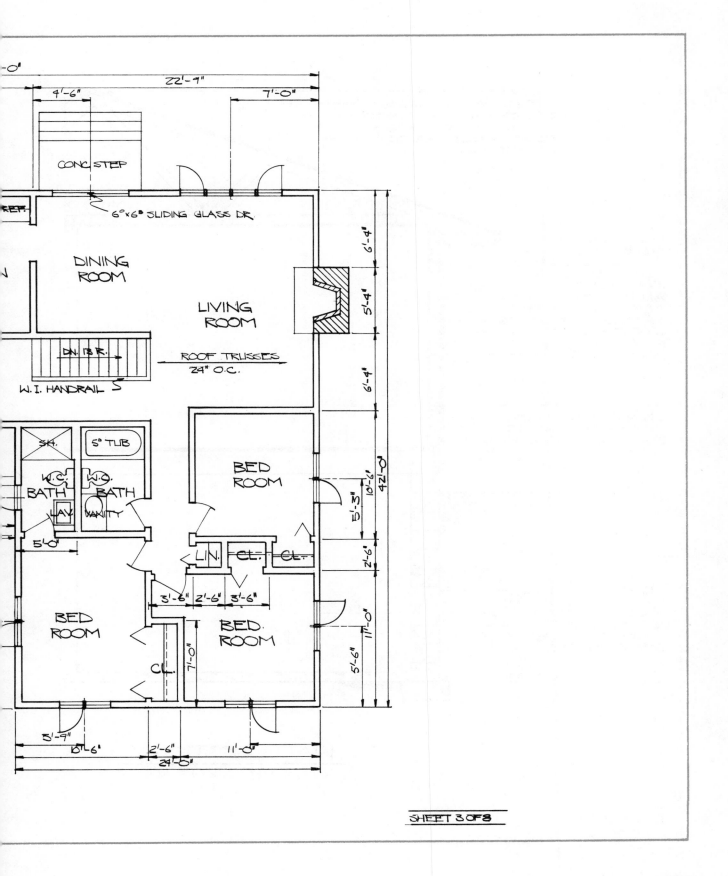

CONC. STEP

6"x6" SLIDING GLASS DR.

DINING
ROOM

LIVING
ROOM

DN. 13 R.

W. I. HANDRAIL

ROOF TRUSSES
24" O.C.

SH.

5° TUB

W.C.
BATH

W.O.
BATH

LAV.

VANITY

5'-0"

BED
ROOM

BED
ROOM

LIN.

CL.

CL.

3'-6"   2'-6"   3'-6"

BED
ROOM

CL.

7'-0"

22'-9"

4'-6"

7'-0"

6'-4"

5'-4"

6'-4"

42'-0"

10'-6"

5'-3"

2'-6"

11'-0"

5'-6"

3'-9"   10'-6"   2'-6"   11'-0"

24'-0"

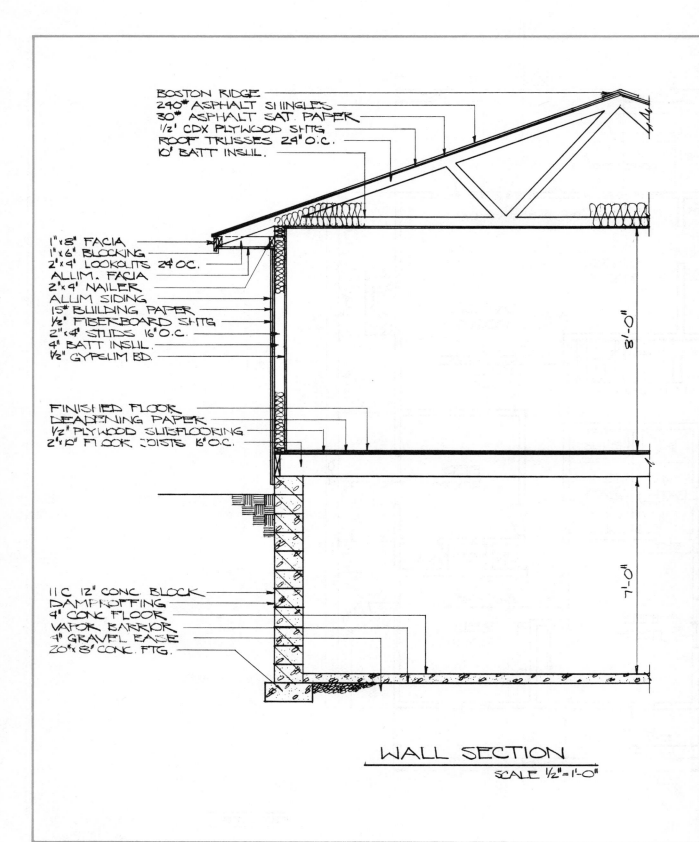

BOSTON RIDGE
240# ASPHALT SHINGLES
30# ASPHALT SAT. PAPER
1/2' CDX PLYWOOD SHTG
ROOF TRUSSES 24" O.C.
10' BATT INSUL.

1"x 8" FACIA
1"x 6" BLOCKING
2"x 4" LOOKOUTS 24 O.C.
ALUM. FACIA
2"x 4" NAILER
ALUM SIDING
15# BUILDING PAPER
1/2" FIBERBOARD SHTG
2"x 4" STUDS 16" O.C.
4" BATT INSUL.
1/2" GYPSUM BD.

FINISHED FLOOR
DEADENING PAPER
1/2" PLYWOOD SUBFLOORING
2"x 10" FLOOR JOISTS 16' O.C.

11 C 12" CONC. BLOCK
DAMPROOFING
4" CONC FLOOR
VAPOR BARRIER
4" GRAVEL BASE
20"x 8' CONC. FTG.

8'-0"

7'-0"

WALL SECTION

SCALE 1/2"= 1'-0"

320

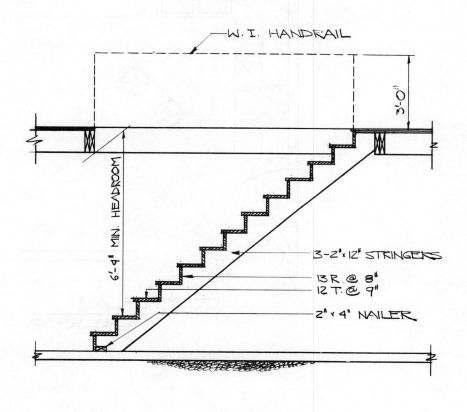

W. I. HANDRAIL

3'-0"

6'-4" MIN. HEADROOM

3-2"x12" STRINGERS

13 R. @ 8"

12 T. @ 9"

2" x 4" NAILER

STAIR SECTION

SCALE ½"=1'-0"

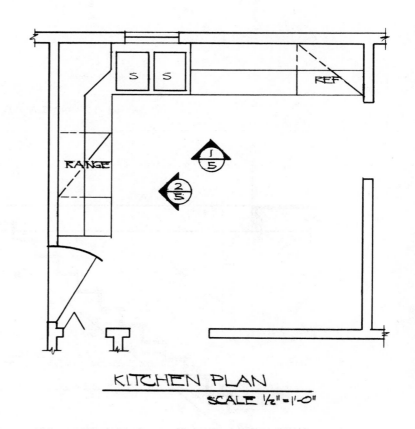

KITCHEN PLAN
SCALE 1/2" = 1'-0"

322

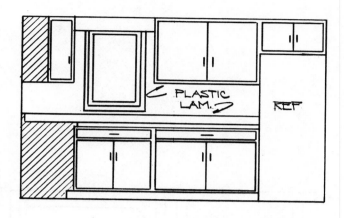

①/5 KITCHEN ELE.

SCALE ½"=1'-0"

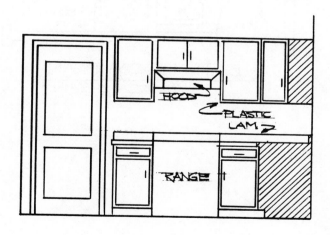

②/5 KITCHEN ELE.

SCALE ½"=1'-0"

SHEET 5 OF 5

323

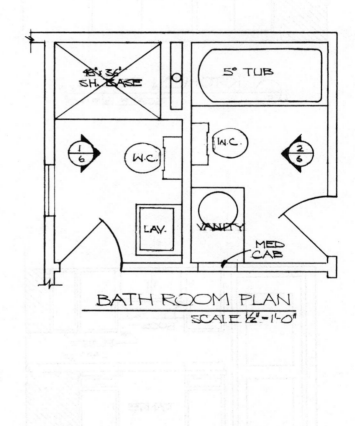

BATH ROOM PLAN

SCALE ½"-1'-0"

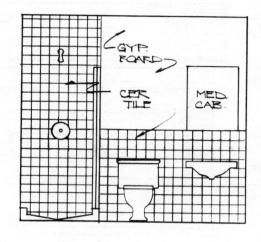

GYP.
BOARD

CER
TILE

MED
CAB.

1 / 6   BATH ROOM ELE
SCALE ½" = 1'-0"

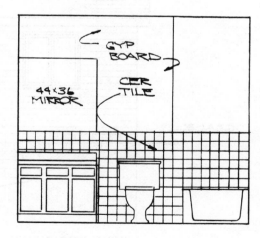

GYP
BOARD

CER
TILE

44 × 36
MIRROR

2 / 6   BATH ROOM ELE.
SCALE ½" = 1'-0"

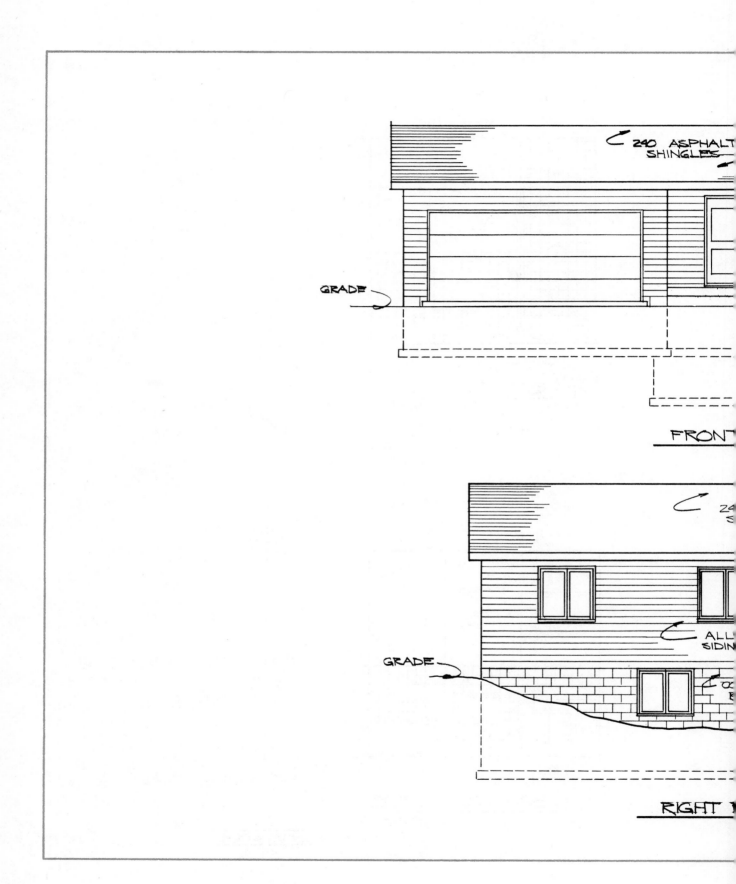

240 ASPHALT SHINGLES

GRADE

FRONT

24
S

ALL
SIDIN

GRADE

RIGHT

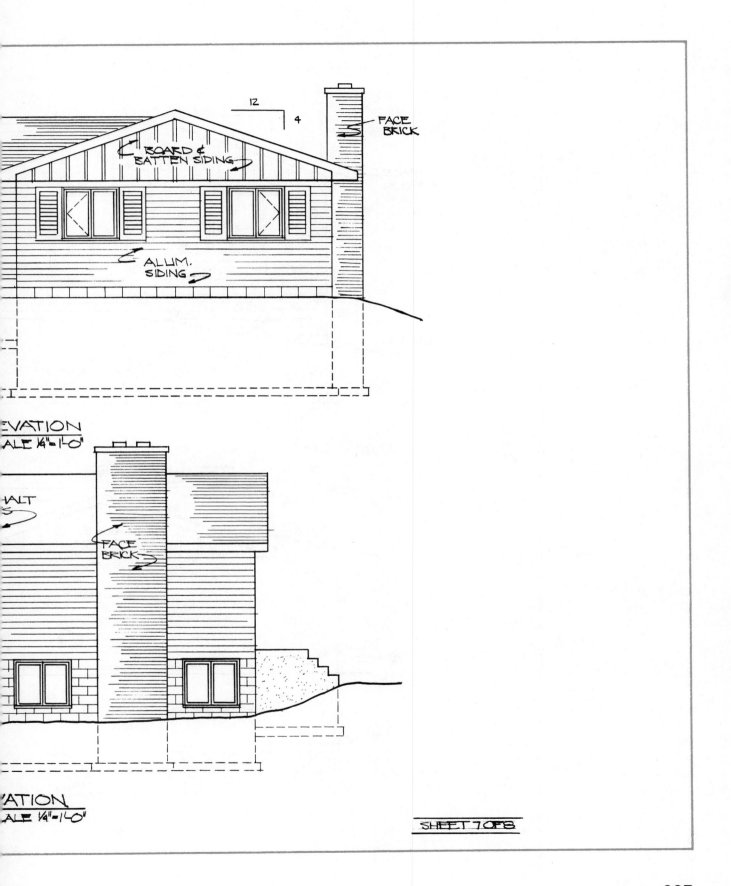

12

4

FACE
BRICK

BOARD &
BATTEN SIDING

ALUM.
SIDING

EVATION
ALE ¼"=1'-0"

HALT
S

FACE
BRICK

ATION
ALE ¼"=1'-0"

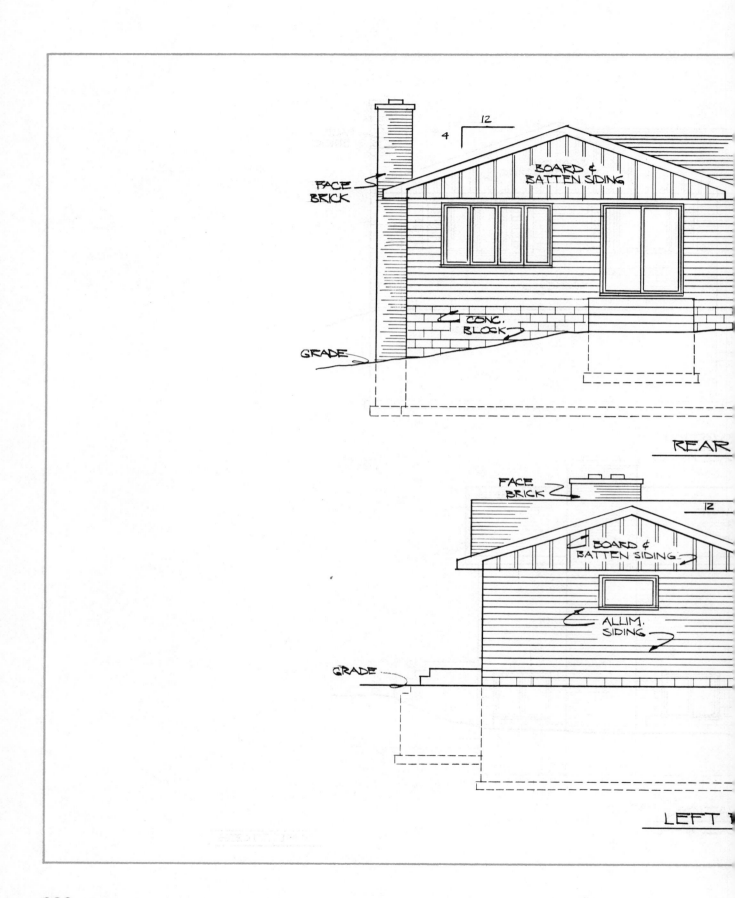

FACE
BRICK

12

4

BOARD &
BATTEN SIDING

CONC.
BLOCK

GRADE

REAR

FACE
BRICK

12

BOARD &
BATTEN SIDING

ALUM.
SIDING

GRADE

LEFT

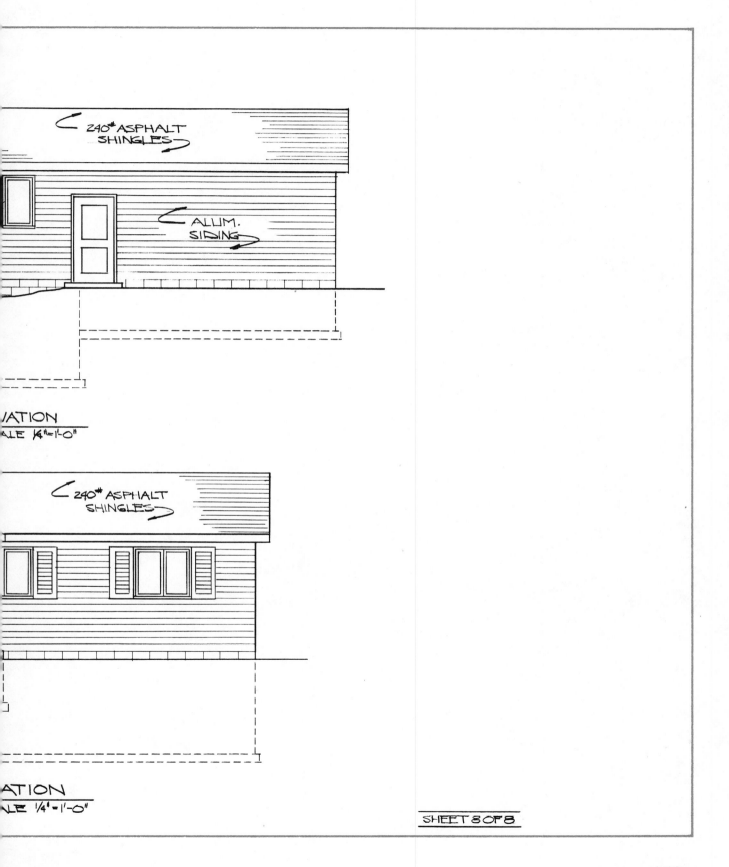

240# ASPHALT SHINGLES

ALUM. SIDING

~ATION
~LE 1/4"=1'-0"

240# ASPHALT SHINGLES

~ATION
~LE 1/4"=1'-0"

Glossary of Terms

Glossary

Air gap (Drainage System)—The unobstructed vertical distance through the free atmosphere between the outlet of a waste pipe and the flood level rim of the fixture or receptacle into which it is discharging.

Angle valve—A globe valve in which the inlet and outlet openings are at 90° angles to one another.

Anti-syphon trap—A trap which is designed to prevent the syphonage of its water seal by increasing the diameter of the outlet leg of the trap so that it contains a sufficient volume of water to prevent a syphoning action.

Apprentice plumber—One who is learning by practical experience the plumbing trade.

Area drain—A receptacle designed to collect surface or storm water from an open area.

Anchors—See *Supports.*

Backflow—The flow of water in pipes in a reverse direction from that normally intended.

Backing—Wood or other supports placed in the building walls to which plumbing fixtures can be attached.

Back pressure—See *Positive pressure.*

Back vent—See *Individual vent.*

Backwater valve—A type of check valve installed to prevent the backflow of sewage from flooding the basement or lower levels of a building.

Bag trap—A water seal trap which has a shape resembling an inflated bag.

Ball cock—A faucet valve that is opened and closed by the fall or rise of a ball floating on the surface of a container of water whose elevation is controlled wholly or in part by the faucet valve.

Ball valves—A valve in which the flow of fluid is controlled by a rotating drilled ball that fits tightly against a resilient (flexible) seat in the valve body.

Basket strainer—A kitchen sink drain fitting consisting of a strainer body attached to the drain opening at the bottom of the sink compartment and a removeable basket (also called a crumb cup). The strainer body has a strainer across the drain outlet. Also called a duo-strainer.

Bathtub—A receptacle for water that is shaped to fit a human body in which a person bathes.

Battery of fixtures—Any group of two or more similar adjacent fixtures which discharge into a common horizontal waste or soil branch.

Bell trap—A trap consisting of an inverted bell with a water seal.

Bidet—A low set bowl equipped with cold and hot running water which is used especially for bathing the external genitals and posterior parts of the body.

Bell or Hub—That portion of a pipe which, for a short distance, is sufficiently enlarged to receive the end of another pipe of the same diameter for the purpose of making a joint.

Branch—Any part of the piping system other than a riser, main, or stack.

Branch interval—A vertical length of stack at least eight-feet high (it corresponds in general to a story height), within which the horizontal branches from one story or floor of the building are connected to the stack. (That is, one branch interval equals one floor of plumbing fixture drains.)

Branch vent—A vent pipe connecting two or more individual vents with either a vent stack or a stack vent.

Building sanitary drain—A building drain which conveys sewage only.

Building drain branch—A soil or waste pipe that extends horizontally from the building drain and receives only the discharge from fixtures on the same floor as the branch.

Building drain—That part of the lowest piping of the drainage system that receives the discharge from soil, waste, and other drainage pipes inside the walls of the building and conveys it to the building sewer.

Building sewer—That part of the drainage system that extends from the end of the building drain and conveys its discharge to the public sewer, private sewer, individual sewage-disposal system, or other point of disposal.

Building sanitary sewer—A building sewer which conveys sewage only.

Building storm drain—A building drain that conveys storm water but no sewage.

Building storm sewer—A building sewer that conveys storm water but no sewage.

Building sub-drain—That portion of a drainage system which cannot drain by gravity into the building sewer.

Building supply pipe—The first section of water supply piping in a building after the water meter.

Butterfly valve—A valve with a rotating disk (the butterfly) that fits within a shaft-controlled valve body. In its closed position, the disk seats against a resilient seat. A butterfly valve is opened or closed by only a 90° turn of the handle.

Change in direction—The term applied to the various turns that may be required in drainage pipes.

Check valve—A valve that permits the flow of water within the pipe in only one direction and closes automatically to prevent backflow (the flow in a reverse direction).

Cleanout—A fitting with a removable plate or plug that is placed in plumbing drainage pipe

Glossary

lines to afford access to the pipes for the purpose of cleaning their interior.

Combination sewer—A sewer which conveys both sewage and storm water.

Common seal trap—A P-trap with a water seal depth of 2 to 4 inches.

Common vent—A vent connecting at the junction of two fixture drains and serving as a vent for both fixture drains. Also called a unit vent.

Compression faucet—A faucet in which the flow of water is shut off by means of a washer that is forced down (or compressed) onto its seat.

Compression stop—A non-rated globe valve.

Conductor—See *Rainwater leader*.

Continuous vent—A continuous vent is a vertical vent that is a continuation of the drain to which it connects.

Continuous waste—A waste from two or more fixtures (or compartments of the same fixture) connected to a single trap.

Core cock—A type of valve through which the flow of water is controlled by a circular core or plug that fits closely in a machined seat. This core has a part bored through it to serve as a water passageway. Also called plug valve.

Corporation cock or corporation stop—A valve placed on the water main to which the building water service is connected.

Crown of a trap—That part of a trap in which the direction of flow is changed from upward to downward.

Crown vent—A vent pipe which is connected at the topmost point in the crown of a trap.

Crown weir—The highest part of the inside portion of the bottom surface at the crown of a trap.

Curb cock or curb stop—A valve placed on the water service usually near the curb line.

Dead end—A branch leading from a soil, waste, or vent pipe, building drain, or building sewer and terminating at a developed length of 2 feet or more by means of a plug, cap, or other fitting.

Deep seal trap—A P-trap with a water seal depth of more than 4 inches.

Developed length—The length of pipe measured along the center line of the pipe and fittings.

Dip of a trap—The lowest portion of the inside top surface of the channel through a trap.

Dishwasher—An electric appliance for washing dishes.

Downspout—See *Rainwater leader*.

Drain—Any pipe which carries waste water or water-born wastes in a building drainage system.

Drainage fixture unit (dfu)—A common measure of the probable discharge into the drainage system by various types of plumbing fixtures on the basis of one dfu being equal to 7.5 gallons per minute discharge. The drainage fixture-unit value for a particular fixture depends on its volume rate of drainage discharge, on the time duration of a single drainage operation, and on the average time between successive operations.

Drinking fountain—A fixture that delivers a stream or jet of drinking water through a nozzle. This stream of water is delivered at an upward angle so that the user of the fixture may drink from the fountain without the use of a cup.

Drum trap—A trap whose main body is a cylinder with its axis vertical. This cylinder is large in diameter than the inlet or outlet pipe.

Duo strainer—See *Basket strainer*.

Durham system—A term sometimes used to describe a soil or waste pipe system which is constructed of threaded pipe, tubing, or other rigid construction using recessed drainage fittings corresponding to the type of piping.

DWV—The abbreviation for drainage, waste, and vent.

Female thread—A thread on the inside of a pipe or fitting.

Finishing—Work done after roughing-in. Usually means the actual installation or setting of the plumbing fixtures and appliances.

Fixture branch—A water supply pipe between the fixture supply pipe and a water distributing pipe.

Fixture drain—The drain from the trap of a fixture to the junction of that drain with any other drain pipe.

Fixture supply—A water supply pipe connecting the fixture with the fixture branch pipe.

Flood level rim—The top edge of a plumbing fixture or receptacle from which water overflows.

Floor drain—A receptacle used to receive water that is to be drained from the floor into the drainage system.

Floor-set—Refers to a plumbing fixture that rests on the floor.

Flow pressure—The pressure in the water supply pipe near the faucet or water outlet while the faucet or water outlet is wide open and flowing.

Glossary

Flow rate—The volume of water used by a plumbing fixture in a given amount of time. Usually expressed in gallons per minute (gpm).

Flush—To wash out by drenching with a large amount of water.

Flush valve—A device located at the bottom of a flush tank for flusing water closets and similar fixtures.

Flushometer valve—A device which discharges a predetermined quantity of water to fixtures for flushing purposes and is actuated by direct water pressure.

Front main cleanout—A plugged fitting located near the front wall of a building where the building drain leaves the building. The front main cleanout may be either inside or directly outside of the building foundation wall.

Full-bath—A bathroom containing a water closet, a lavatory, and a bathtub.

Garbage disposal—An electric grinding device used with water to grind into pulp food wastes and discharge these wastes into the drainage system.

Gate valve—A valve that controls the flow of fluid moving through the valve by means of a gatelike wedge disk that fits against smoothly machined surfaces, called seats, within the valve body.

Globe valve—A compression type valve in which the flow of water is controlled by means of a circular disk that is forced (compressed) onto or withdrawn from a ring—the seat—surrounding the opening through which water flows.

Grade or pitch—The fall (slope) of a line of pipe in reference to a horizontal plane. As applied to plumbing drainage pitch is usually expressed as the fall in a fraction of an inch per foot length of pipe (for example, $1/4$ inch per foot).

Half-bath—A bathroom containing a water closet and a lavatory.

Hangers—See *Supports*.

Header—A pipe of many outlets. The outlets are usually parallel and at $90°$ to the centerline of the header.

Horizontal branch drain—A drain pipe extending horizontally from a soil or waste stack or building drain with or without vertical sections or branches, which receives the discharge from one or more fixture drains on the same floor as the horizontal branch and conducts it to the soil or waste stack or to the building drain.

Horizontal pipe—Any pipe or fitting which makes an angle of less than 45 degrees with the horizontal.

Hub—The bell end or enlarged end of a cast iron or vitrified-clay pipe. See *Bell*.

Indirect waste pipe—A waste pipe that does not connect directly with the drainage system but conveys liquid wastes by discharging into a plumbing fixture, interceptor, or receptacle which is directly connected to the drainage system.

Individual vent—A pipe installed to vent an individual fixture trap. It may terminate either into a branch vent, a vent stack, a stack vent, or the open air. Individual vents are also called back vents because they are commonly installed directly in back of the fixtures they serve.

Invert—The lowest portion of the inside of any horizontal pipe.

Journeyman plumber—A plumber who has served his apprenticeship. One who does plumbing work for another for hire.

Kitchen sink—A shallow, flat bottom, fixture that is used in the kitchen for cleaning dishes and in connection with the preparation of food.

Laundry tray—A fixed tub, installed in a laundry room of a home, that is supplied with cold and hot water and a drain connection, which is used for washing clothes and other household items.

Lavatory—A fixture designed for washing of the hands and face which is commonly found in bathrooms and rest rooms.

Length of pipe—See *Developed length*.

Main—The principle pipe artery to which branches may be connected.

Main vent—The principle artery of the venting system to which vent branches may be connected.

Male thread—A thread on the outside of a pipe or fitting.

Master plumber—A person holding a master plumber's license who may be engaged in the business of plumbing as a contractor, installer. or as an employer of journeyman and apprentice plumbers.

Minus pressure—See *Negative pressure*.

Mop basin—A floor set service sink. Also called a mop receptor.

Negative pressure—A pressure within a pipe that is less than atmospheric pressure. Also called a minus pressure.

Offset—A combination of elbows or bends which brings one section of the pipe out of line but into a line parallel with the other section.

Overrim bathtub fitting—A faucet assembly and a mixing spout mounted on the drain end of

Glossary

the bathtub with the spout above the flood level rim of the tub.

Pipe—A cylindrical conduit or conductor, the wall thickness of which is sufficient to receive a standard pipe thread. May be installed plain end or threaded. Compare with tube.

Pitch—See *Grade*.

Plug valve—See *Core cock*.

Plumbing—Plumbing is the art of installing in buildings the pipes, fixtures, and other apparatus for bringing in the water supply and removing waste water and water-carried waste.

Plumbing appliance—A special class of plumbing fixture intended to perform a special function.

Plumbing fixture—A receptacle for wastes which are ultimately discharged into the sanitary drainage system.

Plumbing system—The plumbing system of a building includes the water supply distributing pipes; the fixtures and fixture traps; the soil, waste, and vent pipes; the building drain and building sewer; the storm water drainage; with their devices, appurtenances, and connections within the building and outside the building within the property line.

Plumbing wall—The wall in a building in which the plumbing pipes are installed. Usually the wall directly behind the plumbing fixtures in a bathroom.

Plus pressure—See *Positive pressure*.

Port control faucet—A single handle, noncompression faucet that contains within the faucet body a port for both cold and hot water and some method of opening and closing these ports.

Positive pressure—A pressure within the sanitary drainage or vent piping system that is greater than atmospheric pressure. Also called plus pressure.

Potable water—Water free from impurities present in amounts sufficient to cause disease or harmful physiological effects. Its bacteriological and chemical quality shall conform to the requirements of the state board of health.

Potable water supply system—The water service pipe, the water distributing pipes, and the necessary connecting pipes, fittings, control valves, and all appurtenances within the building or outside the building within the property lines.

Pressure-reducing valve—An automatic device used for converting high, fluctuating inlet water pressure to a lower constant pressure. This valve is also called a pressure-regulating valve.

Private or private use—In the classification of plumbing fixtures, private applies to fixtures in residences and apartments, and to fixtures in private bathrooms of hotels, as well as similar installations in other buildings where fixtures are intended for use of one family or an individual.

P-Trap—A P-shaped commonly used on plumbing fixtures.

Public or public use—In the classification of plumbing fixtures, public applies to fixtures in general toilet rooms of schools, gymnasiums, hotels, railroad stations, bars, public comfort stations, and other installations (whether pay or free) where fixtures are installed so that their use is similarly unrestricted.

Rainwater leader—A pipe inside the building that conveys storm water from the roof to a storm drain. Also called a conductor or downspout.

Regular seal trap—See *Common seal trap*.

Relief valve—A safety device that automatically provides protection against excessive temperatures, excessive pressures, or both. See also *Temperature and pressure relief valve*.

Relief vent—A vent whose primary function is to provide additional circulation of air between drainage and vent systems or to act as an auxiliary vent on a specially designed system.

Riser—A water supply pipe that extends vertically one full story or more to convey water to fixture branches or to a group of fixtures.

Roof drain—A drain installed to receive water collecting on the surface of a roof and to discharge it into a rainwater leader, conductor, or downspout.

Roof jacket or flange—A jacket or flange installed on the roof terminals of vent stacks and stack vents to seal this opening to prevent rainwater from entering into the building around the vent pipe.

Rough-in—The installation of all parts of the plumbing system which can be completed prior to the installation of fixtures. This includes drainage, water supply, and vent piping, and necessary fixture supports. Also called roughing-in.

Run—That portion of a pipe or fitting continuing in a straight line in the direction of flow in the pipe to which it is connected. Sometimes used to describe an appreciable length of straight or nearly straight piping.

Running trap—A trap in which the inlet and outlet are in a straight horizontal line between which the water way is depressed to below the bottom of either the inlet or outlet.

Sanitary drainage pipe—Pipes installed to remove the waste water and waterborne wastes

Glossary

from plumbing fixtures and convey these wastes to the sanitary sewer or other point of disposal.

Sanitary drainage and vent piping system— The sanitary drainage and vent piping systems are installed by the plumber to remove the waste water and waterborne wastes from the plumbing fixtures and appliances, and to provide a circulation of air within the drainage piping.

Sanitary sewer—A sewer that carries sewage and excludes storm, surface, and groundwater.

Seal of a trap—See *Trap seal*.

Self-syphonage—The loss of the seal of a trap as a result of removing the water from the trap that is caused by the discharge of the fixture to which the trap is connected.

Service sink—A sink with a deep basin to accomodate a scrub pail. It is used for the filling and emptying of scrub pails, the rinsing of mops, and the disposal of cleaning water. Also called a slop sink.

Sewage—Any liquid waste containing animal or vegetable matter in suspension or solution. It may include liquids containing chemicals in solution. (Minnesota Plumbing Code.)

Sewer—An artificial conduit, usually underground, for carrying off waste water and refuse.

Sewer gas—The mixture of vapors, odors, and gases found in sewers.

Shower bath—A bath in which the water for bathing is showered upon the body from above.

Sill cock—A type of lawn faucet used on the outside of a building to which a garden hose may be attached.

Single handle faucet—See *Port control faucet*.

Siphonage—See *Syphonage*.

Slop sink—See *Service sink*.

Soil pipe—A pipe that conveys the discharge of water closets or similar fixtures, containing fecal matter with or without the discharge of other fixtures to the building drain or building sewer. Also cast iron pipe with either bell and spigot or hubless (no-hub) ends used in plumbing to convey wastes.

Soil stack—A vertical line of piping that extends one or more floors and receives the discharge of water closets, urinals, and similar fixtures. It may also receive the discharge from other fixtures.

Stack—A general term for any vertical line of soil, waste, or vent piping extending through one or more stories.

Stack cleanout—A plugged fitting located at the base of all soil or waste stacks.

Stack group—A group of fixtures located adjacent to the stack so that by means of proper fittings, vents may be reduced to a minimum.

Stack vent—The extension of a soil or waste stack above the highest horizontal drain connected to the stack.

Stop and waste valve—A gate or compression type valve that has a side opening or port which may be opened to allow water to drain from the piping supplied by the valve.

Stop box or curb box—An adjustable cast iron box that is brought up to grade with a removable iron cover. By inserting a shutoff rod down into the stop box it is possible to turn off the curb cock.

Storm sewer—A sewer used for conveying groundwater, rainwater, surface water, or similar nonpollutional wastes.

Storm water drainage system—The piping system used for conveying rainwater or other precipitation to the storm sewer or other place of disposal.

S-Trap—An S-shaped, water seal trap sometimes used in plumbing. (Most water closet traps are S-traps.)

³/₄ S-Trap—A trap shaped like three-fourths of the letter S.

Supports—Devices for supporting and securing pipe, fixtures, and equipment.

Syphonage—A suction created by the flow of liquid in pipes.

Swing joint—A joint in a threaded pipe line which permits motion in the line in a plane normal to the direction of one part of the line.

Temperature and pressure relief valve—A safety valve designed to protect against the development of a dangerous condition by relieving high temperature and/or high pressure from a water heater.

Three-quarter bath—A bathroom containing a water closet, a lavatory, and a shower bath.

Trap—A fitting or device which provides, when properly vented, a liquid seal to prevent the emission of sewer gases without materially affecting the flow of sewage or waste water through it.

Trap arm—That portion of a plumbing fixture drain between the trap weir and the vent pipe connection.

Trap seal—The column of water between the crown weir and the top dip of a trap. Also called seal of a trap.

Trim—The water supply and drainage fittings which are installed on the fixture to control the flow of water into the fixture and the flow of waste water from the fixture to the sanitary drainage system.

Glossary

Tube—A conduit or conductor of cylindrical shape, the wall thickness of which is less than that needed to receive a standard pipe thread. Compare with pipe.

Unit vent—See *Common vent*.

Yoke vent—A pipe that connects upward from a soil or waste stack to a vent stack and prevents pressure differences in the two stacks. Yoke vents are a type of relief vent.

Wet vent—A vent which also serves as a drain.

Water service—The pipe from the water main or other source of water supply to the water distributing system of the building.

Water main—The pipe that conveys potable water for public or community use from the municipal water supply source.

Water distributing pipe—A pipe that conveys water from the water service pipe to the point of use.

Waste stack—A vertical line of piping that extends one or more floors and receives the discharge of fixtures other than water closets and urinals.

Waste pipe—A pipe that conveys only liquid waste free from fecal material.

Water supply fixture unit—(wsfu)—A common measure of the probable hydraulic demand on the water supply by various types of plumbing fixtures. The supply fixture-unit valve for a particular fixture depends on its volume rate of supply operation, and on the average time between successive operations.

Water softener—An appliance which removes dissolved water hardness minerals (calcium and magnesium) from water by the process of ion exchange.

Water meter—A device used to measure the amount of water in cubic feet or gallons that passes through the water service.

Water heater—An appliance for the heating and supplying the hot water used within a building for purposes other than space heating.

Water hammer—The banging noise in pipes caused by a quick closing valve.

Water cooler—An electric appliance which combines a drinking fountain with a water cooling unit.

Water closet seat—A device installed around the rim of a water closet to support the person using the closet.

Water closet—A water flushed plumbing fixture which is designed to receive human excrement directly from the user of the fixture. (The term is sometimes used to indicate the room or compartment in which the fixture is located.)

Waste and overflow fitting—A bathtub drain fitting which provides both the outlet for the bathtub drain and an overflow to drain excess water from the tub.

Wall-hung—Refers to a plumbing fixture which is supported from the wall.

Vertical pipe—Any pipe or fitting which makes an angle of 45 degrees or less with the vertical.

Vent system—A pipe or pipes installed to provide a flow of air to or from a drainage system or to provide a circulation of air within such system to protect trap seals from syphonage and back pressure.

Vent stack—A vertical pipe installed to provide circulation of air to and from the drainage system.

Vent pipe—The pipe installed to ventilate a building drainage system and to prevent trap syphonage and back pressure.

Vanity—A bathroom fixture consisting of a lavatory set into or onto the top of a cupboard or cabinet.

Valve—A valve is a fitting installed by plumbers on a piping system to control the flow of fluid within that system in one or more of the following ways:

1. To turn the flow on;
2. To turn the flow off;
3. To regulate the flow by permitting flow in one direction only (that is, to prevent backflow), to regulate pressure, or to relieve excess temperature and/or pressure.

Vacuum—Any air pressure less than that exerted by the atmosphere.

Urinal, womens—A fixture designed for use by women so that it may be straddled.

Urinal—A water flushed plumbing fixture designed to receive urine directly from the user of the fixture.

Metric Information for Plumbers

TABLE 1

QUANTITY	NAME OF BASIC METRIC UNIT	SYMBOL
Length	metre	m
area	square metre	m²
volume	cubic metre	m³
liquid measure	litre	*l*
weight	gram	g
pressure	kilopascal	kp$_a$
temperature	degrees celsius	°C

TABLE 2

METRIC PREFIXES

PREFIX	VALUE	MEANING	SYMBOL
Milli	thousandths	$\frac{1}{1000}$	m
Centi	hundredths	$\frac{1}{100}$	c
Deci	tenths	$\frac{1}{10}$	d
Decka	tens	10x	d^a
Hecto	hundreds	100x	h
Kilo	thousands	1000x	k

TABLE 3

DECIMAL AND METRIC EQUIVALENTS

FRACTIONS	DECIMAL INCHES	MILLIMETRES
1/16	.0625	1.58
1/8	.125	3.18
3/16	.1875	4.76
1/4	.250	6.35
5/16	.3125	7.97
3/8	.375	9.52
7/16	.4375	11.11
1/2	.500	12.70
9/16	.5625	14.29
5/8	.625	15.88
11/16	.6875	17.46
3/4	.750	19.05
13/16	.8125	20.64
7/8	.875	22.22
1	1.00	25.40

TABLE 4

ENGLISH TO METRIC CONVERSION FACTORS

QUANTITY	TO CONVERT	TO	MULTIPLY BY
Length	inches	millimetres	25.4
	inches	centimetres	2.54
	feet	centimetres	30.48
	feet	metres	.3048
	yards	centimetres	91.44
	yards	metres	.9144
Area	square inches	square millimetres	645.2
	square inches	square centimetres	6.452
	square feet	square centimetres	929.0
	square feet	square metres	.0929
	square yards	square metres	.8361
Volume	cubic inches	cubic millimetres	1639
	cubic inches	cubic centimetres	16.39
	cubic feet	cubic centimetres	2.832
	cubic feet	cubic metres	.02832
	cubic yards	cubic metres	.7646
Liquid Measure	pints	cubic centimetres	473.2
	pints	litres	.4732
	quarts	cubic centimetres	946.3
	quarts	litres	.9463
	gallons	cubic centimetres	3785
	gallons	litres	3.785
Weight	ounces	grams	28.35
	ounces	kilograms	.02835
	pounds	grams	453.6
	pounds	kilograms	.4536
	short tons (2000 lbs.)	kilograms	907.2
	short tons (2000 lbs.)	metric tonne (1000 kg)	.9072
Pressure	inches of water column	kilopascals	.2491
	feet of water column	kilopascals	2.989
	pounds per square inch	kilopascals	6.895
Temperature	degrees fahrenheit (°F)	degrees celsius (°C)	5/9 (°F – 32)

TABLE 5

METRIC TO ENGLISH CONVERSION FACTORS

QUANTITY	TO CONVERT	TO	MULTIPLY BY
Length	millimetres	inches	.03937
	centimetres	inches	.3937
	metres	feet	3.281
	metres	yards	1.0937
Area	square millimetres	square inches	.00155
	square centimetres	square inches	.1550
	square centimetres	square feet	.0010
	square metres	square feet	10.76
	square metres	square yards	1.196
Volume	cubic centimetres	cubic inches	.06102
	cubic metres	cubic feet	35.31
	cubic metres	cubic yards	1.308
Liquid Measure	litres	pints	2.113
	litres	quarts	1.057
	litres	gallons	.2642
Weight	grams	ounces	.03527
	kilograms	pounds	2.205
	metric tonne (100 kg)	pounds	2205
Pressure	kilopascals	inches of water column	4.014
	kilopascals	feet of water column	.3346
	kilopascals	pounds per square inch	.1450
Temperature	degrees celsius (°C)	degrees fahrenheit (°F)	(9/5 °C) + 32

Some important features of the Metric System are:
1 cubic centimetre of water weights 1 gram.
Pure water freezes at 0°C and boils at 100 °C.

Numbers appearing in **bold type** refer to illustrations.

A

abrasive saw, 68, **68**, 72, **72**
acrylonitrile-butadiene-styrene
    (ABS) plastic, 43, **43**, 44
adjustable wrench, **77**, 78
air compressor, 245, **245**
air test,
    of distribution piping, 250-251
    of finish plumbing, 311
    of sanitary drainage and vent
        piping, 244-248, **245, 246, 247,
        248**
    of water supply, 250-251
American Standard Taper Pipe
    Thread (NPT), 100, **100**
anchoring devices, 114, **114**
angle of repose, 19, **19**
angle test plug, 246, **246**
angle valves, 49-50, **50**
annealed copper tubing, 38-39
anti-syphon P-traps, 170, **170**
apartment building bathroom
    stack,
    individually vented fixtures,
        160, **160**
    minimum venting, 159-160, **160**
apartment building kitchen sink
    waste stack, 160, **161**, 162
appliance, plumbing, 201
appliances, installing, 240-241
apprentice plumber, 4-5
apprenticeship, 3
architectural blueprints, 7
arm protection, 15
arrestor, water hammer, 239, **239**
assembly finishing tools, 77-78,
    77, 78, 79
automatic thermostat, water
    heater, **237**, 238
automatic water softener, 233-
    235, **233, 234**
available pressure, 181
aviation snips, 87, **88**

B

backing for fixtures, 299
back pressure, 179, **179**
backwater valves, 52, **52**
bag traps, 173-174, **173**
ball valves, 50-51, **51**
band saw, 89, **89**
base, shower, 299
basement fixtures, 289, **290**, 291
    vent installation, 295
basement water supply system,
    308-311
basin, mop, 230-231, **231**
basin wrench, 78, **79**

basket strainer for kitchen sink,
    223-224, **224**
basket strainer wrench, 78, **79**
bathroom stack,
    apartment building, 159-160,
        160, 161
    multistory building, 162-163,
        162
bath-shower module, 219, **219**
bathtub, 218-220, **219**
    installing, 299, 301, **302**
    left-hand, 218
    recessed, 219, **219**
    right-hand, 218
    rough-in sheet, **268**
bathtub drain,
    fitting, 220, **220**
    installation, 295
    rough-in sheet, **269**
bathtub faucets, 219-220, **219**
bathtub fitting,
    installation, 301, **302**
    overrim, 219, **219**
    rough-in sheet, **270**
bathtub vent, installing, 295
bell and spigot cast iron soil pipe,
    30, **31**, 33
bell traps, 173, **173**
bench grinder, 87, 89, **89**
bench setup for threading pipe,
    66, **66**
bending tools, 75, **75**
bent, individual, 11
bent test plug, 246, **246**
bidets, 222, **222**
blade cutter, 76, **76**
blade-type geared pipe cutter, 67
blowout urinals, 211, 213, **213**
blowout water closet, 204, **204,
    209, 209**
blueprints, 257
    architectural, 7
    mechanical, 7
    plumbing, 5, **6**, 7-8, **7, 8**
    structural, 7
boring tools, 82-83, **82, 83**
bottom outlet wall-hung washout
    urinal, 210, **211**
box,
    curb, 9-10
    stop, 9-10
    trench, 21, **21**
box end wrench, 78, **78**
braces, 19
    wall, 231, **231**
bracing trenches, 18
branch,
    building drain, 10
    fixture, 10

    horizontal, 10
branch drains, 123-124
branch vent, 11, 142-143
breathing apparatus, 15-16
brine solution, 233
brine tank of water softener,
    234, **234**
brine well, 234, **234**
builder's level, 81, **81**
building drain, 10
    installing, 292-293
    sizing, 120-122
building drain branch, 10
building drain trench, laying out,
    291, **291**
building height, 185
building sewer, 10
    installing, 303-306, **304, 305**
    sizing, 120-122
    testing, 250
building storm drain, 12, 149-151,
    151
building storm sewer, 11
built-in fixtures, 240
bullnose chisel, 85, **85**
bumping method of gasket
    installation, 96, **96**
burner, main, **236**, 237
bushings, 47
butterfly valves, 51, **51**
by-pass, water softener, 234-235,
    235
by-pass port, **205, 206**, 208

C

cable choker, 89, **90**
cap,
    pipe, 245
    water closet, **206**, 208
capillary action, 180, **180**
cast iron recessed drainage
    fittings, 35-37, **36**
cast iron sink, 222, **223**
cast iron soil pipe, 29-30, **30,
    31, 32**, 33
    bell and spigot, 30, **31**, 33
    joints, 92, **92, 93**, 94-96, **94,
        95, 96, 97**, 98, **98, 99**
    tools, 57-60, **58, 59, 60**
cast iron threaded fittings,
    standard, 35, **35**
caulked soil pipe joint, 92-95,
    92, 93, 94
caulking hammer, **58**, 59
caulk joint tools, 59, **59**
center-set fitting, 218, **218**
ceramic tile shower bath, 221
chain tongs, 70-71, **71**

chain wrench, 70-71, **71**
chalk line, 81, **81**
change in direction, drainage, 128-129
check valves, 51-52, **52**
chisel,
    bullnose, 85, **85**
    cold, **58**, 59, 85, **85**
chlorinated polyvinyl chloride plastic, 43, 44-45, **44**
choker, cable, 89, **90**
cleaning tools, 73-74, **73, 74**
cleanout, 10
    drainage, 129-130, **129**, 130
    front main, 10
    P-trap, 173, **173**
    stack, 10
cleanup of finish plumbing, 311
cleats, 19
closet bend support, 117, **117**
closet seat wrench, 78, **79**
clothing, safety, 15-16
cock,
    core, 50, **50**
    corporation, 9, 50, **50**
    curb, 9, 50, **50**
    gas, 50, **50**
cold chisel, **58**, 59, 85, **85**
cold hard water, 308-309
cold soft water, 309-310
cold water inlet for water heater, 235, **236**
cold water supply fixture unit, 189, **192, 196**
combination compression valve, 221, **221**
combination sewer, 147, **147**
combination wrench, 78, **78**
commercial water closets, 209, **209**
common vents, 141-142, **141, 142, 143**
compound-leverage wrench, 71, **71**
compound water meter, 54, **55**
compressibility, 131
compression faucet, 217, 219, **219**, 222, **223, 227**
compression gasket soil pipe joint, 95-96, **95, 96, 97**, 98, **98**
    tools, 59-60, **60**
compression joints, 107, **107**
compression valve, combination, 221, **221**
compressor, air, 245, **245**
concealed faucet, 218, **218**
concrete drilling tools, 83-85, **84, 85**
conductor, rainwater, 12, 147-148, **147, 148**
contractor, plumbing, 4

control,
    high limit, **237**, 238
    thermostatic, 236-237, **236**
    time clock, 234, **234**
coolers, water, 229-230, **229**
copper, 38-39
copper joints,
    compression, 107, **107**
    flared, 106-107, **107**
    solder, 102-106 , **103, 104, 105, 106**
copper tubing, 38-42, **40, 41, 42**
    annealed, 38-39
    drawn, 38-39
    hard temper, 39
copper tubing fittings, 40-42, **40, 41, 42**
    flared, 41-42, **42**
    solder joint pressure, 41, **41**
copper tubing joints, 102-107, **103, 104, 105, 106, 107**
copper tubing tools,
    bending, 75, **75**
    cleaning, 73-74, **73, 74**
    cutting, 72, **72**
    flaring, 74, **75**
    reaming, 73, **73**
    soldering, 74, **74**
core cocks, 50, **50**
corporation cock, 9, 50, **50**
corporation stop, 9
couplings, reducing, 46, 47
curb box, 9-10
curb cock, 9, 50, **50**
curb stop, 9
cutter,
    blade, 76, **76**
    blade-type geared pipe, 67
    four-wheel pipe, 66-67, **67**
    hydraulic, 58, **58**
    internal, 80, **80**
    ratchet-type, 58, **58**
    single-wheel pipe, 66, **66**
    squeeze type, 58, **58**
    wheel pipe, 61, **61**, 66-67, **66, 67**
cutting tools, 57-59, **58**, 72, **72**, 80, **80**, 82-83, **82, 83**
    abrasive saw, 68, **68**
    pipe vise, 61, **61**
    power, 64-70, **64, 65, 66, 67, 68**
    wheel pipe cutter, 61, **61**

D

defective plumbing systems, 255
diagrammatic piping drawing, 7, **7, 9, 11, 12**
diameter, outside, 38
diaphragm, segment, **205**, 208
diaphragm flushometer valve, **205**, 207-208
dies, threading, 62-63, **62**

dip tube on water heater, 236, **236**
dishwasher, 225-226, **225**
    drain connection, 225-226, **226**
    rough-in sheet, **267**
disks, wedge, 47, **48**
disk-type water meter, 54, **55, 56**
disposal, garbage, 224-225, **225**
distributing water pipe, 10
distribution piping test,
    air, 250-251
    hydrostatic, **250**, 251
diverter, water, 220
domestic dishwasher, 225-226, **225**
domestic garbage disposal, 224-225, **225**
double-molded cap for water closet, **206**, 208
downspout, 12, 147-148, **147, 148**
drain,
    bathtub, 295
    branch, 123-124
    building, 10, 292-293
    dishwasher, 225-226, **226**
    fixture, 10
    floor, 227-229, **228**
    lavatory, 295
    roof, 12, 148, **149**
    shower, 295
    shower bath, 222
    sizing of, 120-122
    storm, 12, 146-151, **146, 147, 149, 151**
drainage,
    sanitary, 35
    storm, 35, 146-151, **146, 147, 148, 149, 151**
drainage fixture unit system, 120
drainage pipe,
    grade of, 128, **128**
    horizontal, 122-123
    pitch of, 128, **128**
    sanitary, 10, 119-127, **120, 122, 125, 126, 127**
drainage system,
    sanitary, 10-11, **11**, 128-130, **128, 129, 130**
    storm, 11-12, **12**
    venting of, 131-145, **134, 135, 136, 137, 138, 139, 140, 141, 142, 143, 144, 145**
drain branch, building, 10
drain fitting,
    bathtub, 220, **220**
    kitchen sink, 223-224, **224**
    laundry tray, 226-227, **227**
    lavatory, 217-218, **218**
    mop basin, 231
    service sink, 231

drain rough-in sheet, bathtub, **269**
drain trench, building, 291, **291**
drawings, rough-in, 263, **264-280**
drawn copper tubing, 38-39
drill, 82-83, **82, 83**
   hammer, 83, **84**
   star, 85, **85**
   wet core, 84-85, **85**
drilling tools for concrete, 83-85, **84, 85**
drinking fountains, 229-230, **229**
   trim, 230
   waste connections, 230, **230**
   drive, power, 68, **69**, 70
drive shaft, universal, 68, **69**, 70
drophead threading dies, 62, **62**
dry pan floor drain, 228, **228**
drum traps, 171, **171**
duo strainer for kitchen sink, 223-224, **224**
duplex residence, plumbing installation, **158,** 159
Durham fittings, 36-37, **36**
DWV copper fittings, **40,** 41

E

ear protection, 15
elbows, reducing, **46,** 47
electrical safety, 17, **17**
electric storage tank water heater, 237-238, **237**
enclosure, shower, 220-221, **220**
end pipe wrench, 70, **70**
equipment, protective, 15-16
estimator, plumbing, 4
evaporation, 179-180
excavations, 19
   trench, 306
expeller, *See* xpelor
eye protection, 15

F

face protection, 15
faucet,
   bathtub, 219-220, **219**
   compression, 217, 219, **219,** 222, **223, 227**
   concealed, 218, **218**
   kitchen sink, 222, **223**
   laundry tray, 226, **227**
   port control, 217, **217,** 222, **223**
   single-handle, 218, **218,** 219, **219**
faucet rough-in sheet, **266**
   laundry tray, **278**
   lavatory, **274**
   shower, **276**
file, half-round, 67, **67**

final air test, 251-254, **251, 252, 253,** 311
   251, 252, 253
finishing plumbing, 240, 308
finishing tools, 77-80, **77, 78, 79, 80**
finish plumbing,
   basement water supply system, 308-311
   cleanup of, 311
   final air test, 311
   installing, 308-311
first-floor bathroom, installing, 293, **294,** 295
first-floor fixtures, **285,** 286-289, **286, 287**
fit, interference, 108, **108**
fitting rough-in sheet, bathtub, **270**
fittings,
   bathtub, 301, **302**
   bathtub drain, 220, **220**
   bathtub faucet, 219-220, **219**
   bushings, 47
   cast iron soil pipe, 29-30, **30, 31, 32,** 33
   cast iron threaded, 35, **35**
   center-set, 218, **218**
   copper tubing, 40-42, **40, 41, 42**
   drain, 223-224, **224,** 231
   Durham, 36-37, **36**
   DWV copper tubing, **40,** 41
   flared copper tubing, 41-42, **42**
   galvanized malleable iron, 37, **37**
   galvanized steel (threaded), 35-37, **35, 36, 37**
   identification of plumbing, 46-47, **46**
   installation of, 98, **98**
   laundry tray drain, 226-227, **227**
   lavatory drain, 217-218, **218**
   no-hub soil pipe, **32,** 33
   overrim, 219, **219**
   recessed drainage, 35-37, **36**
   reducing couplings, **46,** 47
   reducing elbows, **46,** 47
   shower, 299
   solder joint pressure, 41, **41**
   tees, **46,** 47
   trap, 174, **175**
   water supply, 231
   wyes, **46,** 47
fixture backing, 299
fixture branch, 10
fixture drain, 10
fixtures, plumbing, 201
   basement, 289, **290,** 291
   built-in, 240
   first floor, **285,** 286-289, **286, 287**
   installing, 240-241

   location of, **285,** 286-289, **286, 287, 290,** 291
   venting of, 139-141, **139, 140, 141**
fixture supply, 10
fixture trap, 10-11
flange, roof, 11
flared copper tubing fittings, 41-42, **42**
flared joints, 106-107, **107**
   fittings, 41-42, **42**
flare fitting joint, 111, **111**
flaring tools, 74, **75,** 77, **77**
floor drain, 227-229, **228**
floor-set syphon jet water closet, 209, **209**
floor-set water closet, 207, 208
flow pressure, 185-186
flow rate, 181-182
flush devices for urinals, 213-214, **213, 214**
flush devices for water closets, 205-208, **205, 206**
flushometer valves, 205, 206, **206,** 207-208, 209, 213, **213,** 214
   diaphragm, **205,** 207-208
   piston, **206,** 207-208
flush tanks, 205-207, **205**
flush valve, syphon, 214

folded gasket method of installation, 95, **95**
foot protection, 15
foreman plumber, 4
fountains, drinking, 229-230, **229**
four-unit apartment building, water pipe sizing, 191, **192, 193,** 194-195
four-wheel pipe cutter, 66-67, **67**
freon gas leak detection, 252-253, **252**
front main cleanout, 10
full-way valves, 47

G

galvanized malleable iron fittings, 37, **37**
galvanized steel pipe, 33-37, **35, 36, 37**
galvanized steel pipe tools, assembly, 70-71, **70, 71**
   hand cutting, 61-63, **61, 62, 63**
   hand threading, 61-63, **61, 62, 63**
galvanized steel threaded joints, 100-102, **100, 101, 102**
garbage disposal, 224-225, **225**
   rough-in sheet, **266**
   waste piping, 225, **225**
gas, sewer, 10

gas cock, 50, **50**
gasket, 95, **95**, 96, **96**
gasket installation
    bumping method, 96, **96**
    folded method, 95, **95**
gas removal, 131-132
gas storage tank water heater,
    235-237, **236**
gate valves, 47, **48**, 49
geared pipe cutter, blade-type, 67
geared ratchet-type bender, 75, **75**
geared threader, 68, **68**
globe valves, 49, **49**
grade of horizontal drainage pipe,
    128, **128**
grade of vent, 133
grinder, bench, 87, 89, **89**

H

hacksaw, 76-77, **76**
    electric, 82, **83**
half-round file, 67, **67**
halide torch, 87, **87**, 252, **252**
hammer,
    caulking, **58**, 59
    heavy, 85, **85**
    rotary, 84, **84**
    water, 239
hammer drill, 83, **84**
hand hoist, 89, **90**
handle on water closet, **205, 206,**
    208
hand protection, 15
hand tool safety, 16-17
hangers,
    above-ground piping, 113-114,
    113
    closet bends, 117, **117**
    horizontal piping, 116, **116, 117**
    stack base, 117, **117**
    vertical piping, 115, **115**
hard copper, 38-39
hard temper copper tubing, 39
hard water, 232
    cold, 308-309
head protection, 15
heaters, water, 235-240, **236, 237**
heavy hammer, 85, **85**
height of building, 185
high limit control, water heater,
    237, 238
high water service pressure, 239
hinged four-wheel pipe cutter,
    66-67, **67**
hoist,
    hand, 89, **90**
    ratchet lever, 89, **90**
holes, location of, **285,** 286-289,
    286, 287
horizontal branch, 10
horizontal branch drains, sizing,
    123-124

horizontal drainage pipe,
    grade of, 128, **128**
    pitch of, 128, **128**
    sizing of, 122-123
horizontal pipes, 120, **120**
horizontal piping support, 116,
    116, 117
hot water, 310-311
hot water outlet for water heater,
    235, **236**
hot water supply fixture unit, 189,
    193, 197
house plans, 313-330
hydraulic cutter, 58, **58**
hydrostatic test of distribution
    piping, **250,** 251
hydrostatic test of water supply,
    250, 251

I

immersion elements, water
    heater, **237,** 238
indirect waste receptors, 229
individual vent, 11, 138-139, **139**
inflatable rubber test plugs, **246,**
    247, **247**
inlet, cold water, 235, **236**
insert fitting joint, 111, **111**
inspecting plumbing system,
    243-244, 255
inspecting rough plumbing, 299
installation,
    appliances, 240-241
installation,
    basement fixture vent, 295
    bathtub, 299, 301, **302**
    bathtub drain, 295
    bathtub fitting, 301, **302**
    bathtub vent, 295
    building drain, 292-293
    building sewer, 303-306, **304,**
    305
    finish plumbing, 308-311
    first-floor bathroom, 293, **294,**
    295
    fixtures, 240-241
    kitchen sink stack, 293, **293**
    laundry tray vent, 297, **298**
    laundry tray waste, 297, **298**
    lavatory drains, 295
    lavatory water piping, 301, **303**
    pipe, 112-113, **112**
    roof jacket, 301
    rough plumbing, 281-302
    shower base, 299
    shower valve, **300**
    three-quarter bathroom vents,
    296, 297
    water closet water piping, 301
    water service, 306, **307**
insulation of water heater, 235,
    236

integral trap floor drain, 228, **228**
interference fit, 108, **108**
internal cutter, 80, **80**
internal partition traps, 174, **174**
ion exchange, 232
isometric piping drawing, 8, **8**

J

jab saw, 80, **80**
jacket,
    roof, 11
    water heater, 235, **236**
jam-proof threading dies, 62, **62**
joining pipe, 91-111
joint,
    cast iron soil pipe, 92, **92, 93,**
    94-96, **94, 95, 96, 97, 98, 98,**
    99
    caulked soil pipe, 92-95, **92, 93,**
    94
    compression, 107, **107**
    compression gasket soil pipe,
    95-96, **95, 96, 97,** 98, **98**
    copper tubing, 102-107, **103,**
    104, 105, 106, 107
    flared, 106-107, **107**
    flare fitting, 111, **111**
    galvanized steel, 100, 102, **100,**
    101, 102
    insert fitting, 111, **111**
    no-hub soil pipe, 98, **99**
    plastic pipe, 108-111, **108, 109,**
    110, 111
    solder, 102-106, **103, 104, 105,**
    106
    solvent weld, 108-110, **108, 109,**
    110
    threaded, 100-102, **100, 101, 102**
Joint Apprenticeship and Training
    Committee, 3-5
joint tools, 59-60, **59, 60**
journeyman plumber, 4

K

keyhole saw, 83, **83**
kitchen sink, 222-224, **223**
    drain fittings, 223-224, **224**
    faucet, 222, **223**
    faucet rough-in sheet, **265**
    rough-in sheet, **264**
kitchen sink stack, installing,
    293, **293**
kitchen sink waste stack, 160,
    161, 162

L

ladder, 89, **90**
ladder safety, 22-24, **22, 23**
laundry tray, 226-227, **227**
    drain fittings, 226-227, **227**

faucets, 226, **227**
faucet rough-in sheet, **278**
rough-in sheet, **277**
vent installation, 297, **298**
waste installation, 297, **298**
lavatory, 214-218, **215, 216, 217, 218**
rim type, 216, **216**
self-trimming, 216, **216**
undercounter, 216, **216**
vanity, 215-216, **216, 217**
wall-hung, 214-215, **215**
wheelchair, 215, **215**
lavatory drain installing, 295
lavatory drain fittings, 217-218, **218**
lavatory faucet rough-in sheet, **274**
lavatory rough-in sheet, **272**
lavatory trim, 216-218, **217, 218**
lavatory water piping, 301, **303**
layout tools, 81, **81**
leaders, rainwater, 12, 147-148, **147, 148,** 149-151
leak, 180
leak detection,
freon gas, 252-253, **252**
peppermint, 254
smoke, 253-254, **253**
left-hand bathtub, 218
leg protection, 15
length of water supply pipe, 182-185
level, 87, **88**
builder's 81, **81**
lever-type bender, 75, **75**
lifting slings, nylon, 89, **90**
lumber for trench shoring, 20
list of materials, 281, **284, 308**

M

main, water, 9, 10
main burner of water heater, **236,** 237
main cleanout, front, 10
main vents, sizing, 135-137, **136, 137**
malleable iron fittings, 37, **37**
manometer, 87, **87,** 251, **251**
materials, piping, 45
materials list, 281, **284, 308**
measuring tape, steel, 81, **81**
measuring tools, 81, **81**
mechanical blueprints, 7
mechanically sealed traps, 173-174, **174**
mechanical test plug, 246, **246**
meters, water, 54, **55,** 56
compound, 54, **55**
disk-type water, 54, **55, 56**
turbine, 54, **55**

mineral tank of water softener, 234, **234**
module, bath-shower, 219, **219**
mop basins, 230-231, **231**
multiple port valve, 234, **234**
multistory building bathroom stack, 162-163, **162**

N

no-hub soil pipe, **32,** 33
no-hub soil pipe joint, 98, **99**
no-hub soil pipe joint tools, 60, **60**
nonratcheting threading dies, 62, **62,**
nylon lifting slings, 89, **90**

O

Occupational Safety and Health Act (OSHA), 13-14
offset hex wrench, **77,** 78
offset stacks, sizing, 126-127, **127**
oil for threading iron pipe, 63, **63**
one-piece water closet, **207,** 208
outdoor register, 56, **56**
outlet,
hot water, 235, **236**
valve, **205, 206,** 208
outside diameter, 38
outside register, 56, **56**
overrim bathtub fitting, 219, **219**
oxacetylene welding safety, 26-27, **26, 27**

P

pedestal syphon jet urinal, 211, **212**
peppermint leak detection, 254
permits, plumbing, 263
pick, 87, **88**
pilot, safety, **236,** 237
pipe,
ABS plastic, 43, **43,** 44
bell and spigot cast iron, 30, **31,** 33
cast iron soil, 29-30, **30, 31, 32,** 33
CPVC plastic, 43, 44-45, **44**
drainage, 128, **128**
galvanized steel, 33-37, **35, 36, 37**
horizontal, 120, **120,** 122-123
installing, 112-113, **112**
joining, 91-111
no-hub soil, **32,** 33
PE plastic, 43, **44,** 45
plastic, 42-45, **43, 44**
PVC plastic, 43, **43,** 44
sanitary drainage, 10
seamless, 33
sizing of vent, 133-145

soil, 10, 119-120
supporting, 113-117, **113, 114, 115, 116, 117**
vent, 10, 131
vertical, 120, **120**
waste, 10, 119-120
water distributing, 10
water supply, 186
water in four-unit apartment building, 191, **192, 193,** 194-195
water in larger installations, 189-190
water in public building, 195, **196, 197,** 198-199
welded, 33
pipe caps, 245
pipe cutter,
blade-type geared, 67
hinged four-wheel, 66-67, **67**
single-wheel, 66, **66**
wheeled, 61, **61,** 66-67, **66, 67**
pipe joints,
cast iron soil, 92, **92, 93,** 94-96, **94, 95, 96, 97,** 98, **98, 99**
caulked soil, 92-95, **92, 93, 94**
compression gasket, 95-96, **95, 96, 97, 98, 98**
no-hub soil, 98, **99**
plastic, 108-111, **108, 109, 110, 111**
pipe support, **64,** 65
pipe testing, 311
pipe vise, 61, **61**
pipe wrench,
chain, 70-71, **71**
compound-leverage, 71, **71**
end, 70, **70**
straight, 70, **70**
piping,
horizontal, 116, **116, 117**
lavatory water , 301, **303**
rough water, 299
sanitary drainage, 119-127, **120, 122, 125, 126, 127**
toilet room, 163, **163, 164,** 165-166, **166**
underground, 112-113, **112**
vent, 35, 281, **282, 283**
vertical, 115, **115**
waste, 225, **225,** 281, **282, 283**
water closet, 301
piping drawing,
isometric, 8, **8**
schematic, 7, **7, 9, 11, 12**
piping length, water supply, 182-185
piping materials, 45
piping symbols, **6,** 7
piston flushometer valve, **206,** 207-208
pitch of horizontal drainage pipe, 128, **128**

plans, house, 313-330
plan views of blueprints, 7, **7**
plaster of Paris, 247-248
plastic,
    ABS (acrylonitrile-butadiene-styrene), 43, **43**, 44
    chlorinated polyvinyl chloride, 43, 44-45, **44**
    flare fitting joint, 111, **111**
    polyethylene (PE), 43, **44**, 45
    polyvinyl chloride (PVC), 43, **43**, 44
    thermoplastic, 42
    thermosetting, 42
plastic pipe, 42-45, **43, 44**
plastic pipe joints, 108-111, **108, 109, 110, 111**
    flare fitting, 111, **111**
    insert fitting, 111, **111**
    solvent weld, 108-110, **108, 109, 110**
plastic pipe tools, 76-77, **76, 77**
plastic primer, 108, 109, **109**
plastic solvent, 108-110, **110**
pliers, smooth-jawed, **77**, 78
plug,
    angle, 246, **246**
    bent, 246, **246**
    inflatable rubber, **246**, 247, **247**
    mechanical, 246, **246**
    rubber, **246**, 247, **247**
    test, 86-87, **86**
plug valves, 50, **50**
plumb bob, 81, **81**
plumber,
    foreman, 4
    journeyman, 4
    qualifications of, 5
    responsibilities of, 1
    plumber apprentice, 4-5
plumber apprenticeship, 3
plumber superintendent, 4
plumbing,
    finish, 308-311
    rough, 281-302
plumbing appliance, 201
    installing, 240-241
plumbing code, 1, **2**, 3
plumbing contractor, 4
plumbing estimator, 4
plumbing fittings, 46-47, **46**
plumbing fixture, 201
    installing, 240-241
    venting, 136-141, **139, 140, 141**
plumbing history, 1, 3
plumbing inspection, 255
plumbing permits, 263
plumbing symbols, **6**, 7
plumbing system, 1-3, 8-12, **9, 11, 12**
    defects in, 245

inspecting, 243-244
potable water supply, 9-10, **9**
    sanitary drainage and vent piping, 10-11, **11**
    storm water drainage, 11-12, **12**
    testing, 243, 244-254
plumbing test,
    air, 244-248, **245, 246, 247, 248**
    building sewer, 250
    distribution piping, 250-251, **250**
    finished, 251-254, **251, 252, 253**
    sanitary drainage, 244-249, **245, 246, 247, 248, 249**
    storm drainage piping, 250
    vent piping, 244-249, **245, 246, 247, 248, 249**
    water supply, 250-251, **250**
plumbing testing apparatus, 254-255
plumbing tools, 57-90
    assembly, 70-71, **70, 71**, 77-78, **77, 78, 79**
    basic, 57
    bending, 75, **75**
    boring, 82-83, **82, 83**
    for cast iron soil pipe, 57-60, **58, 59, 60**
    caulk joint, 59, **59**
    cleaning, 73-74, **73, 74**
    compression gasket soil pipe joint, 59-60, **60**
    copper tubing, 72-75, **72, 73, 74, 75**
    cutting, 57-59, **58**, 72, **72**, 76, **76**, 82-83, **82, 83**
    drilling, 83-85, **84**, 85
    flaring, 74, **75**, 77, **77**
    galvanized steel pipe, 61-72
    hand cutting, 61-63, **61, 62, 63**
    hand threading, 61-63, **61, 62, 63**
    joint, 59-60, **59, 60**
    layout, 81, **81**
    measuring, 81, **81**
    miscellaneous, 87, **88, 89, 89, 90**
    no-hub soil pipe joint, 60, **60**
    pipe cutters, 66-70, **66, 67, 68**
    pipe support, **64**, 65
    pipe vise, 61, **61**
    power cutting, 64-70, **64, 65, 66, 67, 68**
    for plastics, 76-77, **76, 77**
    power cutting, 64-70, **64, 65, 66, 67, 68,**
    power threading, 64-70, **64, 65, 66, 67, 68, 69**
    reaming, 61-62, **61**, 73, **73**
    testing, 86-87, **86, 87**
    threading dies, 62-63, **62**
    wheel pipe cutter, 61, **61**

plumbing traps, 168-173, **168, 169, 170, 171, 172, 173**
plunger on watercloset, **205, 206**, 208
polyethylene plastic piping, 43, **44**, 45
port, by-pass, **205, 206**, 208
port control faucet, 217, **217**, 222, **223**
port control valve, 221, **221**
potable water supply system, 9-10
pot-scrubber dishwasher rough-in sheet, **267**
power drive, 68, **69**, 70
power threading tools, 64-70, **64, 65, 66, 68, 69**
pressure,
    available, 181
    back, 179, **179**
    flow, 185-186
    high water service, 239
pressure loss by friction, 182, 184
pressure-reducing valve, 53, **53**, 181, **182**
pressure-regulating valve, 53, **53**
pressure relief valve, **239**, 240
primer, plastic, 108, 109, **109**
private use of plumbing fixtures, 182
prohibited traps, 173-174, **173, 174, 175**
protective equipment, 15-16
P-traps, 168-170, **168, 169, 170**
    floor drain, 228, **228**
public building, water piping sizing, 195, **196, 197**, 198-199
public use of plumbing fixtures, 182

R

ratchet, 78, **78**
ratchet lever hoist, 89, **90**
ratchet threading dies, 62, **62**
ratchet-type bender, 75, **75**
ratchet-type cutter, 58, **58**
rated valves, 47
rainwater leaders, 12, 147-148, **147, 148**
    sizing of, 149-151
reamer, 61-62, **61**
    for copper tubing, 73, **73**
    for plastic, **76**, 77
    spiral, 67, **67**
recessed bathtub, 219, **219**
recessed drainage fittings, cast iron, 35-37, **36**
reducing couplings, **46**, 47
reducing elbows, **46**, 47
reducing valve, 239
regeneration of water softener, 233, **233**

register, outdoor, 56, **56**
relief capacity, 240
relief valve, 53, **53**, 205, **206**, 208
    pressure, **239**, 240
    temperature, **239**, 240
    water heater, 238-240, **239**
relief valve opening, water heater, 235-236, **236**
repose, angle of, 19, **19**
residential water closets, **207**, 208-209
reverse trap water closet, 204, **204**
right-hand bathtub, 218
rim type lavatory, 216, **216**
rim wrench, 78, **79**
riser, 10
roof drain, 12, 148, **149**
roof flange, 11
roof jacket, 11
    installing, 301
rotary hammer, 84, **84**
rough-in drawings, 263, **264-280**
rough-in plumbing, 240
rough-in sheet,
    bathtub, **268**
    bathtub drain, **269**
    bathtub fitting, **270**
    dishwasher, **267**
    garbage disposal, **266**
    kitchen sink, **264**
    kitchen sink faucet, **265**
    laundry tray, **277**
    laundry tray faucet, **278**
    lavatory, **272**, **273**
    lavatory faucet, **274**
    shower base, **275**
    shower faucet, **276**
    wall-hung lavatory, **273**
    water closet, **271**
    water heater, **279**
    water softener, **280**
rough plumbing,
    basement, three-quarter
        bathroom vents, **296**, 297
    inspecting, 299
    installing building drain, 292-293
    installing first-floor bathrooms, 293, **294**, 295
    installing kitchen sink stack, 293, **293**
    laundry tray waste, 297, **298**
    laying out building drain
        trench, 291, **291**
    location of basement fixtures, 289, **290**, 291
    location of first floor fixtures, **285**, 286-289, **286**, **287**
    sanitary waste piping sketch, 281, **282**, **283**
    testing, 299
    vent piping sketch, 281, **282**, **283**

    waste piping sketch, 281, **282**, **283**
rough plumbing installation, 281-302
rough water piping, 299
rubber test plugs, **246**, 247, **247**
running traps, 171, **172**

S

safety, 13-27
    clothing, 15-16
    equipment, 15-16
    general, 14-15
    hand tools, 16-17
    ladder, 22-24, **22**, **23**
    oxyacetylene welding, 26-27, **26**, **27**
    scaffolding, 24-26, **24**
    welding, 26-27, **26**, **27**
safety pilot, water heater, **236**, 237
sand holes, 35
sanitary drainage piping, 10, 35, 119-127, **120**, **122**, **125**, **126**, **127**
    horizontal pipes, 120, **120**
    soil pipes, 119-120
    symbols, 153
    vertical pipes, 120, **120**
    waste pipes, 119-120
sanitary drainage system, 10-11, **11**
    installing, 128-130, **128**, **129**, **130**
    venting, 131-145, **134**, **135**, **136**, **137**, **138**, **139**, **140**, **141**, **142**, **143**, **144**, **145**
sanitary drainage tests,
    air, 244-248, **245**, **246**, **247**, **248**
    water, 248-249, **249**
sanitary sewer, 10
sanitary waste piping sketch, 281, **282**, **283**
saw,
    abrasive, 68, **68**, 72, **72**
    band, 89, **89**
    jab, 80, **80**
    keyhole, 83, **83**
scaffolding safety, 24-26, **24**
schematic piping drawing, 7, **7**, **9**, **11**, **12**
screwdrivers, 78, **79**
seal, trap, 131, 175-180, **176**, **177**, **178**, **179**, **180**
seamless pipe, 33
seats, water closet, 209, **210**
segment diaphragm, **205**, 208
self-contained water softener, 234, **234**
self-rimming enameled cast iron
    sink, 222, **223**

self-rimming lavatory, 216, **216**
self-syphonage, 175-178, **176**, **177**, **178**
service, water, 9
service pressure, high water, 239
service sinks, 230-231, **231**
sewage, 10
sewer,
    building, 10, 303-306, **304**, **305**
    building storm, 11
    combination, 147, **147**
    sanitary, 10
    sizing of, 120-122
    storm, 11
sewer gas, 10
sewer test, 250
sheeting, 19
sheet metal snips, 87, **88**
shield, trench, 19
shoring of trenches, 18, **18**
shovel, 87, **88**
shower, 219, **219**
shower base, 221, **221**, 299
    installing, 299
    rough-in sheet, **275**
shower bath, 220-222, **220**, **221**
    drain, 222
    water supply valve, 221, **221**
shower drain, installing, 295
shower enclosure, 220-221, **220**
shower faucet rough-in sheet, **276**
shower fitting, 299
shower valve, installing, 300
single family home,
    individually vented fixtures, 155, **155**
    minimum venting requirements, 154-155, **154**
    water piping sizing, 187, **188**, 189
single-handle faucet, 218, **218**, 219, **219**
sink,
    kitchen, 222-224, **223**
    service, 230-231, **231**
    slop, 230-231, **231**
sink drain fittings, kitchen, 223-224, **224**
sink stack, installing, 293, **293**
siphon, See syphon
sizes of pipe, water supply, 186
sizing
    branch vents, 142-143
    building drain, 120-122
    building sewer, 120-122
    building storm drains, 149-151,
    common vents, 141-142, **151**
    horizontal branch drains, 123-124
    horizontal drainage pipe, 122-123
    individual vents, 139
    main vents, 135-137, **136**, **137**

sizing (cont.),
 offset stacks, 126-127, **127**
 rainwater leaders, 149-151
 soil stacks, 124-125
 stack vents, 134-135, **135**
 vent pipes, 133-145
 vent stacks, 135-137, **136, 137**
 waste stacks, 124-125
 water pipe in four-unit
  apartment building, 191, **192,
  193,** 194-195
 water pipe in larger installations, 189-190
 water pipe in public building,
  195, **196, 197,** 198-199
 water pipe in single-family
  home, 187, **188,** 189
 wet vents, 143, **144,** 145
sketches, 281, **282, 283**
sledgehammer, 85, **85**
slings, lifting, 89, **90**
slop sink, 230-231, **231**
smoke leak detection, 253-254,
 253
smooth-jawed pliers, **77,** 78
snips, aviation (sheet-metal),
 87, **88**
soap solution, testing for leaks,
 248
socket wrench, 78, **78**
soft copper, 38-39
softeners, water, 232-235, **233,
 234, 235**
soft water, cold, 309-310
soil pipe, 10, 119-120
 bell and spigot, 30, **31,** 33
 cast iron, 29-30, **30, 31, 32,** 33
 caulked joint, 92-95, **92, 93, 94**
 compression gasket joint,
  95-96, **95, 96, 97, 98, 98**
 no-hub, **32,** 33, 98, **99**
soil stack, 10
 sizing, 124-125
solder joints, 102-106, **103, 104,
 105, 106,**
 pressure fittings, 41, **41**
solvent, plastic, 108-110, **110**
solvent weld joint, 108-110, **108,
 109, 110**
specifications, 8, 258, **259-262**
spiral reamer, 67, **67**
spoil, 19
spring bender, 75, **75**
spud wrench, **77,** 78
squeeze-type cutter, 58, **58**
stack, 10
 bathroom, 159-160, **160, 161,**
  162-163, **162**
 kitchen sink waste, 160, **161,**
  162
 offset, 126-127, **127**
 sink, 293, **293**
 soil, 10, 124-125

 vent, 11, 135-137, **136, 137**
 waste, 10, 124-125
stack cleanout, 10
stack group, 154, **154**
stack support, 117, **117**
stack terminals, 137-138
stack vents, sizing, 134-135, **135**
stainless steel sink, 222, **223**
stall urinals, 210, **211**
standard cast iron threaded
 fittings, 35, **35**
 star drill, 85, **85**
 stepladder, 89, **90**
steel pipe, galvanized, 33-37, **35,
 36, 37**
steel tape, 81, **81**
stop,
 corporation, 9
 curb, 9
stop box, 9-10
storage tank, water heater,
 237, 238, 235-236, **236**
storm drain, 146-147, **146, 147,**
 148-151, **149, 151**
 building, 12, 149-151, **151**
storm drainage piping, 35
 testing, 250
storm sewer, 11
storm water drainage, 11-12, **12,**
 146-151, **146, 147, 148, 149,
 151**
straight pipe wrench, 70, **70**
strainer,
 basket, 223-224, **224**
 duo, 223
 kitchen sink, 223-224, **224**
S-traps, 171-172, **172**
strap wrench, 78, **79**
stringers, 19
structural blueprints, 7
struts, 19
superheating water, 238
superintendent plumber, 4
supervisor plumber, 4
supply, fixture, 10
supply fitting,
 mop basin, 231
 service sink, 231
supply system, basement water,
 308-311
supply valve, 214
 shower bath, 221, **221**
supporting pipe, 113-117, **113, 114,
 115, 116, 117**
supports,
 above-ground piping, 113-114,
  113
 closet bends, 117, **117**
 horizontal piping, 116, **116,
  117**
 pipe, **64,** 65
 stack base, 117, **117**
 vertical piping, 115, **115**

swing-check backwater valve,
 52, **52**
symbols,
 drainage, 153
 piping, **6,** 7, 187
 valves, **6,** 7
syphonage, 175-179, **176, 177, 178**
 direct, 175-178, **176, 177, 178**
 indirect, 178-179, **178**
 momentum, 178-179, **178**
syphon flush valve, 214
syphon jet urinals, 210-211, **212**
 pedestal, 211, **212**
 wall-hung, 211, **212**
syphon jet water closet, 202, **203**
 floor-set, 209, **209**
 wall-hung, 209, **209**
syphon water closet, wall-hung,
 207, 208-209

T

tank,
 brine, 234, **234**
 flush, 205-207, **205,** 213, 214,
  214
 mineral, 234, **234**
 storage, 235-236, **236, 237,** 238
 urinal flush, 213, 214, **214**
tape, measuring, 81, **81**
tees, **46,** 47, 157, **157**
temperature relief valve, **239,** 240
terminals, stack, 137-138
test,
 building sewer, 250
 distribution piping, 250-251, **250**
 finished plumbing, 251-254, **251,
  252, 253**
 plumbing system, 244-254
 sanitary drainage, 244-249,
  245, 246, 247, 248
 storm drainage piping, 250
 vent piping, 244-249, **245, 246,
  247, 249**
 water supply, 250-251, **250**
test gage assembly, 248, **248**
testing apparatus, 254-255
testing finish plumbing with air,
 311
testing plumbing system, 243
testing rough plumbing, 299
testing tools, 86-87, **86, 87**
testing water pipe, 311
test plugs, 86-87, **86**
 angle, 246, **246**
 bent, 246, **246**
 inflatable rubber, **246,** 247, **247**
 mechanical, 246, **246**
 rubber inflatable, **246,** 247, **247**
thermal expansion, 239
thermocouple, water heater, **236,**
 237
thermoplastic plastics, 42

thermosetting plastics, 42
thermostat, water heater, **237**, 238
thermostatic control of water heater, 236-237, **236**
thermostatic-mixing shower valve, 221, **221**
threaded fittings,
    cast iron, 35, **35**
    for galvanized steel pipe, 35-37, **35, 36, 37**
threaded joints, galvanized steel, 100-102 **100, 101, 102**
threader, geared, 68, **68**
threading dies,
    drophead, 62, **62**
    jam-proof, 62, **62**
    nonratcheting, 62, **62**
    ratchet, 62, **62**
threading oil, 63, **63**
threading tools, power, 64-70 **64, 65, 66, 68, 69**
three-quarter bathroom vents, installing, **296**, 297
3/4 S-traps, 171-172, **172**
tight sheeting, 19
tile shower bath, 221
time clock control, 234, **234**
toilet room piping,
    two-story industrial building, 163, **163**, 165
    two-story office building, **164**, 165-166, **166**
tongs, chain, 70-71, **71**
tools,
    assembly, 70-71, **70, 71**, 77-78, **77, 78, 79**
    basic plumbing, 57
    bending, 75, **75**
    boring, 82-83, **82, 83**
    cast iron soil pipe, 57-60, **58, 59, 60**
    caulk joint, 59, **59**
    cleaning, 73-74, **73, 74**
    compression gasket soil pipe joint, 59-60, **60**
    copper tubing, 72-75, **72, 73, 74, 75**
    cutting, 57-59, **58**, 64-70, **64, 65, 66, 67, 68**, 72, **72**, 76, **76**, 80, **80**, 82-83, **82, 83**
    drilling, 83-85, **84, 85**
    flaring, 74, **75**, 77, **77**
    hand, 16-17
    hand cutting, 61-63, **61, 62, 63**
    hand threading, 61-63, **61, 62, 63**
    layout, 81, **81**
    measuring, 81, **81**
    miscellaneous, 87, **88**, 89, **89**, 90
    no-hub soil pipe joint, 60, **60**
    pipe support, **64, 65**

pipe vise, 61, **61**
    for plastics, 76-77, **76, 77**
    power cutting, 64-70, **64, 65, 66, 67, 68**
    power threading, 64-70, **64, 65, 66, 68, 69**
    reaming, 61-62, **61**, 73, **73, 76, 77**
    testing, 86-87, **86, 87**
    threading, 64-70, **64, 65, 66, 68, 69**
    wheeled pipe cutter, 61, **61**, 66-67, **66, 67**
top cover, water heater, 235, **236**
torch, halide, 87, **87**
trade, plumbing, 3-4
traps,
    anti-syphon P-, 170, **170**
    bell, 173, **173**
    drum, 171, **171**
    from fittings, 174, **175**
    fixture, 10-11
    internal partition   , 174
    material for, 173
    mechanically se   173-174, **174**
    P-, 168-170, **168**,   **170**
    prohibited, 173-174, **173, 174, 175**
    running, 171, **172**
    S-, 171-172, **172**
    size of, 173
    3/4 S-, 171-172, **172**
    storm drain, 148-149, **149**
    types of, 168-173, **168, 169, 170, 171, 172, 173**
trap seal loss, 131, 175-180, **176, 177, 178, 179, 180**
trays, laundry, 226-227, **227**
trench, 19
    building drain, 291, **291**
    excavation, 306
    failure, 18, **18**
    safety, 17-21, **18, 19, 20, 21**
trench box, 21, **21**
trench shield, 19
trench shoring lumber, 20
trim,
    drinking fountain, 230
    lavatory, 216-218, **217, 218**
    water cooler, 230
    trough urinals, 210, **211**
    tube, dip, 236, **236**
tubing, copper, 38-42, **40, 41, 42**
turbine water meter, 54, **55**
two-story industrial building toilet room piping, 163, **163**, 165
two-story office building toilet room piping, 164, 165-166, **166**
two-story single family home, individually vented fixtures, **156**, 157, 159

two-tank water softener, 233-234, **233**

U

undercounter lavatory, 216, **216**
underground piping, 112-113, **112**
universal drive shaft, 68, **69**, 70
urinal, 210-211, **211, 212**, 213-214, **213, 214**
    blowout, 211, 213, **213**
    bottom outlet, 210, **211**
    pedestal syphon jet 211, **212**
    stall, 210, **211**
    syphon jet, 210-211, **212**
    trough, 210, **211**
    wall hung syphon jet, 211, **212**
    wall-hung washout, 210, **211**
    washout, 210, **211**
    women's, 211, **212**
urinal flush devices, 213-214, **213, 214**
urinal flush tank, 213, 214, **214**

V

valves, 47-54
    angle, 49-50, **50**
    backwater, 52, **52**
    ball, 50-51, **51**
    butterfly, 51, **51**
    check, 51-52, **52**
    combination compression, 221, **221**
    flushometer, 205, 206, **206**, 207-209, 213, **213**, 214
    full-way, 47
    gate, 47, **48**, 49
    globe, 49, **49**
    multiple port, 234, **234**
    plug, 50, **50**
    port control, 221, **221**
    pressure-reducing, 53, **53**, 181, **182**
    pressure-regulating, 53, **53**
    pressure relief, **239**, 240
    rated, 47
    reducing, 239
    relief, 53, **53**, 205, **206**, 208, 235-236, **236**, 238-240, **239**
    shower, **300**
    supply, 214, 221, **221**
    symbols, **6**, 7
    syphon flush, 214
    temperature relief, **239**, 240
    thermostatic-mixing shower, 221, **221**
    water supply for shower bath, 221, **221**
    uses of, 53-54
valve outlet, 205, **206**, 208
vanity lavatory, 215-216, **216, 217**
    rough-in sheet, **272**

vent,
  basement fixture, 295
  bathtub, 295
  branch, 11, 142-143
  common, 141-142, **141, 142, 143**
  individual, 138-139, **139**
  laundry tray, 297, **298**
  main, 135-137, **136, 137**
  stack, 11, 134-135, **135**
  three-quarter bathroom, **296,**
    297
  wet, 143, **144,** 145
vented closet tees, 157, **157**
vented fixtures,
  bathroom stack, 160, **160**
  for single family home, 155,
    155
  for two-story single family
    home, **156,** 157, 159
vent grades, 133
venting methods, 132-133
venting requirements, single
    family home, 154-155, **154**
vent piping, 10-11, **11,** 35, 131,
    281, **282, 283**
  sizing, 133-145
  testing, 244-249, **245, 246,**
    247, 248, 249
vent piping sketch, 281, **282, 283**
vent stack, 11
  sizing, 135-137, **136, 137**
vertical pipes, 120, **120**
vertical piping support, 115, **115**
views, plan, 7, **7**
vise, pipe, 61, **61**

W

wales, 19
wall brace, 231, **231**
wall-hung blowout water closet,
    209, **209**
wall-hung drinking fountains, 229,
    229
wall-hung lavatory, 214-215, **215**
wall-hung lavatory rough-in sheet,
    273
wall-hung syphon jet urinal, 211,
    212
wall-hung syphon jet water closet,
    209, **209**
wall-hung syphon water closet,
    207, 208-209
wall-hung washout urinal, 210,
    211
wall-hung water cooler, 229, **229**
washdown water closet, 204, **204**
washout urinal, 210, **211**
waste, laundry tray, 297, **298**
waste connections,
  drinking fountain, 230, **230**
  water cooler, 230, **230**

waste piping, 10, 119-120, 281,
    282, 283
  garbage disposal, 225, **225**
waste piping sketch, 281, **282,**
    283
waste receptors, indirect, 229
waste stack, 10
  kitchen sink, 160, **161,** 162
  sizing of, 124-125
water,
  cold hard, 308-309
  cold soft, 309-310
  hot, 310-311
  potable, 9
water closet, 201-209, **203, 204,**
    205, 206, 207, 209
  blowout, 204, **204**
  commercial, 209, **209**
  floor-set, **207,** 208
  floor-set syphon jet, 209, **209**
  flush devices for, 205-208, **205,**
    206
  one-piece, **207,** 208
  operation of, 202, **203,** 204,
    204
  residential, **207,** 208-209
  reverse trap, 204, **204**
  rough-in sheet, **271**
  seats, 209, **210**
  syphon jet, 202, **203**
  wall-hung blowout, 209, **209**
  wall-hung syphon, **207,** 208-209
  wall-hung syphon jet, 209, **209**
  washdown, 204, **204**
water closet piping, 301
water coolers, 229-230, **229**
  trim, 230
  waste connections, 230, **230**
water demand, 181-182
water distributing pipe, 10
water hammer, 239
water hammer arrestor, 239, **239**
water heater, 235-240, **236, 237**
  automatic electric storage
    tank, 237-238, **237**
  automatic gas storage tank,
    235-237, **236**
  electric, 237-238, **237**
  gas, 235-237, **236**
  insulation for, 235, **236**
  jacket, 235, **236**
  rough-in sheet, **279**
  top cover, 235, **236**
water main, 9
water meter, 10, 54, **55,** 56
  compound, 54, **55**
  disk-type, 54, **55, 56**
  turbine, 54, **55**
water pipe,
  in four-unit apartment building,
    191, **192, 193,** 194-195
  in larger installations, 189-190
  lavatory, 301, **303**

minimum sizes, 186
in public building, 195, **196,**
    197, 198-199
rough, 299
symbols, 187
testing, 311
water seal traps, 168-173, **168,**
    169, 170, 171, 172, 173
water service, 9, 306, **307**
water softeners, 232-235, **233, 234,**
    235
  automatic, 233-235, **233, 234**
  rough-in sheet, **280**
water supply fitting,
  mop basin, 231
  service sink, 231
water supply fixture unit system,
    182, 183
water supply system, basement,
    308-311
water supply test, air, 250-251
water supply test, hydrostatic,
    250, 251
water supply valve, shower bath,
    221, **221**
water test,
  of sanitary drainage, 248-249,
    249
  of vent piping, 248-249, **249**
wedge disks, 47, **48**
welded pipe, 33
welding safety, 26-27, **26, 27**
well, brine, 234, **234**
wet core drilling outfit, 84-85, **85**
wet vents, sizing, 143, **144,** 145
wheelchair lavatory, 215, **215**
wheeled pipe cutters, 61, **61**
    66-67, **66, 67**
wind effect on trap seal, 180
women's urinals, 211, **212**
wrench,
  adjustable, **77,** 78
  basin, 78, **79**
  basket strainer, 78, **79**
  box, 78, **78**
  chain, 70-71, **71**
  closet seat, 78, **79**
  combination, 78, **78**
  compound-leverage, 71, **71**
  end pipe, 70, **70**
  finishing, 77-78, **77, 78**
  offset hex, **77,** 78
  pipe, 70-71, **70, 71**
  rim, 78, **79**
  socket, 78, **78**
  spud, **77,** 78
  straightpipe, 70, **70**
  strap, 78, **79**
wyes, **46,** 47

X

xpelor, **206,** 208